Free! Online PDF version of the

BNi HOME REMODELER'S 2014 COSTBOOK

that you can download, print, share and *customize*!

Just go to www.ConstructionWorkZone.com/bnicostbooks

On ConstructionWorkZone.com you can quickly create a PDF version of this entire Costbook. You can even customize it! Just enter your own labor and markup rates and every cost item contained in this book will be instantly recalculated.

Your purchase of this BNi Costbook not only entitles you to full access to the interactive online PDF version, it also brings you a wide range of additional estimating tools available on ConstructionWorkZone.com, including a FREE Davis-Bacon wage rate database you can use to pinpoint the prevailing labor costs for over 400 metro areas throughout the country!

ConstructionWorkZone.com ...
your source for "all things construction"

In addition to a wide range of online estimating tools, ConstructionWorkZone.com is your go-to source for current industry-specific news articles, search tools, manufacturer data, 3-part specifications, marketing resources and much more. And because you are a valued BNi customer, registration is free!

Questions?

Call us toll-free: 1.888.BNI.BOOK

BNI Building News

HOME REMODELER'S

2014 COSTBOOK

TWENTIETH EDITION

BNi® Building News

EDITOR-IN-CHIEF

William D. Mahoney, P.E.

TECHNICAL SERVICES

Brett Crouse
Tony De Augustine
Anthony Jackson
Eric Mahoney, AIA
Ana Varela
Mathew Woolsey

GRAPHIC DESIGN

Robert O. Wright Jr.

***BNi* Publications, Inc.**

VISTA
990 PARK CENTER DRIVE, SUITE E
VISTA, CA 92081

1-888-BNI-BOOK (1-888-264-2665)
www.bnibooks.com

ISBN 978-1-55701-796-3

Copyright © 2013 by BNI Publications, Inc. All rights reserved. Printed in the United States of America. Except as permitted under the United States Copyright Act of 1976, no part of this publication may be reproduced or distributed in any form or by any means, or stored in a data base or retrieval system, without the prior written permission of the publisher.

While diligent effort is made to provide reliable, accurate and up-to-date information, neither BNI Publications Inc., nor its authors or editors, can place a guarantee on the correctness of the data or information contained in this book. BNI Publications Inc., and its authors and editors, do hereby disclaim any responsibility or liability in connection with the use of this book or of any data or other information contained therein.

Features of this Book *(Continued)*

Adjustments to Costs

The costs as presented in this book attempt to represent national averages. Costs, however, vary among regions, states and even between adjacent localities.

In order to more closely approximate the probable costs for specific locations throughout the U.S., a table of Geographic Multipliers is provided. These adjustment factors are used to modify costs obtained from this book to help account for regional variations of construction costs. Whenever local current costs are known, whether material or equipment prices or labor rates, they should be used if more accuracy is required.

Hours (Man-Hours)

These productivities represent typical installation labor for thousands of construction items. The data takes into account all activities involved in normal construction under commonly experienced working conditions such as site movement, material handling, start-up, etc.

Editor's Note: This **2014 Costbook** is intended to provide accurate, reliable, average costs and typical productivities for thousands of common construction components. The data is developed and compiled from various industry sources, including government, manufacturers, suppliers and working professionals. The intent of the information is to provide assistance and guidelines to construction professionals in estimating. The user should be aware that local conditions, material and labor availability and cost variations, economic considerations, weather, local codes and regulations, etc., all affect the actual cost of construction. These and other such factors must be considered and incorporated into any and all construction estimates.

Features of this Book

The construction estimating information in this book is divided into two main sections: Costbook Pages and Man-Hour Tables. Each section is organized according to the 16 division format. In addition there are extensive Supporting References.

Sample pages with graphic explanations are included before the Costbook pages. These explanations, along with the discussions below, will provide a good understanding of what is included in this book and how it can best be used in construction estimating.

Material Costs

The material costs used in this book represent national averages for prices that a contractor would expect to pay plus an allowance for freight (if applicable) and handling and storage. These costs reflect neither the lowest or highest prices, but rather a typical average cost over time. Periodic fluctuations in availability and in certain commodities (e.g. copper, conduit) can significantly affect local material pricing. In the final estimating and bidding stages of a project when the highest degree of accuracy is required, it is best to check local, current prices.

Labor Costs

Labor costs include the basic wage, plus commonly applicable taxes, insurance and markups for overhead and profit. The labor rates used here to develop the costs are typical average prevailing wage rates. Rates for different trades are used where appropriate for each type of work.

Fixed government rates and average allowances for taxes and insurance are included in the labor costs. These include employer-paid Social Security/Medicare taxes (FICA), Worker's Compensation insurance, state and federal unemployment taxes, and business insurance.

Please note, however, most of these items vary significantly from state to state and within states. For more specific data, local agencies and sources should be consulted.

Equipment Costs

Costs for various types and pieces of equipment are included in Division 1 - General Requirements and can be included in an estimate when required either as a total "Equipment" category or with specific appropriate trades. Costs for equipment are included when appropriate in the installation costs in the Costbook pages.

Overhead and Profit

Included in the labor costs are allowances for overhead and profit for the contractor/employer whose workers are performing the specific tasks. No cost allowances or fees are included for management of subcontractors by the general contractor or construction manager. These costs, where appropriate, must be added to the costs as listed in the book.

The allowance for overhead is included to account for office overhead, the contractors' typical costs of doing business. These costs normally include in-house office staff salaries and benefits, office rent and operating expenses, professional fees, vehicle costs and other operating costs which are not directly applicable to specific jobs. It should be noted for this book that office overhead as included should be distinguished from project overhead, the General Requirements (Division 1) which are specific to particular projects. Project overhead should be included on an item by item basis for each job.

Depending on the trade, an allowance of 10-15 percent is incorporated into the labor/installation costs to account for typical profit of the installing contractor. See Division 1, General Requirements, for a more detailed review of typical profit allowances.

Format *(Continued)*

SPECIALTIES .... DIVISION 10

VISUAL DISPLAY BOARDS 10100
COMPARTMENTS AND CUBICLES 10150
LOUVERS AND VENTS 10200
GRILLES AND SCREENS 10240
SERVICE WALLS 10250
WALL AND CORNER GUARDS 10260
ACCESS FLOORING 10270
PEST CONTROL 10290
FIREPLACES AND STOVES 10300
MANUFACTURED EXTERIOR SPECIALTIES 10340
FLAGPOLES 10350
IDENTIFICATION DEVICES 10400
PEDESTRIAN CONTROL DEVICES 10450
LOCKERS 10500
FIRE PROTECTION SPECIALTIES 10520
PROTECTIVE COVERS 10530
POSTAL SPECIALTIES 10550
PARTITIONS 10600
STORAGE SHELVING 10670
EXTERIOR PROTECTION 10700
TELEPHONE SPECIALTIES 10750
TOILET, BATH, AND LAUNDRY ACCESSORIES 10800
SCALES 10880
WARDROBE AND CLOSET SPECIALTIES 10900

EQUIPMENT .... DIVISION 11

MAINTENANCE EQUIPMENT 11010
SECURITY AND VAULT EQUIPMENT 11020
TELLER AND SERVICE EQUIPMENT 11030
ECCLESIASTICAL EQUIPMENT 11040
LIBRARY EQUIPMENT 11050
THEATER AND STAGE EQUIPMENT 11060
INSTRUMENTAL EQUIPMENT 11070
REGISTRATION EQUIPMENT 11080
CHECKROOM EQUIPMENT 11090
MERCANTILE EQUIPMENT 11100
COMMERCIAL LAUNDRY AND DRY CLEANING EQUIPMENT 11110
VENDING EQUIPMENT 11120
AUDIO-VISUAL EQUIPMENT 11130
VEHICLE SERVICE EQUIPMENT 11140
PARKING CONTROL EQUIPMENT 11150
LOADING DOCK EQUIPMENT 11160
SOLID WASTE HANDLING EQUIPMENT 11170
DETENTION EQUIPMENT 11190
WATER SUPPLY AND TREATMENT EQUIPMENT 11200
HYDRAULIC GATES AND VALVES 11280
FLUID WASTE TREATMENT AND DISPOSAL EQUIPMENT 11300
FOOD SERVICE EQUIPMENT 11400
RESIDENTIAL EQUIPMENT 11450
UNIT KITCHENS 11460
DARKROOM EQUIPMENT 11470
ATHLETIC, RECREATIONAL, AND THERAPEUTIC EQUIPMENT 11480
INDUSTRIAL AND PROCESS EQUIPMENT 11500
LABORATORY EQUIPMENT 11600
PLANETARIUM EQUIPMENT 11650
OBSERVATORY EQUIPMENT 11660
OFFICE EQUIPMENT 11680
MEDICAL EQUIPMENT 11700
MORTUARY EQUIPMENT 11780
NAVIGATION EQUIPMENT 11850
AGRICULTURAL EQUIPMENT 11870
EXHIBIT EQUIPMENT 11900

FURNISHINGS .... DIVISION 12

FABRICS 12050
ART 12100
MANUFACTURED CASEWORK 12300
FURNISHINGS AND ACCESSORIES 12400
FURNITURE 12500
MULTIPLE SEATING 12600
SYSTEMS FURNITURE 12700
INTERIOR PLANTS AND PLANTERS 12800
FURNISHINGS RESTORATION AND REPAIR 12900

SPECIAL CONSTRUCTION .... DIVISION 13

AIR-SUPPORTED STRUCTURES 13010
BUILDING MODULES 13020
SPECIAL PURPOSE ROOMS 13030
SOUND, VIBRATION, AND SEISMIC CONTROL 13080
RADIATION PROTECTION 13090
LIGHTING PROTECTION 13100
CATHODIC PROTECTION 13110
PRE-ENGINEERED STRUCTURES 13120
SWIMMING POOLS 13150
AQUARIUMS 13160
AQUATIC PARK FACILITIES 13165
TUBS AND POOLS 13170
ICE RINKS 13175
KENNELS AND ANIMAL SHELTERS 13185
SITE-CONSTRUCTED INCINERATORS 13190
STORAGE TANKS 13200
FILTER UNDERDRAINS AND MEDIA 13220
DIGESTER COVERS AND APPURTENANCES 13230
OXYGENATION SYSTEMS 13240
SLUDGE CONDITIONING SYSTEMS 13260
HAZARDOUS MATERIAL REMEDIATION 13280
MEASUREMENT AND CONTROL INSTRUMENTATION 13400
RECORDING INSTRUMENTATION 13500
TRANSPORTATION CONTROL INSTRUMENTATION 13550
SOLAR AND WIND ENERGY EQUIPMENT 13600
SECURITY ACCESS AND SURVEILLANCE 13700
BUILDING AUTOMATION AND CONTROL 13800
DETECTION AND ALARM 13850
FIRE SUPPRESSION 13900

CONVEYING SYSTEMS .... DIVISION 14

DUMBWAITERS 14100
ELEVATORS 14200
ESCALATORS AND MOVING WALKS 14300
LIFTS 14400
MATERIAL HANDLING 14500
HOISTS AND CRANES 14600
TURNTABLES 14700
SCAFFOLDING 14800
TRANSPORTATION 14900

MECHANICAL .... DIVISION 15

BASIC MECHANICAL MATERIALS AND METHODS 15050
BUILDING SERVICES PIPING 15100
PROCESS PIPING 15200
FIRE PROTECTION PIPING 15300
PLUMBING FIXTURES AND EQUIPMENT 15400
HEAT-GENERATION EQUIPMENT 15500
REFRIGERATION EQUIPMENT 15600
HEATING, VENTILATING, AND AIR CONDITIONING EQUIPMENT 15700
AIR DISTRIBUTION 15800
HVAC INSTRUMENTATION AND CONTROLS 15900
TESTING, ADJUSTING, AND BALANCING 15950

ELECTRICAL .... DIVISION 16

BASIC ELECTRICAL MATERIALS AND METHODS 16050
WIRING METHODS 16100
ELECTRICAL POWER 16200
TRANSMISSION AND DISTRIBUTION 16300
LOW-VOLTAGE DISTRIBUTION 16400
LIGHTING 16500
COMMUNICATIONS 16700
SOUND AND VIDEO 16800

Format

All data is categorized according to the 16 division format. This industry standard provides an all-inclusive checklist to ensure that no element of a project is overlooked.

GENERAL REQUIREMENTS	DIVISION 1
SUMMARY	1100
PRICE AND PAYMENT PROCEDURES	1200
ADMINISTRATIVE REQUIREMENTS	1300
QUALITY REQUIREMENTS	1400
TEMPORARY FACILITIES AND CONTROLS	1500
PRODUCT REQUIREMENTS	1600
EXECUTION REQUIREMENTS	1700
FACILITY OPERATION	1800
FACILITY DECOMMISSIONING	1900

SITE CONSTRUCTION	DIVISION 2
BASIC SITE MATERIALS AND METHODS	2050
SITE REMEDIATION	2100
SITE PREPARATION	2200
EARTHWORK	2300
TUNNELING, BORING, AND JACKING	2400
FOUNDATION AND LOAD-BEARING ELEMENTS	2450
UTILITY SERVICES	2500
DRAINAGE AND CONTAINMENT	2600
BASES, BALLASTS, PAVEMENTS, AND APPURTENANCES	2700
SITE IMPROVEMENTS AND AMENITIES	2800
PLANTING	2900
SITE RESTORATION AND REHABILITATION	2950

CONCRETE	DIVISION 3
BASIC CONCRETE MATERIALS AND METHODS	3050
CONCRETE FORMS AND ACCESSORIES	3100
CONCRETE REINFORCEMENT	3200
CAST-IN-PLACE CONCRETE	3300
PRECAST CONCRETE	3400
CEMENTITIOUS DECKS AND UNDERLAYMENT	3500
GROUTS	3600
MASS CONCRETE	3700
CONCRETE RESTORATION AND CLEANING	3900

MASONRY	DIVISION 4
BASIC MASONRY MATERIALS AND METHODS	4050
MASONRY UNITS	4200
STONE	4400
REFRACTORIES	4500
CORROSION-RESISTANT MASONRY	4600
SIMULATED MASONRY	4700
MASONRY ASSEMBLIES	4800
MASONRY RESTORATION AND CLEANING	4900

METALS	DIVISION 5
BASIC METAL MATERIALS AND METHODS	5050
STRUCTURAL METAL FRAMING	5100
METAL JOISTS	5200
METAL DECK	5300
COLD-FORMED METAL FRAMING	5400
METAL FABRICATIONS	5500
HYDRAULIC FABRICATIONS	5600
RAILROAD TRACK AND ACCESSORIES	5650
ORNAMENTAL METAL	5700
EXPANSION CONTROL	5800
METAL RESTORATION AND CLEANING	5900

WOOD AND PLASTICS	DIVISION 6
BASIC WOOD AND PLASTIC MATERIALS AND METHODS	6050
ROUGH CARPENTRY	6100
FINISH CARPENTRY	6200
ARCHITECTURAL WOODWORK	6400
STRUCTURAL PLASTICS	6500
PLASTIC FABRICATIONS	6600
WOOD AND PLASTIC RESTORATION AND CLEANING	6900

THERMAL AND MOISTURE PROTECTION	DIVISION 7
BASIC THERMAL AND MOISTURE PROTECTION MATERIALS AND METHODS	7050
DAMPPROOFING AND WATERPROOFING	7100
THERMAL PROTECTION	7200
SHINGLES, ROOF TILES, AND ROOF COVERINGS	7300
ROOFING AND SIDING PANELS	7400
MEMBRANE ROOFING	7500
FLASHING AND SHEET METAL	7600
ROOF SPECIALTIES AND ACCESSORIES	7700
FIRE AND SMOKE PROTECTION	7800
JOINT SEALERS	7900

DOORS AND WINDOWS	DIVISION 8
BASIC DOOR AND WINDOW MATERIALS AND METHODS	8050
METAL DOORS AND FRAMES	8100
WOOD AND PLASTIC DOORS	8200
SPECIALTY DOORS	8300
ENTRANCES AND STOREFRONTS	8400
WINDOWS	8500
SKYLIGHTS	8600
HARDWARE	8700
GLAZING	8800
GLAZED CURTAIN WALL	8900

FINISHES	DIVISION 9
BASIC FINISH MATERIALS AND METHODS	9050
METAL SUPPORT ASSEMBLIES	9100
PLASTER AND GYPSUM BOARD	9200
TILE	9300
TERRAZZO	9400
CEILINGS	9500
FLOORING	9600
WALL FINISHES	9700
ACOUSTICAL TREATMENT	9800
PAINTS AND COATINGS	9900

Preface

For over 65 years, BNi Building News has been dedicated to providing construction professionals with timely and reliable information. Based on this experience, our staff has researched and compiled thousands of up-to-the-minute costs for the **BNi Costbooks**. This book is an essential reference for contractors, engineers, architects, facility managers — any construction professional who must provide an estimate for any type of building project.

Whether working up a preliminary estimate or submitting a formal bid, the costs listed here can be quickly and easily tailored to your needs. All costs are based on prevailing labor rates. Overhead and profit should be included in all costs. Man-hours are also provided.

All data is categorized according to the 16 division format. This industry standard provides an all-inclusive checklist to ensure that no element of a project is overlooked. In addition, to make specific items even easier to locate, there is a complete alphabetical index.

The "Features of this Book" section presents a clear overview of the many features of this book. Included is an explanation of the data, sample page layout and discussion of how to best use the information in the book.

Of course, all buildings and construction projects are unique. The information provided in this book is based on averages from well-managed projects with good labor productivity under normal working conditions (eight hours a day). Other circumstances affecting costs such as overtime, unusual working conditions, savings from buying bulk quantities for large projects, and unusual or hidden costs must be factored in as they arise.

The data provided in this book is for estimating purposes only. Check all applicable federal, state and local codes and regulations for local requirements.

Table of Contents

Sample Costbook Page

In order to best use the information in this book, please review this sample page and read the "Features In This Book" section.

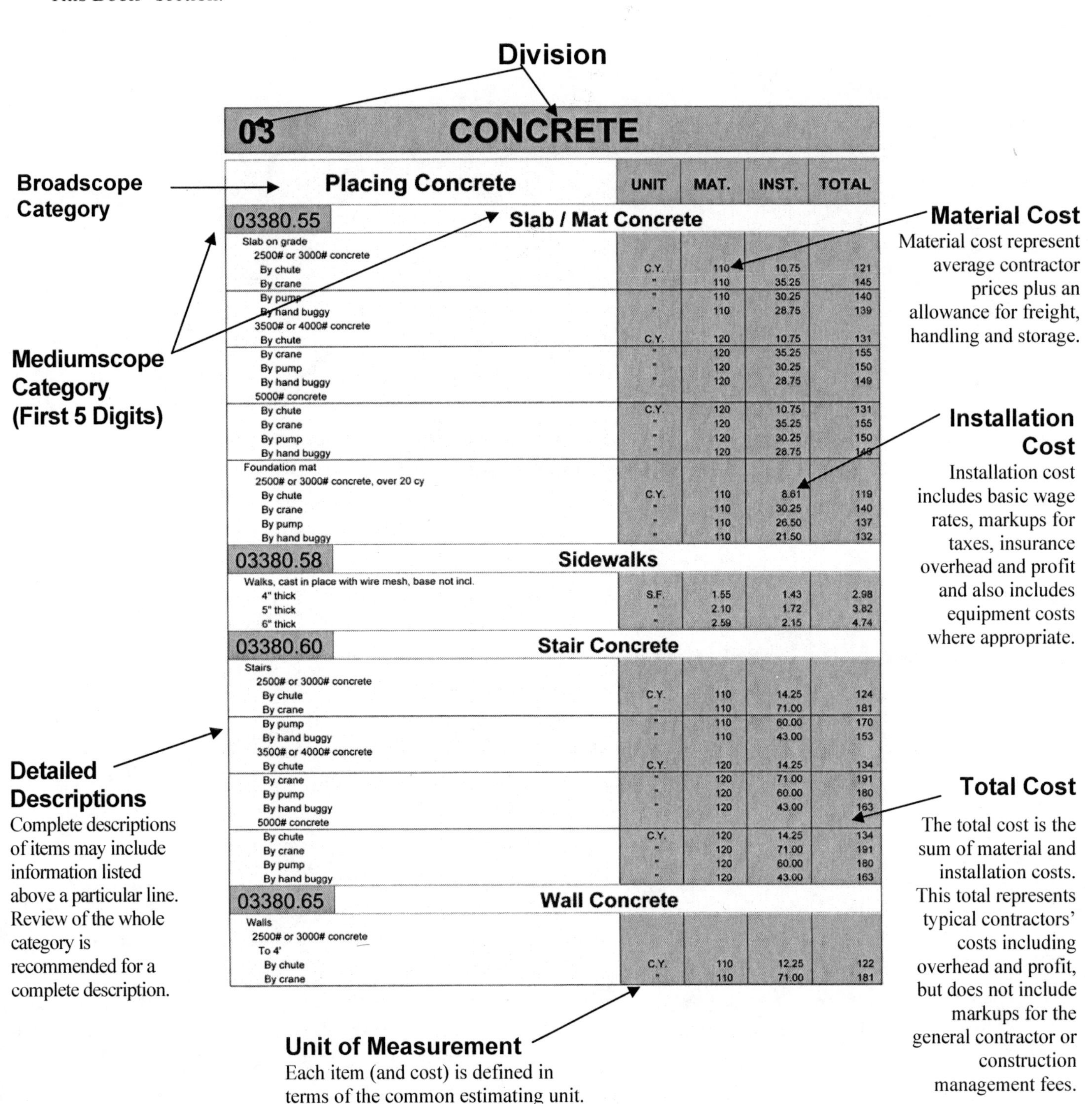

03 CONCRETE

Placing Concrete	UNIT	MAT.	INST.	TOTAL
03380.55 Slab / Mat Concrete				
Slab on grade				
2500# or 3000# concrete				
By chute	C.Y.	110	10.75	121
By crane	"	110	35.25	145
By pump	"	110	30.25	140
By hand buggy	"	110	28.75	139
3500# or 4000# concrete				
By chute	C.Y.	120	10.75	131
By crane	"	120	35.25	155
By pump	"	120	30.25	150
By hand buggy	"	120	28.75	149
5000# concrete				
By chute	C.Y.	120	10.75	131
By crane	"	120	35.25	155
By pump	"	120	30.25	150
By hand buggy	"	120	28.75	149
Foundation mat				
2500# or 3000# concrete, over 20 cy				
By chute	C.Y.	110	8.61	119
By crane	"	110	30.25	140
By pump	"	110	26.50	137
By hand buggy	"	110	21.50	132
03380.58 Sidewalks				
Walks, cast in place with wire mesh, base not incl.				
4" thick	S.F.	1.55	1.43	2.98
5" thick	"	2.10	1.72	3.82
6" thick	"	2.59	2.15	4.74
03380.60 Stair Concrete				
Stairs				
2500# or 3000# concrete				
By chute	C.Y.	110	14.25	124
By crane	"	110	71.00	181
By pump	"	110	60.00	170
By hand buggy	"	110	43.00	153
3500# or 4000# concrete				
By chute	C.Y.	120	14.25	134
By crane	"	120	71.00	191
By pump	"	120	60.00	180
By hand buggy	"	120	43.00	163
5000# concrete				
By chute	C.Y.	120	14.25	134
By crane	"	120	71.00	191
By pump	"	120	60.00	180
By hand buggy	"	120	43.00	163
03380.65 Wall Concrete				
Walls				
2500# or 3000# concrete				
To 4'				
By chute	C.Y.	110	12.25	122
By crane	"	110	71.00	181

BNi® Building News

01 GENERAL

Requirements

	UNIT	MAT.	INST.	TOTAL
01020.10 Allowances				
Overhead				
$20,000 project				
Minimum	PCT.			15.00
Average	"			20.00
Maximum	"			40.00
$100,000 project				
Minimum	PCT.			12.00
Average	"			15.00
Maximum	"			25.00
Profit				
$20,000 project				
Minimum	PCT.			10.00
Average	"			15.00
Maximum	"			25.00
$100,000 project				
Minimum	PCT.			10.00
Average	"			12.00
Maximum	"			20.00
Professional fees				
Architectural				
Minimum	PCT.			5.00
Average	"			10.00
Maximum	"			20.00
Taxes				
Sales tax				
Minimum	PCT.			4.00
Average	"			5.00
Maximum	"			10.00
Unemployment				
Minimum	PCT.			3.00
Average	"			6.50
Maximum	"			8.00
Social security (FICA)	"			7.85
01050.10 Field Staff				
Foreman				
Minimum	YEAR			52,900
Average	"			84,600
Maximum	"			99,050
Watchman				
Minimum	YEAR			22,800
Average	"			30,500
Maximum	"			38,500
01500.10 Temporary Facilities				
Trailers, general office type, per month				
Minimum	EA.			210
Average	"			340
Maximum	"			690

01 GENERAL

Requirements	UNIT	MAT.	INST.	TOTAL
01505.10 Mobilization				
Equipment mobilization				
Backhoe/front-end loader				
Minimum	EA.			120
Average	"			200
Maximum	"			450
Truck crane				
Minimum	EA.			490
Average	"			760
Maximum	"			1,310
01525.10 Construction Aids				
Scaffolding/staging, rent per month				
Measured by lineal feet of base				
10' high	L.F.			12.50
20' high	"			22.50
30' high	"			31.50
Measured by square foot of surface				
Minimum	S.F.			0.55
Average	"			0.94
Maximum	"			1.69
Tarpaulins, fabric, per job				
Minimum	S.F.			0.25
Average	"			0.42
Maximum	"			1.11
01600.10 Equipment				
Air compressor				
60 cfm				
By day	EA.			96.00
By week	"			290
By month	"			870
Air tools, per compressor, per day				
Minimum	EA.			39.50
Average	"			49.50
Maximum	"			69.00
Generators, 5 kw				
By day	EA.			99.00
By week	"			290
By month	"			910
Heaters, salamander type, per week				
Minimum	EA.			120
Average	"			170
Maximum	"			350
Pumps, submersible				
50 gpm				
By day	EA.			79.00
By week	"			240
By month	"			710
Diaphragm pump, by week				
Minimum	EA.			140
Average	"			240
Maximum	"			490
Pickup truck				

Requirements

01600.10 Equipment *(Cont.)*

Requirements	UNIT	MAT.	INST.	TOTAL
By day	EA.			150
By week	"			430
By month	"			1,340
Dump truck				
6 cy truck				
By day	EA.			390
By week	"			1,180
By month	"			3,550
10 cy truck				
By day	EA.			490
By week	"			1,480
By month	"			4,440
Backhoe/loader, rubber tired				
1/2 cy capacity				
By day	EA.			490
By week	"			1,480
By month	"			4,440
Cranes, crawler type				
Truck mounted, hydraulic				
15 ton capacity				
By day	EA.			840
By week	"			2,520
By month	"			7,270

02 SITE CONSTRUCTION

Site Remediation

02115.60 Underground Storage Tank Removal	UNIT	MAT.	INST.	TOTAL
Remove underground storage tank, and backfill				
50 to 250 gals	EA.		800	800
600 gals	"		800	800
1000 gals	"		1,200	1,200

02115.66 Septic Tank Removal	UNIT	MAT.	INST.	TOTAL
Remove septic tank				
1000 gals	EA.		200	200
2000 gals	"		240	240

Site Preparation

02210.10 Soil Boring	UNIT	MAT.	INST.	TOTAL
Borings, uncased, stable earth				
2-1/2" dia.	L.F.		30.00	30.00

Demolition

02220.10 Complete Building Demolition	UNIT	MAT.	INST.	TOTAL
Wood frame	C.F.		0.34	0.34
Concrete	"		0.51	0.51
Steel frame	"		0.69	0.69

02220.15 Selective Building Demolition	UNIT	MAT.	INST.	TOTAL
Partition removal				
Concrete block partitions				
4" thick	S.F.		2.19	2.19
8" thick	"		2.92	2.92
12" thick	"		3.98	3.98
Brick masonry partitions				
4" thick	S.F.		2.19	2.19
8" thick	"		2.74	2.74
12" thick	"		3.65	3.65
16" thick	"		5.48	5.48
Cast in place concrete partitions				
Unreinforced				
6" thick	S.F.		16.00	16.00
8" thick	"		17.25	17.25
10" thick	"		20.00	20.00

Demolition	UNIT	MAT.	INST.	TOTAL
02220.15 Selective Building Demolition *(Cont.)*				
12" thick	S.F.		24.00	24.00
Reinforced				
6" thick	S.F.		18.50	18.50
8" thick	"		24.00	24.00
10" thick	"		26.75	26.75
12" thick	"		32.00	32.00
Terra cotta				
To 6" thick	S.F.		2.19	2.19
Stud partitions				
Metal or wood, with drywall both sides	S.F.		2.19	2.19
Metal studs, both sides, lath and plaster	"		2.92	2.92
Door and frame removal				
Hollow metal in masonry wall				
Single				
2'6"x6'8"	EA.		55.00	55.00
3'x7'	"		73.00	73.00
Double				
3'x7'	EA.		88.00	88.00
4'x8'	"		88.00	88.00
Wood in framed wall				
Single				
2'6"x6'8"	EA.		31.25	31.25
3'x6'8"	"		36.50	36.50
Double				
2'6"x6'8"	EA.		43.75	43.75
3'x6'8"	"		48.75	48.75
Remove for re-use				
Hollow metal	EA.		110	110
Wood	"		73.00	73.00
Floor removal				
Brick flooring	S.F.		1.75	1.75
Ceramic or quarry tile	"		0.97	0.97
Terrazzo	"		1.94	1.94
Heavy wood	"		1.16	1.16
Residential wood	"		1.25	1.25
Resilient tile or linoleum	"		0.43	0.43
Ceiling removal				
Acoustical tile ceiling				
Adhesive fastened	S.F.		0.43	0.43
Furred and glued	"		0.36	0.36
Suspended grid	"		0.27	0.27
Drywall ceiling				
Furred and nailed	S.F.		0.48	0.48
Nailed to framing	"		0.43	0.43
Plastered ceiling				
Furred on framing	S.F.		1.09	1.09
Suspended system	"		1.46	1.46
Roofing removal				
Steel frame				
Corrugated metal roofing	S.F.		0.87	0.87
Built-up roof on metal deck	"		1.46	1.46
Wood frame				
Built up roof on wood deck	S.F.		1.34	1.34

Demolition

02220.15 Selective Building Demolition *(Cont.)*

Demolition	UNIT	MAT.	INST.	TOTAL
Roof shingles	S.F.		0.73	0.73
Roof tiles	"		1.46	1.46
Concrete frame	C.F.		2.92	2.92
Concrete plank	S.F.		2.19	2.19
Built-up roof on concrete	"		1.25	1.25
Cut-outs				
Concrete, elevated slabs, mesh reinforcing				
Under 5 cf	C.F.		43.75	43.75
Over 5 cf	"		36.50	36.50
Bar reinforcing				
Under 5 cf	C.F.		73.00	73.00
Over 5 cf	"		55.00	55.00
Window removal				
Metal windows, trim included				
2'x3'	EA.		43.75	43.75
2'x4'	"		48.75	48.75
2'x6'	"		55.00	55.00
3'x4'	"		55.00	55.00
3'x6'	"		63.00	63.00
3'x8'	"		73.00	73.00
4'x4'	"		73.00	73.00
4'x6'	"		88.00	88.00
4'x8'	"		110	110
Wood windows, trim included				
2'x3'	EA.		24.25	24.25
2'x4'	"		25.75	25.75
2'x6'	"		27.50	27.50
3'x4'	"		29.25	29.25
3'x6'	"		31.25	31.25
3'x8'	"		33.75	33.75
6'x4'	"		36.50	36.50
6'x6'	"		39.75	39.75
6'x8'	"		43.75	43.75
Walls, concrete, bar reinforcing				
Small jobs	C.F.		29.25	29.25
Large jobs	"		24.25	24.25
Brick walls, not including toothing				
4" thick	S.F.		2.19	2.19
8" thick	"		2.74	2.74
12" thick	"		3.65	3.65
16" thick	"		5.48	5.48
Concrete block walls, not including toothing				
4" thick	S.F.		2.43	2.43
6" thick	"		2.57	2.57
8" thick	"		2.74	2.74
10" thick	"		3.13	3.13
12" thick	"		3.65	3.65
Rubbish handling				
Load in dumpster or truck				
Minimum	C.F.		0.97	0.97
Maximum	"		1.46	1.46
Rubbish hauling				
Hand loaded on trucks, 2 mile trip	C.Y.		37.00	37.00

02 SITE CONSTRUCTION

Demolition

Demolition	UNIT	MAT.	INST.	TOTAL
02220.15 Selective Building Demolition *(Cont.)*				
Machine loaded on trucks, 2 mile trip	C.Y.		24.00	24.00

Selective Site Demolition

Selective Site Demolition	UNIT	MAT.	INST.	TOTAL
02225.13 Core Drilling				
Concrete				
6" thick				
3" dia.	EA.		41.00	41.00
8" thick				
3" dia.	EA.		57.00	57.00
02225.20 Fence Demolition				
Remove fencing				
Chain link, 8' high				
For disposal	L.F.		2.19	2.19
For reuse	"		5.48	5.48
Wood				
4' high	S.F.		1.46	1.46
6' high	"		1.75	1.75
8' high	"		2.19	2.19
Masonry				
8" thick				
4' high	S.F.		4.38	4.38
6' high	"		5.48	5.48
8' high	"		6.26	6.26
02225.42 Drainage Piping Demolition				
Remove drainage pipe, not including excavation				
12" dia.	L.F.		10.00	10.00
18" dia.	"		12.75	12.75
02225.43 Gas Piping Demolition				
Remove welded steel pipe, not including excavation				
4" dia.	L.F.		15.00	15.00
5" dia.	"		24.00	24.00
02225.45 Sanitary Piping Demolition				
Remove sewer pipe, not including excavation				
4" dia.	L.F.		9.61	9.61
6" dia.	"		11.00	11.00

02 SITE CONSTRUCTION

Selective Site Demolition

02225.48 Water Piping Demolition

Selective Site Demolition	UNIT	MAT.	INST.	TOTAL
Remove water pipe, not including excavation				
4" dia.	L.F.		11.00	11.00
6" dia.	"		11.50	11.50

02225.50 Saw Cutting Pavement

Selective Site Demolition	UNIT	MAT.	INST.	TOTAL
Pavement, bituminous				
2" thick	L.F.		1.84	1.84
3" thick	"		2.31	2.31
Concrete pavement, with wire mesh				
4" thick	L.F.		3.55	3.55
5" thick	"		3.85	3.85
Plain concrete, unreinforced				
4" thick	L.F.		3.08	3.08
5" thick	"		3.55	3.55

02225.80 Wall, Exterior, Demolition

Selective Site Demolition	UNIT	MAT.	INST.	TOTAL
Concrete wall				
Light reinforcing				
6" thick	S.F.		12.00	12.00
8" thick	"		12.75	12.75
Medium reinforcing				
6" thick	S.F.		12.75	12.75
8" thick	"		13.25	13.25
Heavy reinforcing				
6" thick	S.F.		14.25	14.25
8" thick	"		15.00	15.00
Masonry				
No reinforcing				
8" thick	S.F.		5.34	5.34
12" thick	"		6.00	6.00
Horizontal reinforcing				
8" thick	S.F.		6.00	6.00
12" thick	"		6.49	6.49
Vertical reinforcing				
8" thick	S.F.		7.75	7.75
12" thick	"		8.90	8.90

Site Clearing

02230.50 Tree Cutting & Clearing

Site Clearing	UNIT	MAT.	INST.	TOTAL
Cut trees and clear out stumps				
9" to 12" dia.	EA.		480	480
Loading and trucking				
For machine load, per load, round trip				
1 mile	EA.		96.00	96.00
10 mile	"		160	160

02 SITE CONSTRUCTION

Site Clearing

	UNIT	MAT.	INST.	TOTAL
02230.50 Tree Cutting & Clearing *(Cont.)*				
Hand loaded, round trip				
1 mile	EA.		230	230
10 mile	"		370	370

Earthwork, Excavation & Fill

	UNIT	MAT.	INST.	TOTAL
02315.10 Base Course				
Base course, crushed stone				
3" thick	S.Y.	3.68	0.65	4.33
4" thick	"	4.89	0.70	5.59
6" thick	"	7.31	0.77	8.08
Base course, bank run gravel				
4" deep	S.Y.	3.02	0.68	3.70
6" deep	"	4.62	0.74	5.36
Prepare and roll sub base				
Minimum	S.Y.		0.65	0.65
Average	"		0.81	0.81
Maximum	"		1.09	1.09
02315.20 Borrow				
Borrow fill, F.O.B. at pit				
Sand, haul to site, round trip				
10 mile	C.Y.	20.50	13.00	33.50
02315.30 Bulk Excavation				
Excavation, by small dozer				
Small areas	C.Y.		3.08	3.08
Trim banks	"		4.62	4.62
Hydraulic excavator				
1 cy capacity				
Light material	C.Y.		4.00	4.00
Medium material	"		4.80	4.80
Wet material	"		6.00	6.00
Blasted rock	"		6.86	6.86
Wheel mounted front-end loader				
7/8 cy capacity				
Light material	C.Y.		3.27	3.27
Medium material	"		3.74	3.74
Wet material	"		4.36	4.36
02315.40 Building Excavation				
Structural excavation, unclassified earth				
3/8 cy backhoe	C.Y.		17.50	17.50
3/4 cy backhoe	"		13.00	13.00
1 cy backhoe	"		11.00	11.00
Foundation backfill and compaction by machine	"		26.25	26.25

Earthwork, Excavation & Fill

02315.45 Hand Excavation

Earthwork, Excavation & Fill	UNIT	MAT.	INST.	TOTAL
Excavation				
To 2' deep				
Normal soil	C.Y.		48.75	48.75
Sand and gravel	"		43.75	43.75
Medium clay	"		55.00	55.00
Heavy clay	"		63.00	63.00
Loose rock	"		73.00	73.00
To 6' deep				
Normal soil	C.Y.		63.00	63.00
Sand and gravel	"		55.00	55.00
Medium clay	"		73.00	73.00
Heavy clay	"		88.00	88.00
Loose rock	"		110	110
Backfilling foundation without compaction, 6" lifts	"		27.50	27.50
Compaction of backfill around structures or in trench				
By hand with air tamper	C.Y.		31.25	31.25
By hand with vibrating plate tamper	"		29.25	29.25
1 ton roller	"		46.25	46.25
Miscellaneous hand labor				
Excavation around obstructions and services	C.Y.		150	150

02315.60 Trenching

Earthwork, Excavation & Fill	UNIT	MAT.	INST.	TOTAL
Trenching and continuous footing excavation				
By gradall				
1 cy capacity				
Light soil	C.Y.		3.74	3.74
Medium soil	"		4.02	4.02
Heavy/wet soil	"		4.36	4.36
Loose rock	"		4.76	4.76
Blasted rock	"		5.03	5.03
By hydraulic excavator				
1/2 cy capacity				
Light soil	C.Y.		4.36	4.36
Medium soil	"		4.76	4.76
Heavy/wet soil	"		5.23	5.23
Loose rock	"		5.81	5.81
Blasted rock	"		6.54	6.54
Hand excavation				
Bulk, wheeled 100'				
Normal soil	C.Y.		48.75	48.75
Sand or gravel	"		43.75	43.75
Medium clay	"		63.00	63.00
Heavy clay	"		88.00	88.00
Loose rock	"		110	110
Trenches, up to 2' deep				
Normal soil	C.Y.		55.00	55.00
Sand or gravel	"		48.75	48.75
Medium clay	"		73.00	73.00
Heavy clay	"		110	110
Loose rock	"		150	150
Backfill trenches				
With compaction				
By hand	C.Y.		36.50	36.50

Earthwork, Excavation & Fill

	UNIT	MAT.	INST.	TOTAL
02315.60 Trenching (Cont.)				
By 60 hp tracked dozer	C.Y.		2.31	2.31
By small front-end loader	"		2.64	2.64
Spread dumped fill or gravel, no compaction				
6" layers	S.Y.		1.54	1.54
Compaction in 6" layers				
By hand with air tamper	S.Y.		1.14	1.14
02315.70 Utility Excavation				
Trencher, sandy clay, 8" wide trench				
18" deep	L.F.		2.05	2.05
24" deep	"		2.31	2.31
36" deep	"		2.64	2.64
Trench backfill, 95% compaction				
Tamp by hand	C.Y.		27.50	27.50
Vibratory compaction	"		22.00	22.00
Trench backfilling, with borrow sand, place & compact	"	20.50	22.00	42.50
02315.75 Gravel And Stone				
F.O.B. PLANT				
No. 21 crusher run stone	C.Y.			44.00
No. 68 stone	"			44.00
No. 78 stone	"			44.00
No. 78 gravel, (pea gravel)	"			33.00
02315.80 Hauling Material				
Haul material by 10 cy dump truck, round trip distance				
1 mile	C.Y.		5.13	5.13
5 mile	"		8.40	8.40
Spread topsoil by equipment on site	"		4.10	4.10

Soil Stabilization & Treatment

	UNIT	MAT.	INST.	TOTAL
02340.05 Soil Stabilization				
Straw bale secured with rebar	L.F.	7.54	1.46	9.00
Filter barrier, 18" high filter fabric	"	1.82	4.38	6.20
Sediment fence, 36" fabric with 6" mesh	"	4.32	5.48	9.80
Soil stabilization with tar paper, burlap, straw and stakes	S.F.	0.36	0.06	0.42
02360.20 Soil Treatment				
Soil treatment, termite control pretreatment				
Under slabs	S.F.	0.38	0.24	0.62
By walls	"	0.38	0.29	0.67

Soil Stabilization & Treatment

02445.10 Pipe Jacking

Soil Stabilization & Treatment	UNIT	MAT.	INST.	TOTAL
Pipe casing, horizontal jacking				
18" dia.	L.F.	120	92.00	212

Piles And Caissons

02455.65 Steel Pipe Piles

Piles And Caissons	UNIT	MAT.	INST.	TOTAL
Concrete filled, 3000# concrete, up to 40'				
8" dia.	L.F.	24.25	10.25	34.50
10" dia.	"	31.25	10.50	41.75
12" dia.	"	36.00	11.00	47.00
Pipe piles, non-filled				
8" dia.	L.F.	22.75	7.95	30.70
10" dia.	"	28.75	8.18	36.93
12" dia.	"	35.00	8.42	43.42

02455.80 Wood And Timber Piles

Piles And Caissons	UNIT	MAT.	INST.	TOTAL
Treated wood piles, 12" butt, 8" tip				
25' long	L.F.	12.00	14.25	26.25
30' long	"	12.75	12.00	24.75

02465.50 Prestressed Piling

Piles And Caissons	UNIT	MAT.	INST.	TOTAL
Prestressed concrete piling, less than 60' long				
10" sq.	L.F.	18.00	5.96	23.96
12" sq.	"	25.25	6.22	31.47
Straight cylinder, less than 60' long				
12" dia.	L.F.	23.50	6.50	30.00
14" dia.	"	31.75	6.65	38.40

02475.10 Caissons, Includes Casing

Piles And Caissons	UNIT	MAT.	INST.	TOTAL
Caisson, 3000# conc., 60 # reinf./CY, stable ground				
18" dia., 0.065 CY/ LF	L.F.	12.25	28.75	41.00
24" dia., 0.116 CY/ LF	"	19.50	29.75	49.25
Wet ground, casing required but pulled				
18" dia.	L.F.	12.25	35.75	48.00
24" dia.	"	19.50	39.75	59.25
Soft rock				
18" dia.	L.F.	12.25	100	112
24" dia.	"	19.50	180	200

02 SITE CONSTRUCTION

Utility Services	UNIT	MAT.	INST.	TOTAL
02510.10 Wells				
Domestic water, drilled and cased				
4" dia.	L.F.	27.75	72.00	99.75
6" dia.	"	30.50	80.00	111
02510.15 Water Meters				
Water meter, displacement type				
1"	EA.	200	61.00	261
02510.20 Tapping Saddles & Sleeves				
Tapping saddle, tap size to 2"				
4" saddle	EA.	71.00	22.00	93.00
Tap hole in pipe				
4" hole	EA.		55.00	55.00
02510.25 Valve Boxes				
Valve box, adjustable, for valves up to 20"				
3' deep	EA.	280	14.50	295
4' deep	"	330	17.50	348
5' deep	"	390	22.00	412
02510.40 Ductile Iron Pipe				
Ductile iron pipe, cement lined, slip-on joints				
4"	L.F.	15.50	6.67	22.17
6"	"	18.75	7.06	25.81
Mechanical joint pipe				
4"	L.F.	17.50	9.24	26.74
6"	"	20.75	10.00	30.75
02510.60 Plastic Pipe				
PVC, class 150 pipe				
4" dia.	L.F.	4.60	6.00	10.60
6" dia.	"	8.71	6.49	15.20
Schedule 40 pipe				
1-1/2" dia.	L.F.	1.15	2.57	3.72
2" dia.	"	1.71	2.74	4.45
2-1/2" dia.	"	2.60	2.92	5.52
3" dia.	"	3.54	3.13	6.67
4" dia.	"	5.00	3.65	8.65
6" dia.	"	9.42	4.38	13.80
Drainage pipe				
PVC schedule 80				
1" dia.	L.F.	2.26	2.57	4.83
1-1/2" dia.	"	2.73	2.57	5.30
ABS, 2" dia.	"	3.50	2.74	6.24
2-1/2" dia.	"	4.97	2.92	7.89
3" dia.	"	5.85	3.13	8.98
4" dia.	"	7.96	3.65	11.61
6" dia.	"	13.25	4.38	17.63
90 degree elbows				
1"	EA.	3.76	7.30	11.06
1-1/2"	"	4.69	7.30	11.99
2"	"	5.65	7.97	13.62

02 SITE CONSTRUCTION

Utility Services

02510.60 Plastic Pipe *(Cont.)*	UNIT	MAT.	INST.	TOTAL
2-1/2"	EA.	13.50	8.76	22.26
3"	"	14.00	9.74	23.74
4"	"	24.75	11.00	35.75
6"	"	54.00	14.50	68.50
45 degree elbows				
1"	EA.	6.11	7.30	13.41
1-1/2"	"	7.77	7.30	15.07
2"	"	9.63	7.97	17.60
2-1/2"	"	18.00	8.76	26.76
3"	"	19.25	9.74	28.99
4"	"	36.50	11.00	47.50
6"	"	85.00	14.50	99.50
Tees				
1"	EA.	3.97	8.76	12.73
1-1/2"	"	12.75	8.76	21.51
2"	"	15.50	9.74	25.24
2-1/2"	"	18.00	11.00	29.00
3"	"	19.75	12.50	32.25
4"	"	37.50	14.50	52.00
6"	"	74.00	17.50	91.50
Couplings				
1"	EA.	3.30	7.30	10.60
1-1/2"	"	5.65	7.30	12.95
2"	"	8.33	7.97	16.30
2-1/2"	"	17.50	8.76	26.26
3"	"	18.25	9.74	27.99
4"	"	18.75	11.00	29.75
6"	"	31.75	14.50	46.25

Sanitary Sewer

02530.20 Vitrified Clay Pipe	UNIT	MAT.	INST.	TOTAL
Vitrified clay pipe, extra strength				
6" dia.	L.F.	5.21	11.00	16.21

02530.40 Sanitary Sewers	UNIT	MAT.	INST.	TOTAL
Clay				
6" pipe	L.F.	6.88	8.01	14.89
PVC				
4" pipe	L.F.	3.47	6.00	9.47
6" pipe	"	6.94	6.32	13.26
Connect new sewer line				
To existing manhole	EA.	95.00	150	245
To new manhole	"	68.00	88.00	156

Sanitary Sewer

Sanitary Sewer	UNIT	MAT.	INST.	TOTAL
02540.10 Drainage Fields				
Perforated PVC pipe, for drain field				
4" pipe	L.F.	2.05	5.34	7.39
6" pipe	"	3.85	5.72	9.57
02540.50 Septic Tanks				
Septic tank, precast concrete				
1000 gals	EA.	750	400	1,150
2000 gals	"	2,420	600	3,020
Leaching pit, precast concrete, 72" diameter				
3' deep	EA.	690	300	990
6' deep	"	1,210	340	1,550
8' deep	"	1,540	400	1,940
02630.70 Underdrain				
Drain tile, clay				
6" pipe	L.F.	4.52	5.34	9.86
8" pipe	"	7.21	5.58	12.79
Porous concrete, standard strength				
6" pipe	L.F.	4.07	5.34	9.41
8" pipe	"	4.40	5.58	9.98
Corrugated metal pipe, perforated type				
6" pipe	L.F.	5.64	6.00	11.64
8" pipe	"	6.66	6.32	12.98
Perforated clay pipe				
6" pipe	L.F.	5.44	6.86	12.30
8" pipe	"	7.29	7.06	14.35
Drain tile, concrete				
6" pipe	L.F.	3.23	5.34	8.57
8" pipe	"	5.02	5.58	10.60
Perforated rigid PVC underdrain pipe				
4" pipe	L.F.	1.91	4.00	5.91
6" pipe	"	3.66	4.80	8.46
8" pipe	"	5.60	5.34	10.94
Underslab drainage, crushed stone				
3" thick	S.F.	0.25	0.80	1.05
4" thick	"	0.34	0.92	1.26
6" thick	"	0.51	1.00	1.51

Flexible Surfaces

Flexible Surfaces	UNIT	MAT.	INST.	TOTAL
02740.10 Asphalt Repair				
Coal tar seal coat, rubber add., fuel resist.	S.Y.	2.33	0.62	2.95
Bituminous surface treatment, single	"	2.14	0.43	2.57
Double	"	2.84	0.04	2.88
Bituminous prime coat	"	1.44	0.05	1.49
Tack coat	"	0.70	0.04	0.74

Flexible Surfaces

	UNIT	MAT.	INST.	TOTAL
02740.10 Asphalt Repair (Cont.)				
Crack sealing, concrete paving	L.F.	1.19	0.29	1.48
Rubberized asphalt	S.Y.	2.84	3.98	6.82
Asphalt slurry seal	"	7.02	2.57	9.59
02740.20 Asphalt Surfaces				
Asphalt wearing surface, flexible pavement				
1" thick	S.Y.	4.30	2.38	6.68
1-1/2" thick	"	6.49	2.86	9.35
Binder course				
1-1/2" thick	S.Y.	6.14	2.65	8.79
2" thick	"	8.17	3.25	11.42
Bituminous sidewalk, no base				
2" thick	S.Y.	9.37	2.82	12.19
3" thick	"	14.00	3.00	17.00

Rigid Pavement

	UNIT	MAT.	INST.	TOTAL
02750.10 Concrete Paving				
Concrete paving, reinforced, 5000 psi concrete				
6" thick	S.Y.	23.75	22.25	46.00
8" thick	"	31.75	25.50	57.25
02760.10 Pavement Markings				
Pavement line marking, paint				
4" wide	L.F.	0.25	0.10	0.35
Directional arrows, reflective preformed tape	EA.	140	43.75	184
Messages, reflective preformed tape (per letter)	"	69.00	22.00	91.00
Handicap symbol, preformed tape	"	28.50	43.75	72.25
Parking stall painting	"	6.60	8.76	15.36

Site Improvements

	UNIT	MAT.	INST.	TOTAL
02810.40 Lawn Irrigation				
Residential system, complete				
Minimum	ACRE			15,940
Maximum	"			30,340

Site Improvements

02820.10 Chain Link Fence

Site Improvements	UNIT	MAT.	INST.	TOTAL
Chain link fence, 9 ga., galvanized, with posts 10' o.c.				
4' high	L.F.	6.71	3.13	9.84
5' high	"	8.97	3.98	12.95
6' high	"	10.25	5.48	15.73
Corner or gate post, 3" post				
4' high	EA.	78.00	14.50	92.50
5' high	"	86.00	16.25	102
6' high	"	96.00	19.00	115
Gate with gate posts, galvanized, 3' wide				
4' high	EA.	87.00	110	197
5' high	"	110	150	260
6' high	"	130	150	280
Fabric, galvanized chain link, 2" mesh, 9 ga.				
4' high	L.F.	3.65	1.46	5.11
5' high	"	4.46	1.75	6.21
6' high	"	6.24	2.19	8.43
Line post, no rail fitting, galvanized, 2-1/2" dia.				
4' high	EA.	25.50	12.50	38.00
5' high	"	27.75	13.75	41.50
6' high	"	30.25	14.50	44.75
Vinyl coated, 9 ga., with posts 10' o.c.				
4' high	L.F.	7.26	3.13	10.39
5' high	"	8.63	3.98	12.61
6' high	"	10.25	5.48	15.73
Gate, with posts, 3' wide				
4' high	EA.	110	110	220
5' high	"	120	150	270
6' high	"	140	150	290
Fabric, vinyl, chain link, 2" mesh, 9 ga.				
4' high	L.F.	3.65	1.46	5.11
5' high	"	4.46	1.75	6.21
6' high	"	6.24	2.19	8.43
Swing gates, galvanized, 4' high				
Single gate				
3' wide	EA.	200	110	310
4' wide	"	220	110	330
6' high				
Single gate				
3' wide	EA.	260	150	410
4' wide	"	280	150	430

02880.70 Recreational Courts

Site Improvements	UNIT	MAT.	INST.	TOTAL
Walls, galvanized steel				
8' high	L.F.	14.50	8.76	23.26
10' high	"	17.25	9.74	26.99
12' high	"	19.75	11.50	31.25
Vinyl coated				
8' high	L.F.	13.75	8.76	22.51
10' high	"	17.00	9.74	26.74
12' high	"	18.75	11.50	30.25
Gates, galvanized steel				
Single, 3' transom				
3'x7'	EA.	340	220	560

Site Improvements

02880.70 Recreational Courts (Cont.)

Site Improvements	UNIT	MAT.	INST.	TOTAL
4'x7'	EA.	360	250	610
5'x7'	"	490	290	780
6'x7'	"	530	350	880
Vinyl coated				
Single, 3' transom				
3'x7'	EA.	660	220	880
4'x7'	"	720	250	970
5'x7'	"	720	290	1,010
6'x7'	"	740	350	1,090

Planting

02910.10 Topsoil

Planting	UNIT	MAT.	INST.	TOTAL
Spread topsoil, with equipment				
Minimum	C.Y.		13.00	13.00
Maximum	"		16.25	16.25
By hand				
Minimum	C.Y.		43.75	43.75
Maximum	"		55.00	55.00
Area prep. seeding (grade, rake and clean)				
Square yard	S.Y.		0.35	0.35
Remove topsoil and stockpile on site				
4" deep	C.Y.		11.00	11.00
6" deep	"		10.00	10.00
Spreading topsoil from stock pile				
By loader	C.Y.		12.00	12.00
By hand	"		130	130
Top dress by hand	S.Y.		1.30	1.30
Place imported top soil				
By loader				
4" deep	S.Y.		1.30	1.30
6" deep	"		1.45	1.45
By hand				
4" deep	S.Y.		4.87	4.87
6" deep	"		5.48	5.48
Plant bed preparation, 18" deep				
With backhoe/loader	S.Y.		3.27	3.27
By hand	"		7.30	7.30

02920.10 Fertilizing

Planting	UNIT	MAT.	INST.	TOTAL
Fertilizing (23#/1000 sf)				
By square yard	S.Y.	0.03	0.14	0.17
Liming (70#/1000 sf)				
By square yard	S.Y.	0.03	0.19	0.22

Planting

Planting	UNIT	MAT.	INST.	TOTAL
02920.30 Seeding				
Seeding by hand, 10 lb per 100 s.y.				
By square yard	S.Y.	0.55	0.14	0.69
Reseed disturbed areas	S.F.	0.05	0.21	0.26
02935.10 Shrub & Tree Maintenance				
Moving shrubs on site				
12" ball	EA.		55.00	55.00
24" ball	"		73.00	73.00
3' high	"		43.75	43.75
4' high	"		48.75	48.75
5' high	"		55.00	55.00
Moving trees on site				
24" ball	EA.		120	120
48" ball	"		160	160
Trees				
3' high	EA.		48.00	48.00
6' high	"		53.00	53.00
8' high	"		60.00	60.00
10' high	"		80.00	80.00
Palm trees				
7' high	EA.		60.00	60.00
10' high	"		80.00	80.00
20' high	"		240	240
40' high	"		480	480
Guying trees				
4" dia.	EA.	8.29	22.00	30.29
8" dia.	"	8.29	27.50	35.79
02935.30 Weed Control				
Weed control, bromicil, 15 lb./acre, wettable powder	ACRE	290	220	510
Vegetation control, by application of plant killer	S.Y.	0.02	0.17	0.19
Weed killer, lawns and fields	"	0.24	0.08	0.32
02945.10 Prefabricated Planters				
Concrete precast, circular				
24" dia., 18" high	EA.	350	43.75	394
Fiberglass, circular				
36" dia., 27" high	EA.	590	22.00	612
02945.20 Landscape Accessories				
Steel edging, 3/16" x 4"	L.F.	0.64	0.54	1.18
Landscaping stepping stones, 15"x15", white	EA.	5.83	2.19	8.02
Wood chip mulch	C.Y.	40.75	29.25	70.00
2" thick	S.Y.	2.49	0.87	3.36
4" thick	"	4.70	1.25	5.95
6" thick	"	7.03	1.59	8.62
Gravel mulch, 3/4" stone	C.Y.	32.25	43.75	76.00
White marble chips, 1" deep	S.F.	0.64	0.43	1.07
Peat moss				
2" thick	S.Y.	3.46	0.97	4.43
4" thick	"	6.66	1.46	8.12
6" thick	"	10.25	1.82	12.07
Landscaping timbers, treated lumber				

Planting	UNIT	MAT.	INST.	TOTAL
02945.20 Landscape Accessories *(Cont.)*				
4" x 4"	L.F.	1.35	1.46	2.81
6" x 6"	"	2.69	1.56	4.25
8" x 8"	"	4.40	1.82	6.22

03 CONCRETE

Formwork	UNIT	MAT.	INST.	TOTAL
03110.05 Beam Formwork				
Beam forms, job built				
Beam bottoms				
1 use	S.F.	4.09	9.32	13.41
5 uses	"	1.39	7.99	9.38
Beam sides				
1 use	S.F.	2.92	6.21	9.13
5 uses	"	1.24	5.08	6.32
03110.15 Column Formwork				
Column, square forms, job built				
8" x 8" columns				
1 use	S.F.	3.47	11.25	14.72
5 uses	"	1.23	9.64	10.87
12" x 12" columns				
1 use	S.F.	3.16	10.25	13.41
5 uses	"	1.03	8.88	9.91
Round fiber forms, 1 use				
10" dia.	L.F.	4.53	11.25	15.78
12" dia.	"	5.57	11.50	17.07
03110.18 Curb Formwork				
Curb forms				
Straight, 6" high				
1 use	L.F.	2.06	5.59	7.65
5 uses	"	0.75	4.66	5.41
Curved, 6" high				
1 use	L.F.	2.24	6.99	9.23
5 uses	"	0.91	5.71	6.62
03110.25 Equipment Pad Formwork				
Equipment pad, job built				
1 use	S.F.	3.60	6.99	10.59
2 uses	"	2.16	6.58	8.74
3 uses	"	1.73	6.21	7.94
4 uses	"	1.35	5.89	7.24
5 uses	"	1.07	5.59	6.66
03110.35 Footing Formwork				
Wall footings, job built, continuous				
1 use	S.F.	1.70	5.59	7.29
5 uses	"	0.75	4.66	5.41
Column footings, spread				
1 use	S.F.	1.80	6.99	8.79
5 uses	"	0.73	5.59	6.32
03110.50 Grade Beam Formwork				
Grade beams, job built				
1 use	S.F.	2.68	5.59	8.27
5 uses	"	0.81	4.66	5.47

Formwork

Formwork	UNIT	MAT.	INST.	TOTAL
03110.55 Slab / Mat Formwork				
Mat foundations, job built				
1 use	S.F.	2.66	6.99	9.65
5 uses	"	0.77	5.59	6.36
Edge forms				
6" high				
1 use	L.F.	2.67	5.08	7.75
5 uses	"	0.77	4.30	5.07
5 uses	"	0.70	4.66	5.36
Formwork for openings				
1 use	S.F.	3.60	11.25	14.85
5 uses	"	1.12	7.99	9.11
03110.60 Stair Formwork				
Stairway forms, job built				
1 use	S.F.	4.16	11.25	15.41
5 uses	"	1.39	7.99	9.38
Stairs, elevated				
1 use	S.F.	5.02	11.25	16.27
5 uses	"	1.66	6.99	8.65
03110.65 Wall Formwork				
Wall forms, exterior, job built				
Up to 8' high wall				
1 use	S.F.	2.78	5.59	8.37
5 uses	"	1.02	4.66	5.68
Over 8' high wall				
5 uses	S.F.	1.25	5.59	6.84
Radial wall forms				
1 use	S.F.	2.99	8.60	11.59
5 uses	"	1.35	6.58	7.93
Retaining wall forms				
1 use	S.F.	2.58	6.21	8.79
5 uses	"	0.88	5.08	5.96
Radial retaining wall forms				
1 use	S.F.	2.73	9.32	12.05
5 uses	"	1.17	6.99	8.16
Column pier and pilaster				
1 use	S.F.	3.05	11.25	14.30
5 uses	"	1.37	7.99	9.36
Interior wall forms				
Up to 8' high				
1 use	S.F.	2.78	5.08	7.86
5 uses	"	0.99	4.30	5.29
Radial wall forms				
1 use	S.F.	2.99	7.46	10.45
5 uses	"	1.35	5.89	7.24
Curved wall forms, 24" sections				
1 use	S.F.	2.86	11.25	14.11
5 uses	"	1.28	7.99	9.27
PVC form liner, per side, smooth finish				
1 use	S.F.	7.15	4.66	11.81
5 uses	"	2.04	3.73	5.77

Formwork

03110.90 Miscellaneous Formwork

Formwork	UNIT	MAT.	INST.	TOTAL
Keyway forms (5 uses)				
2 x 4	L.F.	0.24	2.79	3.03
2 x 6	"	0.35	3.10	3.45
Bulkheads				
Walls, with keyways				
2 piece	L.F.	3.99	5.08	9.07
3 piece	"	5.04	5.59	10.63
Ground slab, with keyway				
2 piece	L.F.	4.73	3.99	8.72
3 piece	"	5.77	4.30	10.07
Chamfer strips				
Wood				
1/2" wide	L.F.	0.23	1.24	1.47
3/4" wide	"	0.29	1.24	1.53
1" wide	"	0.39	1.24	1.63
PVC				
1/2" wide	L.F.	1.01	1.24	2.25
3/4" wide	"	1.10	1.24	2.34
1" wide	"	1.59	1.24	2.83
Radius				
1"	L.F.	1.18	1.33	2.51
1-1/2"	"	2.14	1.33	3.47
Reglets				
Galvanized steel, 24 ga.	L.F.	1.54	2.23	3.77
Metal formwork				
Straight edge forms				
4" high	L.F.	18.75	3.49	22.24
6" high	"	20.75	3.73	24.48
8" high	"	28.25	3.99	32.24
Curb form, S-shape				
12" x				
1'-6"	L.F.	41.50	7.99	49.49
2'	"	45.50	7.46	52.96

Reinforcement

03210.05 Beam Reinforcing

Reinforcement	UNIT	MAT.	INST.	TOTAL
Beam-girders				
#3 - #4	TON	1,340	1,430	2,770
#5 - #6	"	1,170	1,140	2,310
Galvanized				
#3 - #4	TON	2,270	1,430	3,700
#5 - #6	"	2,150	1,140	3,290
Bond Beams				
#3 - #4	TON	1,340	1,910	3,250
#5 - #6	"	1,170	1,430	2,600

Reinforcement

Reinforcement	UNIT	MAT.	INST.	TOTAL
03210.05 Beam Reinforcing (Cont.)				
Galvanized				
#3 - #4	TON	2,170	1,910	4,080
#5 - #6	"	2,150	1,430	3,580
03210.15 Column Reinforcing				
Columns				
#3 - #4	TON	1,340	1,630	2,970
#5 - #6	"	1,170	1,270	2,440
Galvanized				
#3 - #4	TON	2,270	1,630	3,900
#5 - #6	"	2,150	1,270	3,420
03210.25 Equip. Pad Reinforcing				
Equipment pad				
#3 - #4	TON	1,340	1,140	2,480
#5 - #6	"	1,170	1,040	2,210
03210.35 Footing Reinforcing				
Footings				
#3 - #4	TON	1,340	950	2,290
#5 - #6	"	1,170	820	1,990
#7 - #8	"	1,110	720	1,830
Straight dowels, 24" long				
3/4" dia. (#6)	EA.	3.72	5.72	9.44
5/8" dia. (#5)	"	3.22	4.76	7.98
1/2" dia. (#4)	"	2.43	4.08	6.51
03210.45 Foundation Reinforcing				
Foundations				
#3 - #4	TON	1,340	950	2,290
#5 - #6	"	1,170	820	1,990
Galvanized				
#3 - #4	TON	2,270	950	3,220
#5 - #6	"	2,150	820	2,970
03210.50 Grade Beam Reinforcing				
Grade beams				
#3 - #4	TON	1,340	880	2,220
#5 - #6	"	1,170	760	1,930
Galvanized				
#3 - #4	TON	2,270	880	3,150
#5 - #6	"	2,150	760	2,910
03210.55 Slab / Mat Reinforcing				
Bars, slabs				
#3 - #4	TON	1,340	950	2,290
#5 - #6	"	1,170	820	1,990
Galvanized				
#3 - #4	TON	2,270	950	3,220
#5 - #6	"	2,150	820	2,970
Wire mesh, slabs				
Galvanized				

Reinforcement

03210.55 Slab / Mat Reinforcing *(Cont.)*

	UNIT	MAT.	INST.	TOTAL
4x4				
W1.4xW1.4	S.F.	0.33	0.38	0.71
W4.0xW4.0	"	0.89	0.47	1.36
6x6				
W1.4xW1.4	S.F.	0.30	0.28	0.58
W4.0xW4.0	"	0.62	0.38	1.00

03210.60 Stair Reinforcing

	UNIT	MAT.	INST.	TOTAL
Stairs				
#3 - #4	TON	1,340	1,140	2,480
#5 - #6	"	1,170	950	2,120
Galvanized				
#3 - #4	TON	2,270	1,140	3,410
#5 - #6	"	2,150	950	3,100

03210.65 Wall Reinforcing

	UNIT	MAT.	INST.	TOTAL
Walls				
#3 - #4	TON	1,340	820	2,160
#5 - #6	"	1,170	720	1,890
Galvanized				
#3 - #4	TON	2,270	820	3,090
#5 - #6	"	2,150	720	2,870
Masonry wall (horizontal)				
#3 - #4	TON	1,340	2,290	3,630
#5 - #6	"	1,170	1,910	3,080
Galvanized				
#3 - #4	TON	2,270	2,290	4,560
#5 - #6	"	2,150	1,910	4,060
Masonry wall (vertical)				
#3 - #4	TON	1,340	2,860	4,200
#5 - #6	"	1,170	2,290	3,460
Galvanized				
#3 - #4	TON	2,270	2,860	5,130
#5 - #6	"	2,150	2,290	4,440

Accessories

03250.40 Concrete Accessories

	UNIT	MAT.	INST.	TOTAL
Expansion joint, poured				
Asphalt				
1/2" x 1"	L.F.	0.74	0.87	1.61
1" x 2"	"	2.33	0.95	3.28
Vapor barrier				
4 mil polyethylene	S.F.	0.05	0.14	0.19
6 mil polyethylene	"	0.07	0.14	0.21
Gravel porous fill, under floor slabs, 3/4" stone	C.Y.	20.50	73.00	93.50

03 CONCRETE

Cast-in-place Concrete

03300.10 Concrete Admixtures

Item	UNIT	MAT.	INST.	TOTAL
Concrete admixtures				
Water reducing admixture	GAL			10.75
Set retarder	"			23.00
Air entraining agent	"			10.00

03350.10 Concrete Finishes

Item	UNIT	MAT.	INST.	TOTAL
Floor finishes				
Broom	S.F.		0.62	0.62
Screed	"		0.54	0.54
Darby	"		0.54	0.54
Steel float	"		0.73	0.73
Break ties and patch holes	"		0.87	0.87
Carborundum				
Dry rub	S.F.		1.46	1.46
Wet rub	"		2.19	2.19

03360.10 Pneumatic Concrete

Item	UNIT	MAT.	INST.	TOTAL
Pneumatic applied concrete (gunite)				
2" thick	S.F.	5.46	3.00	8.46
3" thick	"	6.71	4.00	10.71
4" thick	"	8.18	4.80	12.98
Finish surface				
Minimum	S.F.		2.79	2.79
Maximum	"		5.59	5.59

03370.10 Curing Concrete

Item	UNIT	MAT.	INST.	TOTAL
Sprayed membrane				
Slabs	S.F.	0.05	0.08	0.13

Placing Concrete

03380.05 Beam Concrete

Item	UNIT	MAT.	INST.	TOTAL
Beams and girders				
2500# or 3000# concrete				
By pump	C.Y.	110	80.00	190
By hand buggy	"	110	43.75	154
3500# or 4000# concrete				
By pump	C.Y.	120	80.00	200
By hand buggy	"	120	43.75	164

03380.15 Column Concrete

Item	UNIT	MAT.	INST.	TOTAL
Columns				
2500# or 3000# concrete				
By pump	C.Y.	110	73.00	183
3500# or 4000# concrete				
By pump	C.Y.	120	73.00	193

03 CONCRETE

Placing Concrete	UNIT	MAT.	INST.	TOTAL
03380.25 Equipment Pad Concrete				
Equipment pad				
2500# or 3000# concrete				
By chute	C.Y.	110	14.50	125
By pump	"	110	63.00	173
3500# or 4000# concrete				
By chute	C.Y.	110	14.50	125
By pump	"	110	63.00	173
03380.35 Footing Concrete				
Continuous footing				
2500# or 3000# concrete				
By chute	C.Y.	110	14.50	125
By pump	"	110	55.00	165
Spread footing				
2500# or 3000# concrete				
By chute	C.Y.	110	14.50	125
By pump	"	110	58.00	168
03380.50 Grade Beam Concrete				
Grade beam				
2500# or 3000# concrete				
By chute	C.Y.	110	14.50	125
By pump	"	110	55.00	165
By hand buggy	"	110	43.75	154
03380.53 Pile Cap Concrete				
Pile cap				
2500# or 3000 concrete				
By chute	C.Y.	110	14.50	125
By pump	"	110	63.00	173
By hand buggy	"	110	43.75	154
3500# or 4000# concrete				
By chute	C.Y.	120	14.50	135
By pump	"	120	63.00	183
By hand buggy	"	120	43.75	164
03380.55 Slab / Mat Concrete				
Slab on grade				
2500# or 3000# concrete				
By chute	C.Y.	110	11.00	121
By pump	"	110	31.25	141
By hand buggy	"	110	29.25	139
03380.58 Sidewalks				
Walks, cast in place with wire mesh, base not incl.				
4" thick	S.F.	1.59	1.46	3.05
5" thick	"	2.15	1.75	3.90
6" thick	"	2.65	2.19	4.84

Placing Concrete

	UNIT	MAT.	INST.	TOTAL
03380.60 Stair Concrete				
Stairs				
2500# or 3000# concrete				
By chute	C.Y.	110	14.50	125
By pump	"	110	63.00	173
By hand buggy	"	110	43.75	154
3500# or 4000# concrete				
By chute	C.Y.	120	14.50	135
By pump	"	120	63.00	183
By hand buggy	"	120	43.75	164
03380.65 Wall Concrete				
Walls				
2500# or 3000# concrete				
To 4'				
By chute	C.Y.	110	12.50	123
By pump	"	110	67.00	177
To 8'				
By pump	C.Y.	110	73.00	183
Filled block (CMU)				
3000# concrete, by pump				
4" wide	S.F.	0.42	3.12	3.54
6" wide	"	0.95	3.65	4.60
8" wide	"	1.49	4.38	5.87
Pilasters, 3000# concrete	C.F.	5.84	88.00	93.84
Wall cavity, 2" thick, 3000# concrete	S.F.	1.08	2.92	4.00
03400.90 Precast Specialties				
Precast concrete, coping, 4' to 8' long				
12" wide	L.F.	8.82	6.00	14.82
10" wide	"	7.85	6.86	14.71
Splash block, 30"x12"x4"	EA.	13.50	40.00	53.50
Stair unit, per riser	"	86.00	40.00	126
Sun screen and trellis, 8' long, 12" high				
4" thick blades	EA.	94.00	30.00	124

Concrete Restoration

	UNIT	MAT.	INST.	TOTAL
03730.10 Concrete Repair				
Epoxy grout floor patch, 1/4" thick	S.F.	6.68	4.38	11.06
Grout, epoxy, 2 component system	C.F.			320
Grout crack seal, 2 component	"	750	43.75	794
Concrete, epoxy modified				
Sand mix	C.F.	120	17.50	138
Gravel mix	"	86.00	16.25	102
Concrete repair				
Edge repair				

Concrete Restoration

03730.10 Concrete Repair (Cont.)

Concrete Restoration	UNIT	MAT.	INST.	TOTAL
2" spall	L.F.	1.81	11.00	12.81
3" spall	"	1.81	11.50	13.31
8" spall	"	2.12	13.00	15.12
Crack repair, 1/8" crack	"	3.57	4.38	7.95
Reinforcing steel repair				
1 bar, 4 ft				
#4 bar	L.F.	0.56	7.15	7.71
#5 bar	"	0.75	7.15	7.90
#6 bar	"	0.92	7.62	8.54

04 MASONRY

Mortar And Grout

Mortar And Grout	UNIT	MAT.	INST.	TOTAL
04100.10 Masonry Grout				
Grout, non shrink, non-metallic, trowelable	C.F.	5.35	1.60	6.95
Grout door frame, hollow metal				
Single	EA.	13.50	60.00	73.50
Double	"	18.75	63.00	81.75
Grout-filled concrete block (CMU)				
4" wide	S.F.	0.35	2.00	2.35
6" wide	"	0.93	2.18	3.11
8" wide	"	1.38	2.40	3.78
12" wide	"	2.26	2.52	4.78
Grout-filled individual CMU cells				
4" wide	L.F.	0.29	1.20	1.49
6" wide	"	0.40	1.20	1.60
8" wide	"	0.53	1.20	1.73
10" wide	"	0.66	1.37	2.03
12" wide	"	0.80	1.37	2.17
Bond beams or lintels, 8" deep				
6" thick	L.F.	0.80	1.99	2.79
8" thick	"	1.07	2.19	3.26
10" thick	"	1.34	2.43	3.77
12" thick	"	1.60	2.73	4.33
Cavity walls				
2" thick	S.F.	0.89	2.92	3.81
3" thick	"	1.34	2.92	4.26
4" thick	"	1.78	3.12	4.90
6" thick	"	2.67	3.65	6.32
04150.10 Masonry Accessories				
Foundation vents	EA.	32.25	21.25	53.50
Bar reinforcing				
Horizontal				
#3 - #4	Lb.	0.63	2.13	2.76
#5 - #6	"	0.63	1.78	2.41
Vertical				
#3 - #4	Lb.	0.63	2.67	3.30
#5 - #6	"	0.63	2.13	2.76
Horizontal joint reinforcing				
Truss type				
4" wide, 6" wall	L.F.	0.20	0.21	0.41
6" wide, 8" wall	"	0.20	0.22	0.42
8" wide, 10" wall	"	0.25	0.23	0.48
10" wide, 12" wall	"	0.25	0.24	0.49
12" wide, 14" wall	"	0.31	0.25	0.56
Ladder type				
4" wide, 6" wall	L.F.	0.15	0.21	0.36
6" wide, 8" wall	"	0.17	0.22	0.39
8" wide, 10" wall	"	0.18	0.23	0.41
10" wide, 12" wall	"	0.21	0.23	0.44
Rectangular wall ties				
3/16" dia., galvanized				
2" x 6"	EA.	0.38	0.89	1.27
2" x 8"	"	0.41	0.89	1.30
2" x 10"	"	0.47	0.89	1.36
2" x 12"	"	0.53	0.89	1.42

Mortar And Grout

04150.10 Masonry Accessories (Cont.)	UNIT	MAT.	INST.	TOTAL
4" x 6"	EA.	0.43	1.06	1.49
4" x 8"	"	0.49	1.06	1.55
4" x 10"	"	0.64	1.06	1.70
4" x 12"	"	0.73	1.06	1.79
1/4" dia., galvanized				
2" x 6"	EA.	0.71	0.89	1.60
2" x 8"	"	0.80	0.89	1.69
2" x 10"	"	0.90	0.89	1.79
2" x 12"	"	1.04	0.89	1.93
4" x 6"	"	0.82	1.06	1.88
4" x 8"	"	0.90	1.06	1.96
4" x 10"	"	1.04	1.06	2.10
4" x 12"	"	1.09	1.06	2.15
"Z" type wall ties, galvanized				
6" long				
1/8" dia.	EA.	0.33	0.89	1.22
3/16" dia.	"	0.36	0.89	1.25
1/4" dia.	"	0.38	0.89	1.27
8" long				
1/8" dia.	EA.	0.36	0.89	1.25
3/16" dia.	"	0.38	0.89	1.27
1/4" dia.	"	0.41	0.89	1.30
10" long				
1/8" dia.	EA.	0.38	0.89	1.27
3/16" dia.	"	0.43	0.89	1.32
1/4" dia.	"	0.49	0.89	1.38
Dovetail anchor slots				
Galvanized steel, filled				
24 ga.	L.F.	0.88	1.33	2.21
20 ga.	"	1.84	1.33	3.17
16 oz. copper, foam filled	"	2.17	1.33	3.50
Dovetail anchors				
16 ga.				
3-1/2" long	EA.	0.27	0.89	1.16
5-1/2" long	"	0.32	0.89	1.21
12 ga.				
3-1/2" long	EA.	0.35	0.89	1.24
5-1/2" long	"	0.58	0.89	1.47
Dovetail, triangular galvanized ties, 12 ga.				
3" x 3"	EA.	0.60	0.89	1.49
5" x 5"	"	0.64	0.89	1.53
7" x 7"	"	0.72	0.89	1.61
7" x 9"	"	0.77	0.89	1.66
Brick anchors				
Corrugated, 3-1/2" long				
16 ga.	EA.	0.25	0.89	1.14
12 ga.	"	0.44	0.89	1.33
Non-corrugated, 3-1/2" long				
16 ga.	EA.	0.35	0.89	1.24
12 ga.	"	0.63	0.89	1.52
Cavity wall anchors, corrugated, galvanized				
5" long				
16 ga.	EA.	0.78	0.89	1.67

04 MASONRY

Mortar And Grout	UNIT	MAT.	INST.	TOTAL
04150.10 Masonry Accessories *(Cont.)*				
12 ga.	EA.	1.16	0.89	2.05
7" long				
28 ga.	EA.	0.86	0.89	1.75
24 ga.	"	1.09	0.89	1.98
22 ga.	"	1.11	0.89	2.00
16 ga.	"	1.25	0.89	2.14
Mesh ties, 16 ga., 3" wide				
8" long	EA.	1.05	0.89	1.94
12" long	"	1.16	0.89	2.05
20" long	"	1.61	0.89	2.50
24" long	"	1.77	0.89	2.66
04150.20 Masonry Control Joints				
Control joint, cross shaped PVC	L.F.	2.17	1.33	3.50
Closed cell joint filler				
1/2"	L.F.	0.38	1.33	1.71
3/4"	"	0.77	1.33	2.10
Rubber, for				
4" wall	L.F.	2.50	1.33	3.83
6" wall	"	3.10	1.40	4.50
8" wall	"	3.74	1.48	5.22
PVC, for				
4" wall	L.F.	1.30	1.33	2.63
6" wall	"	2.20	1.40	3.60
8" wall	"	3.33	1.48	4.81
04150.50 Masonry Flashing				
Through-wall flashing				
5 oz. coated copper	S.F.	3.69	4.45	8.14
0.030" elastomeric	"	1.21	3.56	4.77

Unit Masonry	UNIT	MAT.	INST.	TOTAL
04210.10 Brick Masonry				
Standard size brick, running bond				
Face brick, red (6.4/sf)				
Veneer	S.F.	5.39	8.90	14.29
Cavity wall	"	5.39	7.62	13.01
9" solid wall	"	10.75	15.25	26.00
Common brick (6.4/sf)				
Select common for veneers	S.F.	3.52	8.90	12.42
Back-up				
4" thick	S.F.	3.16	6.67	9.83
8" thick	"	6.33	10.75	17.08
Firewall				
12" thick	S.F.	10.25	17.75	28.00

Unit Masonry

Unit Masonry	UNIT	MAT.	INST.	TOTAL
04210.10 Brick Masonry *(Cont.)*				
16" thick	S.F.	13.75	24.25	38.00
Glazed brick (7.4/sf)				
Veneer	S.F.	14.75	9.70	24.45
Buff or gray face brick (6.4/sf)				
Veneer	S.F.	6.27	8.90	15.17
Cavity wall	"	6.27	7.62	13.89
Jumbo or oversize brick (3/sf)				
4" veneer	S.F.	4.12	5.34	9.46
4" back-up	"	4.12	4.45	8.57
8" back-up	"	4.78	7.62	12.40
12" firewall	"	6.43	13.25	19.68
16" firewall	"	9.07	17.75	26.82
Norman brick, red face, (4.5/sf)				
4" veneer	S.F.	5.94	6.67	12.61
Cavity wall	"	5.94	5.93	11.87
Chimney, standard brick, including flue				
16" x 16"	L.F.	30.50	53.00	83.50
16" x 20"	"	51.00	53.00	104
16" x 24"	"	55.00	53.00	108
20" x 20"	"	42.75	67.00	110
20" x 24"	"	58.00	67.00	125
20" x 32"	"	65.00	76.00	141
Window sill, face brick on edge	"	2.97	13.25	16.22
04210.60 Pavers, Masonry				
Brick walk laid on sand, sand joints				
Laid flat, (4.5 per sf)	S.F.	3.78	5.93	9.71
Laid on edge, (7.2 per sf)	"	6.05	8.90	14.95
Precast concrete patio blocks				
2" thick				
Natural	S.F.	2.68	1.78	4.46
Colors	"	3.74	1.78	5.52
Exposed aggregates, local aggregate				
Natural	S.F.	8.08	1.78	9.86
Colors	"	8.08	1.78	9.86
Granite or limestone aggregate	"	8.44	1.78	10.22
White tumblestone aggregate	"	6.05	1.78	7.83
Stone pavers, set in mortar				
Bluestone				
1" thick				
Irregular	S.F.	6.14	13.25	19.39
Snapped rectangular	"	9.35	10.75	20.10
1-1/2" thick, random rectangular	"	10.75	13.25	24.00
2" thick, random rectangular	"	12.75	15.25	28.00
Slate				
Natural cleft				
Irregular, 3/4" thick	S.F.	7.71	15.25	22.96
Random rectangular				
1-1/4" thick	S.F.	16.75	13.25	30.00
1-1/2" thick	"	18.75	14.75	33.50
Granite blocks				
3" thick, 3" to 6" wide				
4" to 12" long	S.F.	11.00	17.75	28.75

Unit Masonry	UNIT	MAT.	INST.	TOTAL
04210.60 Pavers, Masonry (Cont.)				
6" to 15" long	S.F.	7.15	15.25	22.40
Crushed stone, white marble, 3" thick	"	1.70	0.87	2.57
04220.10 Concrete Masonry Units				
Hollow, load bearing				
4"	S.F.	1.55	3.95	5.50
6"	"	2.27	4.10	6.37
8"	"	2.60	4.45	7.05
Solid, load bearing				
4"	S.F.	2.43	3.95	6.38
6"	"	2.73	4.10	6.83
8"	"	3.72	4.45	8.17
Back-up block, 8" x 16"				
2"	S.F.	1.63	3.05	4.68
4"	"	1.69	3.14	4.83
6"	"	2.48	3.33	5.81
8"	"	2.85	3.56	6.41
Foundation wall, 8" x 16"				
6"	S.F.	2.48	3.81	6.29
8"	"	2.85	4.10	6.95
10"	"	3.93	4.45	8.38
12"	"	4.52	4.85	9.37
Solid				
6"	S.F.	2.99	4.10	7.09
8"	"	4.08	4.45	8.53
10"	"	4.36	4.85	9.21
12"	"	6.47	5.34	11.81
Filled cavities				
4"	S.F.	5.41	5.93	11.34
6"	"	6.24	6.28	12.52
8"	"	8.00	6.67	14.67
Gypsum unit masonry				
Partition blocks (12"x30")				
Solid				
2"	S.F.	1.16	2.13	3.29
Hollow				
3"	S.F.	1.17	2.13	3.30
4"	"	1.35	2.22	3.57
6"	"	1.44	2.42	3.86
Vertical reinforcing				
4' o.c., add 5% to labor				
2'8" o.c., add 15% to labor				
Interior partitions, add 10% to labor				
04220.90 Bond Beams & Lintels				
Bond beam, no grout or reinforcement				
8" x 16" x				
4" thick	L.F.	1.68	4.10	5.78
6" thick	"	2.57	4.27	6.84
8" thick	"	2.94	4.45	7.39
Beam lintel, no grout or reinforcement				
8" x 16" x				
Precast masonry lintel				

Unit Masonry

	UNIT	MAT.	INST.	TOTAL
04220.90 Bond Beams & Lintels *(Cont.)*				
6 lf, 8" high x				
4" thick	L.F.	6.17	8.90	15.07
6" thick	"	7.88	8.90	16.78
8" thick	"	8.91	9.70	18.61
Steel angles and plates				
Minimum	Lb.	1.09	0.76	1.85
Maximum	"	1.61	1.33	2.94
Various size angle lintels				
1/4" stock				
3" x 3"	L.F.	5.65	3.33	8.98
3" x 3-1/2"	"	6.23	3.33	9.56
3/8" stock				
3" x 4"	L.F.	9.82	3.33	13.15
3-1/2" x 4"	"	10.25	3.33	13.58
4" x 4"	"	11.25	3.33	14.58
5" x 3-1/2"	"	12.00	3.33	15.33
6" x 3-1/2"	"	13.50	3.33	16.83
1/2" stock				
6" x 4"	L.F.	15.00	3.33	18.33
04240.10 Clay Tile				
Hollow clay tile, for back-up, 12" x 12"				
Scored face				
Load bearing				
4" thick	S.F.	6.56	3.81	10.37
6" thick	"	7.65	3.95	11.60
8" thick	"	9.53	4.10	13.63
Non-load bearing				
3" thick	S.F.	12.75	3.68	16.43
4" thick	"	6.16	3.81	9.97
6" thick	"	7.12	3.95	11.07
8" thick	"	9.08	4.10	13.18
Clay tile floors				
4" thick	S.F.	6.16	2.96	9.12
6" thick	"	7.65	3.14	10.79
8" thick	"	9.53	3.33	12.86
Terra cotta				
Coping, 10" or 12" wide, 3" thick	L.F.	14.75	10.75	25.50
04270.10 Glass Block				
Glass block, 4" thick				
6" x 6"	S.F.	30.25	17.75	48.00
8" x 8"	"	19.00	13.25	32.25
12" x 12"	"	24.25	10.75	35.00
Replacement glass blocks, 4" x 8" x 8"				
Minimum	S.F.	21.00	53.00	74.00
Maximum	"	26.75	110	137

04 MASONRY

Unit Masonry

04295.10 Parging / Masonry Plaster

Unit Masonry	UNIT	MAT.	INST.	TOTAL
Parging				
1/2" thick	S.F.	0.29	3.56	3.85
3/4" thick	"	0.37	4.45	4.82
1" thick	"	0.50	5.34	5.84

Stone

04400.10 Stone

Stone	UNIT	MAT.	INST.	TOTAL
Rubble stone				
Walls set in mortar				
8" thick	S.F.	15.00	13.25	28.25
12" thick	"	18.25	21.25	39.50
18" thick	"	24.00	26.75	50.75
24" thick	"	30.25	35.50	65.75
Dry set wall				
8" thick	S.F.	17.00	8.90	25.90
12" thick	"	19.00	13.25	32.25
18" thick	"	26.50	17.75	44.25
24" thick	"	32.25	21.25	53.50
Cut stone				
Imported marble				
Facing panels				
3/4" thick	S.F.	39.25	21.25	60.50
1-1/2" thick	"	61.00	24.25	85.25
Flooring, travertine, minimum	"	19.00	8.21	27.21
Average	"	25.25	10.75	36.00
Maximum	"	46.25	11.75	58.00
Domestic marble				
Stairs				
12" treads	L.F.	28.50	26.75	55.25
6" risers	"	21.00	17.75	38.75
Thresholds, 7/8" thick, 3' long, 4" to 6" wide				
Plain	EA.	27.00	44.50	71.50
Beveled	"	30.00	44.50	74.50
Window sill				
6" wide, 2" thick	L.F.	15.25	21.25	36.50
Stools				
5" wide, 7/8" thick	L.F.	22.25	21.25	43.50
Limestone panels up to 12' x 5', smooth finish				
2" thick	S.F.	29.75	9.61	39.36
Miscellaneous limestone items				
Steps, 14" wide, 6" deep	L.F.	63.00	35.50	98.50
Coping, smooth finish	C.F.	88.00	17.75	106
Sills, lintels, jambs, smooth finish	"	88.00	21.25	109
Granite veneer facing panels, polished				
7/8" thick				

Stone

04400.10 Stone (Cont.)	UNIT	MAT.	INST.	TOTAL
Black	S.F.	51.00	21.25	72.25
Gray	"	40.00	21.25	61.25
Slate, panels				
1" thick	S.F.	25.00	21.25	46.25
Sills or stools				
1" thick				
6" wide	L.F.	11.75	21.25	33.00

Masonry Restoration

04520.10 Restoration And Cleaning	UNIT	MAT.	INST.	TOTAL
Masonry cleaning				
Washing brick				
Smooth surface	S.F.	0.21	0.89	1.10
Rough surface	"	0.29	1.18	1.47
Steam clean masonry				
Smooth face				
Minimum	S.F.		0.71	0.71
Maximum	"		1.04	1.04
Rough face				
Minimum	S.F.		0.95	0.95
Maximum	"		1.43	1.43
Sandblast masonry				
Minimum	S.F.	0.48	1.14	1.62
Maximum	"	0.72	1.91	2.63
Pointing masonry				
Brick	S.F.	1.19	2.13	3.32
Concrete block	"	0.53	1.52	2.05
Cut and repoint				
Brick				
Minimum	S.F.	0.37	2.67	3.04
Maximum	"	0.63	5.34	5.97
Stone work	L.F.	0.98	4.10	5.08
Cut and recaulk				
Oil base caulks	L.F.	1.32	3.56	4.88
Butyl caulks	"	1.15	3.56	4.71
Polysulfides and acrylics	"	2.24	3.56	5.80
Silicones	"	2.61	3.56	6.17
Cement and sand grout on walls, to 1/8" thick				
Minimum	S.F.	0.64	2.13	2.77
Maximum	"	1.63	2.67	4.30
Brick removal and replacement				
Minimum	EA.	0.68	6.67	7.35
Average	"	0.89	8.90	9.79
Maximum	"	1.80	26.75	28.55

Masonry Restoration

04550.10 Refractories

Masonry Restoration	UNIT	MAT.	INST.	TOTAL
Flue liners				
Rectangular				
8" x 12"	L.F.	8.30	8.90	17.20
12" x 12"	"	10.25	9.70	19.95
12" x 18"	"	18.25	10.75	29.00
Round				
18" dia.	L.F.	37.50	12.75	50.25
24" dia.	"	74.00	15.25	89.25

05 METALS

Metal Fastening

	UNIT	MAT.	INST.	TOTAL
05050.10 Structural Welding				
Welding				
Single pass				
1/8"	L.F.	0.33	3.09	3.42
3/16"	"	0.55	4.12	4.67
1/4"	"	0.77	5.15	5.92
05050.90 Metal Anchors				
Anchor bolts				
3/8" x				
8" long	EA.			1.00
10" long	"			1.09
12" long	"			1.20
1/2" x				
8" long	EA.			1.50
10" long	"			1.60
12" long	"			1.75
5/8" x				
8" long	EA.			1.40
10" long	"			1.55
12" long	"			1.65
3/4" x				
8" long	EA.			2.00
12" long	"			2.25
Expansion shield				
1/4"	EA.			0.62
3/8"	"			1.04
1/2"	"			2.04
5/8"	"			2.97
3/4"	"			3.63
Non-drilling anchor				
1/4"	EA.			0.64
3/8"	"			0.80
1/2"	"			1.23
5/8"	"			2.02
3/4"	"			3.46
Self-drilling anchor				
1/4"	EA.			1.62
5/16"	"			2.03
3/8"	"			2.44
1/2"	"			3.25
Add 25% for galvanized anchor bolts				
Channel door frame, with anchors	Lb.	1.78	0.68	2.46
Corner guard angle, with anchors	"	1.59	1.03	2.62
05050.95 Metal Lintels				
Lintels, steel				
Plain	Lb.	1.33	1.54	2.87
Galvanized	"	2.00	1.54	3.54

05 METALS

Metal Fastening

05120.10 Structural Steel	UNIT	MAT.	INST.	TOTAL
Beams and girders, A-36				
Welded	TON	2,870	720	3,590
Bolted	"	2,800	650	3,450
Columns				
Pipe				
6" dia.	Lb.	1.61	0.71	2.32

Cold Formed Framing

05410.10 Metal Framing	UNIT	MAT.	INST.	TOTAL
Furring channel, galvanized				
Beams and columns, 3/4"				
12" o.c.	S.F.	0.43	6.18	6.61
16" o.c.	"	0.33	5.62	5.95
Walls, 3/4"				
12" o.c.	S.F.	0.43	3.09	3.52
16" o.c.	"	0.33	2.57	2.90
24" o.c.	"	0.24	2.06	2.30
1-1/2"				
12" o.c.	S.F.	0.72	3.09	3.81
16" o.c.	"	0.54	2.57	3.11
24" o.c.	"	0.37	2.06	2.43
Stud, load bearing				
16" o.c.				
16 ga.				
2-1/2"	S.F.	1.33	2.74	4.07
3-5/8"	"	1.57	2.74	4.31
4"	"	1.63	2.74	4.37
6"	"	2.05	3.09	5.14
18 ga.				
2-1/2"	S.F.	1.08	2.74	3.82
3-5/8"	"	1.33	2.74	4.07
4"	"	1.39	2.74	4.13
6"	"	1.75	3.09	4.84
8"	"	2.11	3.09	5.20
20 ga.				
2-1/2"	S.F.	0.60	2.74	3.34
3-5/8"	"	0.72	2.74	3.46
4"	"	0.78	2.74	3.52
6"	"	0.96	3.09	4.05
8"	"	1.14	3.09	4.23
24" o.c.				
16 ga.				
2-1/2"	S.F.	0.90	2.37	3.27
3-5/8"	"	1.08	2.37	3.45
4"	"	1.14	2.37	3.51

Cold Formed Framing

05410.10 Metal Framing (Cont.)

	UNIT	MAT.	INST.	TOTAL
6"	S.F.	1.39	2.57	3.96
8"	"	1.75	2.57	4.32
18 ga.				
2-1/2"	S.F.	0.72	2.37	3.09
3-5/8"	"	0.84	2.37	3.21
4"	"	0.90	2.37	3.27
6"	"	1.14	2.57	3.71
8"	"	1.39	2.57	3.96
20 ga.				
2-1/2"	S.F.	0.44	2.37	2.81
3-5/8"	"	0.49	2.37	2.86
4"	"	0.55	2.37	2.92
6"	"	0.71	2.57	3.28
8"	"	0.88	2.57	3.45

Metal Fabrications

05520.10 Railings

	UNIT	MAT.	INST.	TOTAL
Railing, pipe				
1-1/4" diameter, welded steel				
2-rail				
Primed	L.F.	31.75	12.25	44.00
Galvanized	"	40.75	12.25	53.00
3-rail				
Primed	L.F.	40.75	15.50	56.25
Galvanized	"	53.00	15.50	68.50
Wall mounted, single rail, welded steel				
Primed	L.F.	21.25	9.51	30.76
Galvanized	"	27.50	9.51	37.01
1-1/2" diameter, welded steel				
2-rail				
Primed	L.F.	34.50	12.25	46.75
Galvanized	"	44.75	12.25	57.00
3-rail				
Primed	L.F.	43.25	15.50	58.75
Galvanized	"	56.00	15.50	71.50
Wall mounted, single rail, welded steel				
Primed	L.F.	21.75	9.51	31.26
Galvanized	"	28.50	9.51	38.01
2" diameter, welded steel				
2-rail				
Primed	L.F.	41.25	13.75	55.00
Galvanized	"	54.00	13.75	67.75
3-rail				
Primed	L.F.	52.00	17.75	69.75
Galvanized	"	68.00	17.75	85.75

Metal Fabrications

05520.10 Railings (Cont.)

Metal Fabrications	UNIT	MAT.	INST.	TOTAL
Wall mounted, single rail, welded steel				
Primed	L.F.	23.75	10.25	34.00
Galvanized	"	30.75	10.25	41.00

Misc. Fabrications

05700.10 Ornamental Metal

Misc. Fabrications	UNIT	MAT.	INST.	TOTAL
Railings, square bars, 6" o.c., shaped top rails				
Steel	L.F.	92.00	31.00	123
Aluminum	"	110	31.00	141
Bronze	"	190	41.25	231
Stainless steel	"	190	41.25	231
Laminated metal or wood handrails				
2-1/2" round or oval shape	L.F.	280	31.00	311
Aluminum louvers				
Residential use, fixed type, with screen				
8" x 8"	EA.	17.75	31.00	48.75
12" x 12"	"	19.50	31.00	50.50
12" x 18"	"	23.50	31.00	54.50
14" x 24"	"	33.75	31.00	64.75
18" x 24"	"	37.75	31.00	68.75
30" x 24"	"	52.00	34.25	86.25

06 WOOD AND PLASTICS

Fasteners And Adhesives	UNIT	MAT.	INST.	TOTAL
06050.10 Accessories				
Column/post base, cast aluminum				
4" x 4"	EA.	15.75	14.00	29.75
6" x 6"	"	21.75	14.00	35.75
Bridging, metal, per pair				
12" o.c.	EA.	2.13	5.59	7.72
16" o.c.	"	1.96	5.08	7.04
Anchors				
Bolts, threaded two ends, with nuts and washers				
1/2" dia.				
4" long	EA.	2.48	3.49	5.97
7-1/2" long	"	2.89	3.49	6.38
3/4" dia.				
7-1/2" long	EA.	5.48	3.49	8.97
15" long	"	8.26	3.49	11.75
Framing anchors				
10 gauge	EA.	0.97	4.66	5.63
Bolts, carriage				
1/4 x 4	EA.	0.64	5.59	6.23
5/16 x 6	"	1.46	5.89	7.35
3/8 x 6	"	2.95	5.89	8.84
1/2 x 6	"	4.12	5.89	10.01
Joist and beam hangers				
18 ga.				
2 x 4	EA.	1.16	5.59	6.75
2 x 6	"	1.39	5.59	6.98
2 x 8	"	1.62	5.59	7.21
2 x 10	"	1.74	6.21	7.95
2 x 12	"	2.26	6.99	9.25
16 ga.				
3 x 6	EA.	4.07	6.21	10.28
3 x 8	"	4.93	6.21	11.14
3 x 10	"	5.57	6.58	12.15
3 x 12	"	6.28	7.46	13.74
3 x 14	"	6.79	7.99	14.78
4 x 6	"	6.97	6.21	13.18
4 x 8	"	8.14	6.21	14.35
4 x 10	"	9.30	6.58	15.88
4 x 12	"	12.00	7.46	19.46
4 x 14	"	12.50	7.99	20.49
Rafter anchors, 18 ga., 1-1/2" wide				
5-1/4" long	EA.	0.88	4.66	5.54
10-3/4" long	"	1.29	4.66	5.95
Shear plates				
2-5/8" dia.	EA.	3.11	4.30	7.41
4" dia.	"	6.46	4.66	11.12
Sill anchors				
Embedded in concrete	EA.	2.28	5.59	7.87
Split rings				
2-1/2" dia.	EA.	1.88	6.21	8.09
4" dia.	"	3.46	6.99	10.45
Strap ties, 14 ga., 1-3/8" wide				
12" long	EA.	2.35	4.66	7.01

Fasteners And Adhesives

06050.10 Accessories (Cont.)

Description	UNIT	MAT.	INST.	TOTAL
18" long	EA.	2.53	5.08	7.61
24" long	"	3.76	5.59	9.35
36" long	"	5.17	6.21	11.38
Toothed rings				
2-5/8" dia.	EA.	2.17	9.32	11.49
4" dia.	"	2.53	11.25	13.78

Rough Carpentry

06110.10 Blocking

Description	UNIT	MAT.	INST.	TOTAL
Steel construction				
Walls				
2x4	L.F.	0.48	3.73	4.21
2x6	"	0.73	4.30	5.03
2x8	"	0.96	4.66	5.62
2x10	"	1.28	5.08	6.36
2x12	"	1.66	5.59	7.25
Ceilings				
2x4	L.F.	0.48	4.30	4.78
2x6	"	0.73	5.08	5.81
2x8	"	0.96	5.59	6.55
2x10	"	1.28	6.21	7.49
2x12	"	1.66	6.99	8.65
Wood construction				
Walls				
2x4	L.F.	0.48	3.10	3.58
2x6	"	0.73	3.49	4.22
2x8	"	0.96	3.73	4.69
2x10	"	1.28	3.99	5.27
2x12	"	1.66	4.30	5.96
Ceilings				
2x4	L.F.	0.48	3.49	3.97
2x6	"	0.73	3.99	4.72
2x8	"	0.96	4.30	5.26
2x10	"	1.28	4.66	5.94
2x12	"	1.66	5.08	6.74

06110.20 Ceiling Framing

Description	UNIT	MAT.	INST.	TOTAL
Ceiling joists				
12" o.c.				
2x4	S.F.	0.71	1.33	2.04
2x6	"	1.03	1.39	2.42
2x8	"	1.51	1.47	2.98
2x10	"	1.72	1.55	3.27
2x12	"	3.17	1.64	4.81
16" o.c.				

Rough Carpentry

06110.20 Ceiling Framing *(Cont.)*

	UNIT	MAT.	INST.	TOTAL
2x4	S.F.	0.58	1.07	1.65
2x6	"	0.86	1.11	1.97
2x8	"	1.23	1.16	2.39
2x10	"	1.38	1.21	2.59
2x12	"	2.59	1.27	3.86
24" o.c.				
2x4	S.F.	0.41	0.88	1.29
2x6	"	0.69	0.93	1.62
2x8	"	1.03	0.98	2.01
2x10	"	1.23	1.03	2.26
2x12	"	3.11	1.09	4.20
Headers and nailers				
2x4	L.F.	0.48	1.80	2.28
2x6	"	0.73	1.86	2.59
2x8	"	0.96	1.99	2.95
2x10	"	1.28	2.15	3.43
2x12	"	1.58	2.33	3.91
Sister joists for ceilings				
2x4	L.F.	0.48	3.99	4.47
2x6	"	0.73	4.66	5.39
2x8	"	0.96	5.59	6.55
2x10	"	1.28	6.99	8.27
2x12	"	1.58	9.32	10.90

06110.30 Floor Framing

	UNIT	MAT.	INST.	TOTAL
Floor joists				
12" o.c.				
2x6	S.F.	0.88	1.11	1.99
2x8	"	1.29	1.14	2.43
2x10	"	1.79	1.16	2.95
2x12	"	2.62	1.21	3.83
2x14	"	4.00	1.16	5.16
3x6	"	2.97	1.19	4.16
3x8	"	3.87	1.21	5.08
3x10	"	4.84	1.27	6.11
3x12	"	5.80	1.33	7.13
3x14	"	6.63	1.39	8.02
4x6	"	3.87	1.16	5.03
4x8	"	4.97	1.21	6.18
4x10	"	6.35	1.27	7.62
4x12	"	7.73	1.33	9.06
4x14	"	8.97	1.39	10.36
16" o.c.				
2x6	S.F.	0.75	0.93	1.68
2x8	"	1.06	0.94	2.00
2x10	"	1.29	0.96	2.25
2x12	"	1.61	0.99	2.60
2x14	"	3.59	1.03	4.62
3x6	"	2.48	0.96	3.44
3x8	"	3.17	0.99	4.16
3x10	"	4.00	1.03	5.03
3x12	"	4.84	1.07	5.91
3x14	"	5.73	1.11	6.84

Rough Carpentry

06110.30 Floor Framing (Cont.)

Item	UNIT	MAT.	INST.	TOTAL
4x6	S.F.	3.17	0.96	4.13
4x8	"	4.35	0.99	5.34
4x10	"	5.39	1.03	6.42
4x12	"	6.35	1.07	7.42
4x14	"	7.60	1.11	8.71
Sister joists for floors				
2x4	L.F.	0.48	3.49	3.97
2x6	"	0.73	3.99	4.72
2x8	"	0.96	4.66	5.62
2x10	"	1.28	5.59	6.87
2x12	"	1.66	6.99	8.65
3x6	"	2.42	5.59	8.01
3x8	"	2.97	6.21	9.18
3x10	"	3.93	6.99	10.92
3x12	"	4.76	7.99	12.75
4x6	"	3.11	5.59	8.70
4x8	"	4.14	6.21	10.35
4x10	"	5.39	6.99	12.38
4x12	"	6.00	7.99	13.99

06110.40 Furring

Item	UNIT	MAT.	INST.	TOTAL
Furring, wood strips				
Walls				
On masonry or concrete walls				
1x2 furring				
12" o.c.	S.F.	0.38	1.74	2.12
16" o.c.	"	0.33	1.59	1.92
24" o.c.	"	0.31	1.47	1.78
1x3 furring				
12" o.c.	S.F.	0.48	1.74	2.22
16" o.c.	"	0.44	1.59	2.03
24" o.c.	"	0.34	1.47	1.81
On wood walls				
1x2 furring				
12" o.c.	S.F.	0.38	1.24	1.62
16" o.c.	"	0.33	1.11	1.44
24" o.c.	"	0.30	1.01	1.31
1x3 furring				
12" o.c.	S.F.	0.49	1.24	1.73
16" o.c.	"	0.41	1.11	1.52
24" o.c.	"	0.34	1.01	1.35
Ceilings				
On masonry or concrete ceilings				
1x2 furring				
12" o.c.	S.F.	0.38	3.10	3.48
16" o.c.	"	0.33	2.79	3.12
24" o.c.	"	0.30	2.54	2.84
1x3 furring				
12" o.c.	S.F.	0.48	3.10	3.58
16" o.c.	"	0.41	2.79	3.20
24" o.c.	"	0.34	2.54	2.88
On wood ceilings				
1x2 furring				

Rough Carpentry

Rough Carpentry	UNIT	MAT.	INST.	TOTAL
06110.40 Furring *(Cont.)*				
12" o.c.	S.F.	0.38	2.07	2.45
16" o.c.	"	0.33	1.86	2.19
24" o.c.	"	0.30	1.69	1.99
1x3				
12" o.c.	S.F.	0.48	2.07	2.55
16" o.c.	"	0.41	1.86	2.27
24" o.c.	"	0.34	1.69	2.03
06110.50 Roof Framing				
Roof framing				
Rafters, gable end				
0-2 pitch (flat to 2-in-12)				
12" o.c.				
2x4	S.F.	0.69	1.16	1.85
2x6	"	0.96	1.21	2.17
2x8	"	1.38	1.27	2.65
2x10	"	1.72	1.33	3.05
2x12	"	3.17	1.39	4.56
16" o.c.				
2x6	S.F.	0.86	0.99	1.85
2x8	"	1.22	1.03	2.25
2x10	"	1.38	1.07	2.45
2x12	"	2.55	1.11	3.66
24" o.c.				
2x6	S.F.	0.48	0.84	1.32
2x8	"	1.01	0.87	1.88
2x10	"	1.17	0.90	2.07
2x12	"	2.06	0.93	2.99
4-6 pitch (4-in-12 to 6-in-12)				
12" o.c.				
2x4	S.F.	0.69	1.21	1.90
2x6	"	1.03	1.27	2.30
2x8	"	1.58	1.33	2.91
2x10	"	1.79	1.39	3.18
2x12	"	2.76	1.47	4.23
16" o.c.				
2x6	S.F.	0.86	1.03	1.89
2x8	"	1.38	1.07	2.45
2x10	"	1.58	1.11	2.69
2x12	"	2.35	1.16	3.51
24" o.c.				
2x6	S.F.	0.69	0.87	1.56
2x8	"	1.17	0.90	2.07
2x10	"	1.24	0.96	2.20
2x12	"	1.93	1.07	3.00
8-12 pitch (8-in-12 to 12-in-12)				
12" o.c.				
2x4	S.F.	0.75	1.27	2.02
2x6	"	1.17	1.33	2.50
2x8	"	1.66	1.39	3.05
2x10	"	1.93	1.47	3.40
2x12	"	2.97	1.55	4.52
16" o.c.				

Rough Carpentry

06110.50 Roof Framing (Cont.)

	UNIT	MAT.	INST.	TOTAL
2x6	S.F.	0.96	1.07	2.03
2x8	"	1.55	1.11	2.66
2x10	"	1.72	1.16	2.88
2x12	"	2.48	1.21	3.69
24" o.c.				
2x6	S.F.	0.75	0.90	1.65
2x8	"	1.23	0.93	2.16
2x10	"	1.38	0.96	2.34
2x12	"	2.21	0.99	3.20
Ridge boards				
2x6	L.F.	0.73	2.79	3.52
2x8	"	0.96	3.10	4.06
2x10	"	1.28	3.49	4.77
2x12	"	1.66	3.99	5.65
Hip rafters				
2x6	L.F.	0.73	1.99	2.72
2x8	"	0.96	2.07	3.03
2x10	"	1.28	2.15	3.43
2x12	"	1.66	2.23	3.89
Jack rafters				
4-6 pitch (4-in-12 to 6-in-12)				
16" o.c.				
2x6	S.F.	0.90	1.64	2.54
2x8	"	1.38	1.69	3.07
2x10	"	1.58	1.80	3.38
2x12	"	2.35	1.86	4.21
24" o.c.				
2x6	S.F.	0.69	1.27	1.96
2x8	"	1.17	1.30	2.47
2x10	"	1.38	1.36	2.74
2x12	"	2.00	1.39	3.39
8-12 pitch (8-in-12 to 12-in-12)				
16" o.c.				
2x6	S.F.	1.38	1.74	3.12
2x8	"	1.72	1.80	3.52
2x10	"	2.48	1.86	4.34
2x12	"	3.45	1.92	5.37
24" o.c.				
2x6	S.F.	1.10	1.33	2.43
2x8	"	1.38	1.36	2.74
2x10	"	2.21	1.39	3.60
2x12	"	3.17	1.43	4.60
Sister rafters				
2x4	L.F.	0.48	3.99	4.47
2x6	"	0.73	4.66	5.39
2x8	"	0.96	5.59	6.55
2x10	"	1.28	6.99	8.27
2x12	"	1.66	9.32	10.98
Fascia boards				
2x4	L.F.	0.48	2.79	3.27
2x6	"	0.73	2.79	3.52
2x8	"	0.96	3.10	4.06
2x10	"	1.28	3.10	4.38

06 WOOD AND PLASTICS

Rough Carpentry	UNIT	MAT.	INST.	TOTAL
06110.50 Roof Framing (Cont.)				
2x12	L.F.	1.66	3.49	5.15
Cant strips				
Fiber				
3x3	L.F.	0.38	1.59	1.97
4x4	"	0.53	1.69	2.22
Wood				
3x3	L.F.	2.00	1.69	3.69
06110.60 Sleepers				
Sleepers, over concrete				
12" o.c.				
1x2	S.F.	0.26	1.27	1.53
1x3	"	0.39	1.33	1.72
2x4	"	0.85	1.55	2.40
2x6	"	1.25	1.64	2.89
16" o.c.				
1x2	S.F.	0.24	1.11	1.35
1x3	"	0.34	1.11	1.45
2x4	"	0.71	1.33	2.04
2x6	"	1.05	1.39	2.44
06110.65 Soffits				
Soffit framing				
2x3	L.F.	0.37	3.99	4.36
2x4	"	0.46	4.30	4.76
2x6	"	0.68	4.66	5.34
2x8	"	0.95	5.08	6.03
06110.70 Wall Framing				
Framing wall, studs				
12" o.c.				
2x3	S.F.	0.48	1.03	1.51
2x4	"	0.68	1.03	1.71
2x6	"	0.99	1.11	2.10
2x8	"	1.32	1.16	2.48
16" o.c.				
2x3	S.F.	0.39	0.87	1.26
2x4	"	0.55	0.87	1.42
2x6	"	0.79	0.93	1.72
2x8	"	1.24	0.96	2.20
24" o.c.				
2x3	S.F.	0.30	0.75	1.05
2x4	"	0.41	0.75	1.16
2x6	"	0.66	0.79	1.45
2x8	"	0.85	0.82	1.67
Plates, top or bottom				
2x3	L.F.	0.37	1.64	2.01
2x4	"	0.46	1.74	2.20
2x6	"	0.68	1.86	2.54
2x8	"	0.95	1.99	2.94
Headers, door or window				
2x6				
Single				

06 WOOD AND PLASTICS

Rough Carpentry	UNIT	MAT.	INST.	TOTAL
06110.70 Wall Framing *(Cont.)*				
3' long	EA.	2.22	28.00	30.22
6' long	"	4.44	35.00	39.44
Double				
3' long	EA.	4.45	31.00	35.45
6' long	"	8.92	40.00	48.92
2x8				
Single				
4' long	EA.	4.07	35.00	39.07
8' long	"	8.12	43.00	51.12
Double				
4' long	EA.	8.12	40.00	48.12
8' long	"	16.25	51.00	67.25
2x10				
Single				
5' long	EA.	6.14	43.00	49.14
10' long	"	12.25	56.00	68.25
Double				
5' long	EA.	12.25	46.75	59.00
10' long	"	24.75	56.00	80.75
2x12				
Single				
6' long	EA.	8.92	43.00	51.92
12' long	"	17.50	56.00	73.50
Double				
6' long	EA.	17.50	51.00	68.50
12' long	"	35.00	62.00	97.00
06115.10 Floor Sheathing				
Sub-flooring, plywood, CDX				
1/2" thick	S.F.	0.52	0.69	1.21
5/8" thick	"	0.75	0.79	1.54
3/4" thick	"	1.39	0.93	2.32
Structural plywood				
1/2" thick	S.F.	0.83	0.69	1.52
5/8" thick	"	1.33	0.79	2.12
3/4" thick	"	1.39	0.86	2.25
Board type subflooring				
1x6				
Minimum	S.F.	1.26	1.24	2.50
Maximum	"	1.60	1.39	2.99
1x8				
Minimum	S.F.	1.39	1.17	2.56
Maximum	"	1.63	1.31	2.94
1x10				
Minimum	S.F.	1.95	1.11	3.06
Maximum	"	2.10	1.24	3.34
Underlayment				
Hardboard, 1/4" tempered	S.F.	0.78	0.69	1.47
Plywood, CDX				
3/8" thick	S.F.	0.81	0.69	1.50
1/2" thick	"	0.96	0.74	1.70
5/8" thick	"	1.12	0.79	1.91
3/4" thick	"	1.39	0.86	2.25

06 WOOD AND PLASTICS

Rough Carpentry

	UNIT	MAT.	INST.	TOTAL
06115.20 Roof Sheathing				
Sheathing				
Plywood, CDX				
3/8" thick	S.F.	0.81	0.72	1.53
1/2" thick	"	0.96	0.74	1.70
5/8" thick	"	1.12	0.79	1.91
3/4" thick	"	1.39	0.86	2.25
Structural plywood				
3/8" thick	S.F.	0.51	0.72	1.23
1/2" thick	"	0.67	0.74	1.41
5/8" thick	"	0.81	0.79	1.60
3/4" thick	"	0.97	0.86	1.83
06115.30 Wall Sheathing				
Sheathing				
Plywood, CDX				
3/8" thick	S.F.	0.81	0.82	1.63
1/2" thick	"	0.96	0.86	1.82
5/8" thick	"	1.12	0.93	2.05
3/4" thick	"	1.39	1.01	2.40
Waferboard				
3/8" thick	S.F.	0.51	0.82	1.33
1/2" thick	"	0.67	0.86	1.53
5/8" thick	"	0.81	0.93	1.74
3/4" thick	"	0.89	1.01	1.90
Structural plywood				
3/8" thick	S.F.	0.81	0.82	1.63
1/2" thick	"	0.96	0.86	1.82
5/8" thick	"	1.12	0.93	2.05
3/4" thick	"	0.96	1.01	1.97
Gypsum, 1/2" thick	"	0.51	0.86	1.37
Asphalt impregnated fiberboard, 1/2" thick	"	0.89	0.86	1.75
06125.10 Wood Decking				
Decking, T&G solid				
Cedar				
3" thick	S.F.	9.62	1.39	11.01
4" thick	"	12.00	1.49	13.49
Fir				
3" thick	S.F.	4.21	1.39	5.60
4" thick	"	5.11	1.49	6.60
Southern yellow pine				
3" thick	S.F.	4.21	1.59	5.80
4" thick	"	4.45	1.72	6.17
White pine				
3" thick	S.F.	5.11	1.39	6.50
4" thick	"	6.91	1.49	8.40
06130.10 Heavy Timber				
Mill framed structures				
Beams to 20' long				
Douglas fir				
6x8	L.F.	6.69	7.30	13.99
6x10	"	7.90	7.55	15.45

Rough Carpentry

06130.10 Heavy Timber *(Cont.)*	UNIT	MAT.	INST.	TOTAL
6x12	L.F.	9.46	8.11	17.57
6x14	"	11.25	8.42	19.67
6x16	"	12.25	8.76	21.01
8x10	"	10.50	7.55	18.05
8x12	"	12.25	8.11	20.36
8x14	"	14.25	8.42	22.67
8x16	"	16.25	8.76	25.01
Southern yellow pine				
6x8	L.F.	5.31	7.30	12.61
6x10	"	6.45	7.55	14.00
6x12	"	8.26	8.11	16.37
6x14	"	9.46	8.42	17.88
6x16	"	10.50	8.76	19.26
8x10	"	8.75	7.55	16.30
8x12	"	10.75	8.11	18.86
8x14	"	12.25	8.42	20.67
8x16	"	14.00	8.76	22.76
Columns to 12' high				
Douglas fir				
6x6	L.F.	4.81	11.00	15.81
8x8	"	8.26	11.00	19.26
10x10	"	14.50	12.25	26.75
12x12	"	18.00	12.25	30.25
Southern yellow pine				
6x6	L.F.	4.14	11.00	15.14
8x8	"	6.97	11.00	17.97
10x10	"	10.75	12.25	23.00
12x12	"	15.00	12.25	27.25
Posts, treated				
4x4	L.F.	1.66	2.23	3.89
6x6	"	4.81	2.79	7.60

06190.20 Wood Trusses	UNIT	MAT.	INST.	TOTAL
Truss, fink, 2x4 members				
3-in-12 slope				
24' span	EA.	99.00	63.00	162
26' span	"	100	63.00	163
28' span	"	110	66.00	176
30' span	"	120	66.00	186
34' span	"	120	71.00	191
38' span	"	120	71.00	191
5-in-12 slope				
24' span	EA.	100	64.00	164
28' span	"	110	66.00	176
30' span	"	120	68.00	188
32' span	"	130	68.00	198
40' span	"	170	73.00	243
Gable, 2x4 members				
5-in-12 slope				
24' span	EA.	120	64.00	184
26' span	"	130	64.00	194
28' span	"	150	66.00	216
30' span	"	150	68.00	218

06 WOOD AND PLASTICS

Rough Carpentry

06190.20 Wood Trusses *(Cont.)*

	UNIT	MAT.	INST.	TOTAL
32' span	EA.	160	68.00	228
36' span	"	170	71.00	241
40' span	"	180	73.00	253
King post type, 2x4 members				
4-in-12 slope				
16' span	EA.	73.00	59.00	132
18' span	"	79.00	61.00	140
24' span	"	84.00	64.00	148
26' span	"	91.00	64.00	155
30' span	"	110	68.00	178
34' span	"	120	68.00	188
38' span	"	150	71.00	221
42' span	"	170	76.00	246

Finish Carpentry

06200.10 Finish Carpentry

	UNIT	MAT.	INST.	TOTAL
Mouldings and trim				
Apron, flat				
9/16 x 2	L.F.	1.54	2.79	4.33
9/16 x 3-1/2	"	3.55	2.94	6.49
Base				
Colonial				
7/16 x 2-1/4	L.F.	1.83	2.79	4.62
7/16 x 3	"	2.37	2.79	5.16
7/16 x 3-1/4	"	2.43	2.79	5.22
9/16 x 3	"	2.37	2.94	5.31
9/16 x 3-1/4	"	2.48	2.94	5.42
11/16 x 2-1/4	"	2.60	3.10	5.70
Ranch				
7/16 x 2-1/4	L.F.	2.01	2.79	4.80
7/16 x 3-1/4	"	2.37	2.79	5.16
9/16 x 2-1/4	"	2.18	2.94	5.12
9/16 x 3	"	2.37	2.94	5.31
9/16 x 3-1/4	"	2.43	2.94	5.37
Casing				
11/16 x 2-1/2	L.F.	1.89	2.54	4.43
11/16 x 3-1/2	"	2.13	2.66	4.79
Chair rail				
9/16 x 2-1/2	L.F.	2.01	2.79	4.80
9/16 x 3-1/2	"	2.78	2.79	5.57
Closet pole				
1-1/8" dia.	L.F.	1.36	3.73	5.09
1-5/8" dia.	"	2.01	3.73	5.74
Cove				
9/16 x 1-3/4	L.F.	1.54	2.79	4.33

Finish Carpentry

06200.10 Finish Carpentry *(Cont.)*	UNIT	MAT.	INST.	TOTAL
11/16 x 2-3/4	L.F.	2.37	2.79	5.16
Crown				
9/16 x 1-5/8	L.F.	2.01	3.73	5.74
9/16 x 2-5/8	"	2.18	4.30	6.48
11/16 x 3-5/8	"	2.37	4.66	7.03
11/16 x 4-1/4	"	3.55	5.08	8.63
11/16 x 5-1/4	"	3.97	5.59	9.56
Drip cap				
1-1/16 x 1-5/8	L.F.	2.13	2.79	4.92
Glass bead				
3/8 x 3/8	L.F.	0.77	3.49	4.26
1/2 x 9/16	"	0.94	3.49	4.43
5/8 x 5/8	"	1.01	3.49	4.50
3/4 x 3/4	"	1.18	3.49	4.67
Half round				
1/2	L.F.	0.89	2.23	3.12
5/8	"	1.18	2.23	3.41
3/4	"	1.60	2.23	3.83
Lattice				
1/4 x 7/8	L.F.	0.71	2.23	2.94
1/4 x 1-1/8	"	0.77	2.23	3.00
1/4 x 1-3/8	"	0.82	2.23	3.05
1/4 x 1-3/4	"	0.92	2.23	3.15
1/4 x 2	"	1.06	2.23	3.29
Ogee molding				
5/8 x 3/4	L.F.	1.41	2.79	4.20
11/16 x 1-1/8	"	3.32	2.79	6.11
11/16 x 1-3/8	"	2.60	2.79	5.39
Parting bead				
3/8 x 7/8	L.F.	1.18	3.49	4.67
Quarter round				
1/4 x 1/4	L.F.	0.41	2.23	2.64
3/8 x 3/8	"	0.59	2.23	2.82
1/2 x 1/2	"	0.77	2.23	3.00
11/16 x 11/16	"	0.77	2.43	3.20
3/4 x 3/4	"	1.41	2.43	3.84
1-1/16 x 1-1/16	"	1.12	2.54	3.66
Railings, balusters				
1-1/8 x 1-1/8	L.F.	3.79	5.59	9.38
1-1/2 x 1-1/2	"	4.44	5.08	9.52
Screen moldings				
1/4 x 3/4	L.F.	0.94	4.66	5.60
5/8 x 5/16	"	1.18	4.66	5.84
Shoe				
7/16 x 11/16	L.F.	1.18	2.23	3.41
Sash beads				
1/2 x 3/4	L.F.	1.36	4.66	6.02
1/2 x 7/8	"	1.54	4.66	6.20
1/2 x 1-1/8	"	1.66	5.08	6.74
5/8 x 7/8	"	1.66	5.08	6.74
Stop				
5/8 x 1-5/8				
Colonial	L.F.	0.82	3.49	4.31

Finish Carpentry	UNIT	MAT.	INST.	TOTAL
06200.10 Finish Carpentry *(Cont.)*				
Ranch	L.F.	0.82	3.49	4.31
Stools				
11/16 x 2-1/4	L.F.	3.61	6.21	9.82
11/16 x 2-1/2	"	3.79	6.21	10.00
11/16 x 5-1/4	"	3.91	6.99	10.90
Exterior trim, casing, select pine, 1x3	"	2.60	2.79	5.39
Douglas fir				
1x3	L.F.	1.24	2.79	4.03
1x4	"	1.54	2.79	4.33
1x6	"	2.01	3.10	5.11
1x8	"	2.78	3.49	6.27
Cornices, white pine, #2 or better				
1x2	L.F.	0.77	2.79	3.56
1x4	"	0.94	2.79	3.73
1x6	"	1.54	3.10	4.64
1x8	"	1.89	3.29	5.18
1x10	"	2.43	3.49	5.92
1x12	"	3.02	3.73	6.75
Shelving, pine				
1x8	L.F.	1.36	4.30	5.66
1x10	"	1.78	4.47	6.25
1x12	"	2.25	4.66	6.91
Plywood shelf, 3/4", with edge band, 12" wide	"	2.43	5.59	8.02
Adjustable shelf, and rod, 12" wide				
3' to 4' long	EA.	19.50	14.00	33.50
5' to 8' long	"	36.25	18.75	55.00
Prefinished wood shelves with brackets and supports				
8" wide				
3' long	EA.	57.00	14.00	71.00
4' long	"	65.00	14.00	79.00
6' long	"	95.00	14.00	109
10" wide				
3' long	EA.	63.00	14.00	77.00
4' long	"	91.00	14.00	105
6' long	"	100	14.00	114
06220.10 Millwork				
Countertop, laminated plastic				
25" x 7/8" thick				
Minimum	L.F.	15.25	14.00	29.25
Average	"	28.75	18.75	47.50
Maximum	"	42.25	22.50	64.75
25" x 1-1/4" thick				
Minimum	L.F.	18.50	18.75	37.25
Average	"	36.75	22.50	59.25
Maximum	"	55.00	28.00	83.00
Add for cutouts	EA.		35.00	35.00
Backsplash, 4" high, 7/8" thick	L.F.	20.25	11.25	31.50
Plywood, sanded, A-C				
1/4" thick	S.F.	1.30	1.86	3.16
3/8" thick	"	1.41	1.99	3.40
1/2" thick	"	1.60	2.15	3.75
A-D				

Finish Carpentry

06220.10 Millwork (Cont.)

	UNIT	MAT.	INST.	TOTAL
1/4" thick	S.F.	1.24	1.86	3.10
3/8" thick	"	1.41	1.99	3.40
1/2" thick	"	1.54	2.15	3.69
Base cab., 34-1/2" high, 24" deep, hardwood				
Minimum	L.F.	200	22.50	223
Average	"	230	28.00	258
Maximum	"	250	37.25	287
Wall cabinets				
Minimum	L.F.	60.00	18.75	78.75
Average	"	82.00	22.50	105
Maximum	"	100	28.00	128

Wood Treatment

06300.10 Wood Treatment

	UNIT	MAT.	INST.	TOTAL
Creosote preservative treatment				
8 lb/cf	B.F.			0.61
10 lb/cf	"			0.73
Salt preservative treatment				
Oil borne				
Minimum	B.F.			0.56
Maximum	"			0.79
Water borne				
Minimum	B.F.			0.39
Maximum	"			0.61
Fire retardant treatment				
Minimum	B.F.			0.79
Maximum	"			0.95
Kiln dried, softwood, add to framing costs				
1" thick	B.F.			0.28
2" thick	"			0.39
3" thick	"			0.50
4" thick	"			0.61

06 WOOD AND PLASTICS

Architectural Woodwork	UNIT	MAT.	INST.	TOTAL
06420.10 Panel Work				
Hardboard, tempered, 1/4" thick				
Natural faced	S.F.	0.94	1.39	2.33
Plastic faced	"	1.41	1.59	3.00
Pegboard, natural	"	1.18	1.39	2.57
Plastic faced	"	1.41	1.59	3.00
Untempered, 1/4" thick				
Natural faced	S.F.	0.89	1.39	2.28
Plastic faced	"	1.54	1.59	3.13
Pegboard, natural	"	0.94	1.39	2.33
Plastic faced	"	1.36	1.59	2.95
Plywood unfinished, 1/4" thick				
Birch				
Natural	S.F.	1.01	1.86	2.87
Select	"	1.48	1.86	3.34
Knotty pine	"	1.95	1.86	3.81
Cedar (closet lining)				
Standard boards T&G	S.F.	2.43	1.86	4.29
Particle board	"	1.48	1.86	3.34
Plywood, prefinished, 1/4" thick, premium grade				
Birch veneer	S.F.	3.55	2.23	5.78
Cherry veneer	"	4.14	2.23	6.37
Chestnut veneer	"	8.00	2.23	10.23
Lauan veneer	"	1.54	2.23	3.77
Mahogany veneer	"	4.09	2.23	6.32
Oak veneer (red)	"	4.09	2.23	6.32
Pecan veneer	"	5.15	2.23	7.38
Rosewood veneer	"	8.00	2.23	10.23
Teak veneer	"	5.28	2.23	7.51
Walnut veneer	"	4.56	2.23	6.79
06430.10 Stairwork				
Risers, 1x8, 42" wide				
White oak	EA.	42.75	28.00	70.75
Pine	"	38.00	28.00	66.00
Treads, 1-1/16" x 9-1/2" x 42"				
White oak	EA.	51.00	35.00	86.00
06440.10 Columns				
Column, hollow, round wood				
12" diameter				
10' high	EA.	750	80.00	830
12' high	"	920	86.00	1,006
14' high	"	1,100	96.00	1,196
16' high	"	1,360	120	1,480
24" diameter				
16' high	EA.	3,110	120	3,230
18' high	"	3,540	130	3,670
20' high	"	4,350	130	4,480
22' high	"	4,580	130	4,710
24' high	"	5,000	130	5,130

07 THERMAL AND MOISTURE

Moisture Protection

	UNIT	MAT.	INST.	TOTAL
07100.10 Waterproofing				
Membrane waterproofing, elastomeric				
Butyl				
1/32" thick	S.F.	1.21	1.75	2.96
1/16" thick	"	1.58	1.82	3.40
Neoprene				
1/32" thick	S.F.	2.07	1.75	3.82
1/16" thick	"	2.97	1.82	4.79
Plastic vapor barrier (polyethylene)				
4 mil	S.F.	0.05	0.17	0.22
6 mil	"	0.07	0.17	0.24
10 mil	"	0.11	0.21	0.32
Bituminous membrane, asphalt felt, 15 lb.				
One ply	S.F.	0.75	1.09	1.84
Two ply	"	0.89	1.32	2.21
Modified asphalt membrane, fibrous asphalt				
One ply	S.F.	0.52	1.82	2.34
Two ply	"	1.06	2.19	3.25
Three ply	"	1.65	2.43	4.08
Asphalt coated protective board				
1/8" thick	S.F.	0.49	1.09	1.58
1/4" thick	"	0.67	1.09	1.76
3/8" thick	"	0.74	1.09	1.83
1/2" thick	"	0.90	1.15	2.05
Cement protective board				
3/8" thick	S.F.	1.38	1.46	2.84
1/2" thick	"	1.93	1.46	3.39
Fluid applied, neoprene				
50 mil	S.F.	1.93	1.46	3.39
90 mil	"	3.20	1.46	4.66
Bentonite waterproofing, panels				
3/16" thick	S.F.	1.60	1.09	2.69
1/4" thick	"	1.82	1.09	2.91
07150.10 Dampproofing				
Silicone dampproofing, sprayed on				
Concrete surface				
1 coat	S.F.	0.64	0.24	0.88
2 coats	"	1.06	0.33	1.39
Concrete block				
1 coat	S.F.	0.64	0.29	0.93
2 coats	"	1.06	0.39	1.45
Brick				
1 coat	S.F.	0.74	0.33	1.07
2 coats	"	1.15	0.43	1.58
07160.10 Bituminous Dampproofing				
Building paper, asphalt felt				
15 lb	S.F.	0.19	1.75	1.94
30 lb	"	0.36	1.82	2.18
Asphalt, troweled, cold, primer plus				
1 coat	S.F.	0.68	1.46	2.14
2 coats	"	1.43	2.19	3.62
3 coats	"	2.04	2.74	4.78

Moisture Protection

07160.10 Bituminous Dampproofing (Cont.)

Moisture Protection	UNIT	MAT.	INST.	TOTAL
Fibrous asphalt, hot troweled, primer plus				
1 coat	S.F.	0.68	1.75	2.43
2 coats	"	1.43	2.43	3.86
3 coats	"	2.04	3.13	5.17
Asphaltic paint dampproofing, per coat				
Brush on	S.F.	0.35	0.62	0.97
Spray on	"	0.49	0.48	0.97

07190.10 Vapor Barriers

Moisture Protection	UNIT	MAT.	INST.	TOTAL
Vapor barrier, polyethylene				
2 mil	S.F.	0.01	0.21	0.22
6 mil	"	0.06	0.21	0.27
8 mil	"	0.07	0.24	0.31
10 mil	"	0.09	0.24	0.33

Insulation

07210.10 Batt Insulation

Insulation	UNIT	MAT.	INST.	TOTAL
Ceiling, fiberglass, unfaced				
3-1/2" thick, R11	S.F.	0.38	0.51	0.89
6" thick, R19	"	0.50	0.58	1.08
9" thick, R30	"	1.00	0.67	1.67
Suspended ceiling, unfaced				
3-1/2" thick, R11	S.F.	0.38	0.48	0.86
6" thick, R19	"	0.50	0.54	1.04
9" thick, R30	"	1.00	0.62	1.62
Crawl space, unfaced				
3-1/2" thick, R11	S.F.	0.38	0.67	1.05
6" thick, R19	"	0.50	0.73	1.23
9" thick, R30	"	1.00	0.79	1.79
Wall, fiberglass				
Paper backed				
2" thick, R7	S.F.	0.25	0.46	0.71
3" thick, R8	"	0.27	0.48	0.75
4" thick, R11	"	0.45	0.51	0.96
6" thick, R19	"	0.67	0.54	1.21
Foil backed, 1 side				
2" thick, R7	S.F.	0.58	0.46	1.04
3" thick, R11	"	0.61	0.48	1.09
4" thick, R14	"	0.64	0.51	1.15
6" thick, R21	"	0.84	0.54	1.38
Foil backed, 2 sides				
2" thick, R7	S.F.	0.66	0.51	1.17
3" thick, R11	"	0.83	0.54	1.37
4" thick, R14	"	0.99	0.58	1.57
6" thick, R21	"	1.06	0.62	1.68

Insulation	UNIT	MAT.	INST.	TOTAL
07210.10 Batt Insulation (Cont.)				
Unfaced				
2" thick, R7	S.F.	0.37	0.46	0.83
3" thick, R9	"	0.41	0.48	0.89
4" thick, R11	"	0.45	0.51	0.96
6" thick, R19	"	0.58	0.54	1.12
Mineral wool batts				
Paper backed				
2" thick, R6	S.F.	0.24	0.46	0.70
4" thick, R12	"	0.52	0.48	1.00
6" thick, R19	"	0.67	0.54	1.21
Fasteners, self adhering, attached to ceiling deck				
2-1/2" long	EA.	0.19	0.73	0.92
4-1/2" long	"	0.22	0.79	1.01
Capped, self-locking washers	"	0.19	0.43	0.62
07210.20 Board Insulation				
Insulation, rigid				
Fiberglass, roof				
0.75" thick, R2.78	S.F.	0.55	0.39	0.94
1.06" thick, R4.17	"	0.84	0.41	1.25
1.31" thick, R5.26	"	1.13	0.43	1.56
1.63" thick, R6.67	"	1.39	0.46	1.85
2.25" thick, R8.33	"	1.54	0.48	2.02
Perlite board, roof				
1.00" thick, R2.78	S.F.	0.58	0.36	0.94
1.50" thick, R4.17	"	0.90	0.38	1.28
2.00" thick, R5.92	"	1.11	0.39	1.50
2.50" thick, R6.67	"	1.35	0.41	1.76
3.00" thick, R8.33	"	1.70	0.43	2.13
4.00" thick, R10.00	"	1.89	0.46	2.35
5.25" thick, R14.29	"	2.07	0.48	2.55
Rigid urethane				
Roof				
1" thick, R6.67	S.F.	1.06	0.36	1.42
1.20" thick, R8.33	"	1.22	0.37	1.59
1.50" thick, R11.11	"	1.45	0.38	1.83
2" thick, R14.29	"	1.88	0.39	2.27
2.25" thick, R16.67	"	2.45	0.41	2.86
Wall				
1" thick, R6.67	S.F.	1.06	0.46	1.52
1.5" thick, R11.11	"	1.45	0.48	1.93
2" thick, R14.29	"	1.92	0.51	2.43
Polystyrene				
Roof				
1.0" thick, R4.17	S.F.	0.43	0.36	0.79
1.5" thick, R6.26	"	0.65	0.38	1.03
2.0" thick, R8.33	"	0.79	0.39	1.18
Wall				
1.0" thick, R4.17	S.F.	0.43	0.46	0.89
1.5" thick, R6.26	"	0.65	0.48	1.13
2.0" thick, R8.33	"	0.79	0.51	1.30
Rigid board insulation, deck				
Mineral fiberboard				

Insulation	UNIT	MAT.	INST.	TOTAL
07210.20 Board Insulation *(Cont.)*				
1" thick, R3.0	S.F.	0.62	0.36	0.98
2" thick, R5.26	"	1.35	0.39	1.74
Fiberglass				
1" thick, R4.3	S.F.	1.05	0.36	1.41
2" thick, R8.5	"	1.56	0.39	1.95
Polystyrene				
1" thick, R5.4	S.F.	0.43	0.36	0.79
2" thick, R10.8	"	1.07	0.39	1.46
Urethane				
.75" thick, R5.4	S.F.	0.97	0.36	1.33
1" thick, R6.4	"	1.15	0.36	1.51
1.5" thick, R10.7	"	1.39	0.38	1.77
2" thick, R14.3	"	1.58	0.39	1.97
Foamglass				
1" thick, R1.8	S.F.	1.57	0.36	1.93
2" thick, R5.26	"	2.01	0.39	2.40
Wood fiber				
1" thick, R3.85	S.F.	1.79	0.36	2.15
2" thick, R7.7	"	2.15	0.39	2.54
Particle board				
3/4" thick, R2.08	S.F.	0.94	0.36	1.30
1" thick, R2.77	"	0.98	0.36	1.34
2" thick, R5.50	"	1.30	0.39	1.69
07210.60 Loose Fill Insulation				
Blown-in type				
Fiberglass				
5" thick, R11	S.F.	0.36	0.36	0.72
6" thick, R13	"	0.41	0.43	0.84
9" thick, R19	"	0.50	0.62	1.12
Rockwool, attic application				
6" thick, R13	S.F.	0.33	0.43	0.76
8" thick, R19	"	0.39	0.54	0.93
10" thick, R22	"	0.47	0.67	1.14
12" thick, R26	"	0.59	0.73	1.32
15" thick, R30	"	0.71	0.87	1.58
Poured type				
Fiberglass				
1" thick, R4	S.F.	0.39	0.27	0.66
2" thick, R8	"	0.73	0.31	1.04
3" thick, R12	"	1.08	0.36	1.44
4" thick, R16	"	1.42	0.43	1.85
Mineral wool				
1" thick, R3	S.F.	0.43	0.27	0.70
2" thick, R6	"	0.80	0.31	1.11
3" thick, R9	"	1.22	0.36	1.58
4" thick, R12	"	1.42	0.43	1.85
Vermiculite or perlite				
2" thick, R4.8	S.F.	0.77	0.31	1.08
3" thick, R7.2	"	1.10	0.36	1.46
4" thick, R9.6	"	1.43	0.43	1.86

Insulation

07210.60 Loose Fill Insulation *(Cont.)*	UNIT	MAT.	INST.	TOTAL
Masonry, poured vermiculite or perlite				
4" block	S.F.	0.39	0.21	0.60
6" block	"	0.59	0.27	0.86
8" block	"	0.85	0.31	1.16
10" block	"	1.13	0.33	1.46
12" block	"	1.41	0.36	1.77

07210.70 Sprayed Insulation	UNIT	MAT.	INST.	TOTAL
Foam, sprayed on				
Polystyrene				
1" thick, R4	S.F.	0.61	0.43	1.04
2" thick, R8	"	1.19	0.58	1.77
Urethane				
1" thick, R4	S.F.	0.58	0.43	1.01
2" thick, R8	"	1.11	0.58	1.69

Shingles And Tiles

07310.10 Asphalt Shingles	UNIT	MAT.	INST.	TOTAL
Standard asphalt shingles, strip shingles				
210 lb/square	SQ.	90.00	53.00	143
235 lb/square	"	95.00	59.00	154
240 lb/square	"	99.00	67.00	166
260 lb/square	"	140	76.00	216
300 lb/square	"	150	89.00	239
385 lb/square	"	210	110	320
Roll roofing, mineral surface				
90 lb	SQ.	55.00	38.25	93.25
110 lb	"	92.00	44.50	137
140 lb	"	95.00	53.00	148

07310.50 Metal Shingles	UNIT	MAT.	INST.	TOTAL
Aluminum, .020" thick				
Plain	SQ.	260	110	370
Colors	"	280	110	390
Steel, galvanized				
26 ga.				
Plain	SQ.	320	110	430
Colors	"	400	110	510
24 ga.				
Plain	SQ.	370	110	480
Colors	"	470	110	580
Porcelain enamel, 22 ga.				
Minimum	SQ.	800	130	930
Average	"	920	130	1,050
Maximum	"	1,030	130	1,160

Shingles And Tiles

07310.60 Slate Shingles

Description	UNIT	MAT.	INST.	TOTAL
Slate shingles				
Pennsylvania				
Ribbon	SQ.	630	270	900
Clear	"	810	270	1,080
Vermont				
Black	SQ.	680	270	950
Gray	"	750	270	1,020
Green	"	760	270	1,030
Red	"	1,380	270	1,650
Replacement shingles				
Small jobs	EA.	12.00	17.75	29.75
Large jobs	S.F.	9.51	8.90	18.41

07310.70 Wood Shingles

Description	UNIT	MAT.	INST.	TOTAL
Wood shingles, on roofs				
White cedar, #1 shingles				
4" exposure	SQ.	230	180	410
5" exposure	"	210	130	340
#2 shingles				
4" exposure	SQ.	160	180	340
5" exposure	"	140	130	270
Resquared and rebutted				
4" exposure	SQ.	210	180	390
5" exposure	"	170	130	300
On walls				
White cedar, #1 shingles				
4" exposure	SQ.	230	270	500
5" exposure	"	210	210	420
6" exposure	"	170	180	350
#2 shingles				
4" exposure	SQ.	160	270	430
5" exposure	"	140	210	350
6" exposure	"	120	180	300
Add for fire retarding	"			110

07310.80 Wood Shakes

Description	UNIT	MAT.	INST.	TOTAL
Shakes, hand split, 24" red cedar, on roofs				
5" exposure	SQ.	260	270	530
7" exposure	"	250	210	460
9" exposure	"	230	180	410
On walls				
6" exposure	SQ.	250	270	520
8" exposure	"	240	210	450
10" exposure	"	220	180	400
Add for fire retarding	"			75.00

07 THERMAL AND MOISTURE

Roofing And Siding

	UNIT	MAT.	INST.	TOTAL
07410.10 Manufactured Roofs				
Aluminum roof panels, for steel framing				
Corrugated				
Unpainted finish				
.024"	S.F.	1.97	1.33	3.30
.030"	"	2.29	1.33	3.62
Painted finish				
.024"	S.F.	2.49	1.33	3.82
.030"	"	3.04	1.33	4.37
Steel roof panels, for structural steel framing				
Corrugated, painted				
18 ga.	S.F.	4.48	1.33	5.81
20 ga.	"	4.18	1.33	5.51
07460.10 Metal Siding Panels				
Aluminum siding panels				
Corrugated				
Plain finish				
.024"	S.F.	2.35	2.47	4.82
.032"	"	2.77	2.47	5.24
Painted finish				
.024"	S.F.	2.93	2.47	5.40
.032"	"	3.36	2.47	5.83
Steel siding panels				
Corrugated				
22 ga.	S.F.	2.25	4.12	6.37
24 ga.	"	2.05	4.12	6.17
26 ga.	"	1.87	4.12	5.99
Box rib				
20 ga.	S.F.	3.37	4.12	7.49
22 ga.	"	2.82	4.12	6.94
24 ga.	"	2.48	4.12	6.60
26 ga.	"	2.03	4.12	6.15
07460.50 Plastic Siding				
Horizontal vinyl siding, solid				
8" wide				
Standard	S.F.	1.23	2.15	3.38
Insulated	"	1.49	2.15	3.64
10" wide				
Standard	S.F.	1.27	1.99	3.26
Insulated	"	1.52	1.99	3.51
Vinyl moldings for doors and windows	L.F.	0.79	2.23	3.02
07460.60 Plywood Siding				
Rough sawn cedar, 3/8" thick	S.F.	1.79	1.86	3.65
Fir, 3/8" thick	"	0.99	1.86	2.85
Texture 1-11, 5/8" thick				
Cedar	S.F.	2.42	1.99	4.41
Fir	"	1.69	1.99	3.68
Redwood	"	2.59	1.90	4.49
Southern Yellow Pine	"	1.37	1.99	3.36

07 THERMAL AND MOISTURE

Roofing And Siding	UNIT	MAT.	INST.	TOTAL
07460.70 Steel Siding				
Ribbed, sheets, galvanized				
22 ga.	S.F.	2.50	2.47	4.97
24 ga.	"	2.26	2.47	4.73
26 ga.	"	1.75	2.47	4.22
28 ga.	"	1.50	2.47	3.97
Primed				
24 ga.	S.F.	2.95	2.47	5.42
26 ga.	"	2.09	2.47	4.56
28 ga.	"	1.75	2.47	4.22
07460.80 Wood Siding				
Beveled siding, cedar				
A grade				
1/2 x 6	S.F.	4.54	2.79	7.33
1/2 x 8	"	4.64	2.23	6.87
3/4 x 10	"	5.97	1.86	7.83
Clear				
1/2 x 6	S.F.	5.06	2.79	7.85
1/2 x 8	"	5.17	2.23	7.40
3/4 x 10	"	6.93	1.86	8.79
B grade				
1/2 x 6	S.F.	4.89	2.79	7.68
1/2 x 8	"	5.52	22.50	28.02
3/4 x 10	"	5.20	1.86	7.06
Board and batten				
Cedar				
1x6	S.F.	5.37	2.79	8.16
1x8	"	4.89	2.23	7.12
1x10	"	4.42	1.99	6.41
1x12	"	3.96	1.80	5.76
Pine				
1x6	S.F.	1.36	2.79	4.15
1x8	"	1.33	2.23	3.56
1x10	"	1.27	1.99	3.26
1x12	"	1.17	1.80	2.97
Redwood				
1x6	S.F.	5.84	2.79	8.63
1x8	"	5.44	2.23	7.67
1x10	"	5.04	1.99	7.03
1x12	"	4.66	1.80	6.46
Tongue and groove				
Cedar				
1x4	S.F.	5.04	3.10	8.14
1x6	"	4.86	2.94	7.80
1x8	"	4.55	2.79	7.34
1x10	"	4.47	2.66	7.13
Pine				
1x4	S.F.	1.51	3.10	4.61
1x6	"	1.43	2.94	4.37
1x8	"	1.34	2.79	4.13
1x10	"	1.27	2.66	3.93
Redwood				
1x4	S.F.	5.34	3.10	8.44

Roofing And Siding

07460.80 Wood Siding (Cont.)	UNIT	MAT.	INST.	TOTAL
1x6	S.F.	5.14	2.94	8.08
1x8	"	4.97	2.79	7.76
1x10	"	4.74	2.66	7.40

Membrane Roofing

07510.10 Built-up Asphalt Roofing	UNIT	MAT.	INST.	TOTAL
Built-up roofing, asphalt felt, including gravel				
2 ply	SQ.	88.00	130	218
3 ply	"	120	180	300
4 ply	"	170	210	380
Walkway, for built-up roofs				
3' x 3' x				
1/2" thick	S.F.	2.34	1.78	4.12
3/4" thick	"	3.63	1.78	5.41
1" thick	"	3.93	1.78	5.71
Roof bonds				
Asphalt felt				
10 yrs	SQ.			37.25
20 yrs	"			42.50
Cant strip, 4" x 4"				
Treated wood	L.F.	2.32	1.52	3.84
Foamglass	"	1.99	1.33	3.32
Mineral fiber	"	0.39	1.33	1.72
New gravel for built-up roofing, 400 lb/sq	SQ.	35.00	110	145
Roof gravel (ballast)	C.Y.	23.50	270	294
Aluminum coating, top surfacing, for built-up roofing	SQ.	44.50	89.00	134
Remove 4-ply built-up roof (includes gravel)	"		270	270
Remove & replace gravel, includes flood coat	"	53.00	180	233

07530.10 Single-ply Roofing	UNIT	MAT.	INST.	TOTAL
Elastic sheet roofing				
Neoprene, 1/16" thick	S.F.	2.83	0.66	3.49
EPDM rubber				
45 mil	S.F.	1.47	0.66	2.13
60 mil	"	2.02	0.66	2.68
PVC				
45 mil	S.F.	2.03	0.66	2.69
60 mil	"	2.42	0.66	3.08
Flashing				
Pipe flashing, 90 mil thick				
1" pipe	EA.	33.00	13.25	46.25
2" pipe	"	35.25	13.25	48.50
3" pipe	"	35.50	14.00	49.50
4" pipe	"	38.50	14.00	52.50
5" pipe	"	41.25	14.75	56.00

Membrane Roofing

07530.10 Single-ply Roofing (Cont.)

Description	UNIT	MAT.	INST.	TOTAL
6" pipe	EA.	45.00	14.75	59.75
8" pipe	"	51.00	15.75	66.75
10" pipe	"	58.00	17.75	75.75
12" pipe	"	71.00	17.75	88.75
Neoprene flashing, 60 mil thick strip				
6" wide	L.F.	1.91	4.45	6.36
12" wide	"	3.76	6.68	10.44
18" wide	"	5.52	8.90	14.42
24" wide	"	7.27	13.25	20.52
Adhesives				
Mastic sealer, applied at joints only				
1/4" bead	L.F.	0.11	0.26	0.37
Fluid applied roofing				
Urethane, 2 part, elastomeric membrane				
1" thick	S.F.	2.97	0.89	3.86
Vinyl liquid roofing, 2 coats, 2 mils per coat	"	4.65	0.76	5.41
Silicone roofing, 2 coats sprayed, 16 mil per coat	"	3.47	0.89	4.36
Inverted roof system				
Insulated membrane with coarse gravel ballast				
3 ply with 2" polystyrene	S.F.	6.69	0.89	7.58
Ballast, 3/4" through 1-1/2" gravel, 100lb/sf	"	0.40	53.00	53.40
Walkway for membrane roofs, 1/2" thick	"	2.13	1.78	3.91

Flashing And Sheet Metal

07610.10 Metal Roofing

Description	UNIT	MAT.	INST.	TOTAL
Sheet metal roofing, copper, 16 oz, batten seam	SQ.	1,800	360	2,160
Standing seam	"	1,760	330	2,090
Aluminum roofing, natural finish				
Corrugated, on steel frame				
.0175" thick	SQ.	130	150	280
.0215" thick	"	170	150	320
.024" thick	"	200	150	350
.032" thick	"	250	150	400
V-beam, on steel frame				
.032" thick	SQ.	260	150	410
.040" thick	"	280	150	430
.050" thick	"	350	150	500
Ridge cap				
.019" thick	L.F.	4.04	1.78	5.82
Corrugated galvanized steel roofing, on steel frame				
28 ga.	SQ.	220	150	370
26 ga.	"	250	150	400
24 ga.	"	290	150	440
22 ga.	"	320	150	470
26 ga., factory insulated with 1" polystyrene	"	490	210	700

Flashing And Sheet Metal

07610.10 Metal Roofing (Cont.)

Flashing And Sheet Metal	UNIT	MAT.	INST.	TOTAL
Ridge roll				
10" wide	L.F.	2.20	1.78	3.98
20" wide	"	4.47	2.13	6.60

07620.10 Flashing And Trim

Flashing And Sheet Metal	UNIT	MAT.	INST.	TOTAL
Counter flashing				
Aluminum, .032"	S.F.	1.82	5.34	7.16
Stainless steel, .015"	"	5.81	5.34	11.15
Copper				
16 oz.	S.F.	9.36	5.34	14.70
20 oz.	"	11.00	5.34	16.34
24 oz.	"	13.50	5.34	18.84
32 oz.	"	16.50	5.34	21.84
Valley flashing				
Aluminum, .032"	S.F.	1.58	3.34	4.92
Stainless steel, .015	"	5.06	3.34	8.40
Copper				
16 oz.	S.F.	9.36	3.34	12.70
20 oz.	"	11.00	4.45	15.45
24 oz.	"	13.50	3.34	16.84
32 oz.	"	16.50	3.34	19.84
Base flashing				
Aluminum, .040"	S.F.	2.60	4.45	7.05
Stainless steel, .018"	"	6.05	4.45	10.50
Copper				
16 oz.	S.F.	9.36	4.45	13.81
20 oz.	"	11.00	3.34	14.34
24 oz.	"	13.50	4.45	17.95
32 oz.	"	16.50	4.45	20.95
Waterstop, "T" section, 22 ga.				
1-1/2" x 3"	L.F.	3.25	2.67	5.92
2" x 2"	"	3.60	2.67	6.27
4" x 3"	"	4.01	2.67	6.68
6" x 4"	"	4.24	2.67	6.91
8" x 4"	"	5.26	2.67	7.93
Scupper outlets				
10" x 10" x 4"	EA.	34.00	13.25	47.25
22" x 4" x 4"	"	42.00	13.25	55.25
8" x 8" x 5"	"	34.00	13.25	47.25
Flashing and trim, aluminum				
.019" thick	S.F.	1.28	3.81	5.09
.032" thick	"	1.57	3.81	5.38
.040" thick	"	2.69	4.11	6.80
Neoprene sheet flashing, .060" thick	"	2.14	3.34	5.48
Copper, paper backed				
2 oz.	S.F.	2.75	5.34	8.09
5 oz.	"	3.54	5.34	8.88
Drainage boots, roof, cast iron				
2 x 3	L.F.	110	6.68	117
3 x 4	"	140	6.68	147
4 x 5	"	200	7.12	207
4 x 6	"	200	7.12	207
5 x 7	"	230	7.63	238

Flashing And Sheet Metal

07620.10 Flashing And Trim *(Cont.)*

Flashing And Sheet Metal	UNIT	MAT.	INST.	TOTAL
Pitch pocket, copper, 16 oz.				
4 x 4	EA.	170	13.25	183
6 x 6	"	180	13.25	193
8 x 8	"	200	13.25	213
8 x 10	"	210	13.25	223
8 x 12	"	250	13.25	263
Reglets, copper 10 oz.	L.F.	6.71	3.56	10.27
Stainless steel, .020"	"	3.04	3.56	6.60
Gravel stop				
Aluminum, .032"				
4"	L.F.	0.93	1.78	2.71
6"	"	1.37	1.78	3.15
8"	"	1.84	2.05	3.89
10"	"	2.31	2.05	4.36
Copper, 16 oz.				
4"	L.F.	3.91	1.78	5.69
6"	"	5.84	1.78	7.62
8"	"	7.83	2.05	9.88
10"	"	9.76	2.05	11.81

07620.20 Gutters And Downspouts

Gutters And Downspouts	UNIT	MAT.	INST.	TOTAL
Copper gutter and downspout				
Downspouts, 16 oz. copper				
Round				
3" dia.	L.F.	13.00	3.56	16.56
4" dia.	"	16.25	3.56	19.81
Rectangular, corrugated				
2" x 3"	L.F.	12.75	3.34	16.09
3" x 4"	"	15.50	3.34	18.84
Rectangular, flat surface				
2" x 3"	L.F.	14.50	3.56	18.06
3" x 4"	"	20.50	3.56	24.06
Lead-coated copper downspouts				
Round				
3" dia.	L.F.	17.00	3.34	20.34
4" dia.	"	20.75	3.81	24.56
Rectangular, corrugated				
2" x 3"	L.F.	17.25	3.56	20.81
3" x 4"	"	20.50	3.56	24.06
Rectangular, plain				
2" x 3"	L.F.	12.00	3.56	15.56
3" x 4"	"	14.00	3.56	17.56
Gutters, 16 oz. copper				
Half round				
4" wide	L.F.	11.75	5.34	17.09
5" wide	"	14.25	5.93	20.18
Type K				
4" wide	L.F.	13.00	5.34	18.34
5" wide	"	13.75	5.93	19.68
Lead-coated copper gutters				
Half round				
4" wide	L.F.	14.25	5.34	19.59
6" wide	"	19.50	5.93	25.43

Flashing And Sheet Metal

07620.20 Gutters And Downspouts *(Cont.)*

Item	UNIT	MAT.	INST.	TOTAL
Type K				
4" wide	L.F.	15.75	5.34	21.09
5" wide	"	20.25	5.93	26.18
Aluminum gutter and downspout				
Downspouts				
2" x 3"	L.F.	1.32	3.56	4.88
3" x 4"	"	1.81	3.81	5.62
4" x 5"	"	2.01	4.11	6.12
Round				
3" dia.	L.F.	2.23	3.56	5.79
4" dia.	"	2.85	3.81	6.66
Gutters, stock units				
4" wide	L.F.	2.16	5.62	7.78
5" wide	"	2.57	5.93	8.50
Galvanized steel gutter and downspout				
Downspouts, round corrugated				
3" dia.	L.F.	1.99	3.56	5.55
4" dia.	"	2.68	3.56	6.24
5" dia.	"	3.99	3.81	7.80
6" dia.	"	5.30	3.81	9.11
Rectangular				
2" x 3"	L.F.	1.80	3.56	5.36
3" x 4"	"	2.59	3.34	5.93
4" x 4"	"	3.24	3.34	6.58
Gutters, stock units				
5" wide				
Plain	L.F.	1.74	5.93	7.67
Painted	"	1.90	5.93	7.83
6" wide				
Plain	L.F.	2.43	6.28	8.71
Painted	"	2.73	6.28	9.01

Roofing Specialties

07700.10 Manufactured Specialties

Item	UNIT	MAT.	INST.	TOTAL
Moisture relief vent				
Aluminum	EA.	19.25	7.63	26.88
Copper	"	52.00	7.63	59.63
Smoke vent, 48" x 48"				
Aluminum	EA.	2,100	130	2,230
Galvanized steel	"	1,850	130	1,980
Heat/smoke vent, 48" x 96"				
Aluminum	EA.	2,950	180	3,130
Galvanized steel	"	2,510	180	2,690
Ridge vent strips				
Mill finish	L.F.	3.90	3.56	7.46

07 THERMAL AND MOISTURE

Roofing Specialties

07700.10 Manufactured Specialties (Cont.)

Roofing Specialties	UNIT	MAT.	INST.	TOTAL
Soffit vents				
Mill finish				
2-1/2" wide	L.F.	0.49	2.13	2.62

Skylights

07810.10 Plastic Skylights

Skylights	UNIT	MAT.	INST.	TOTAL
Single thickness, not including mounting curb				
2' x 4'	EA.	370	67.00	437
4' x 4'	"	500	89.00	589
5' x 5'	"	670	130	800
6' x 8'	"	1,410	180	1,590
Double thickness, not including mounting curb				
2' x 4'	EA.	490	67.00	557
4' x 4'	"	610	89.00	699
5' x 5'	"	900	130	1,030
6' x 8'	"	1,580	180	1,760
Metal framed skylights				
Translucent panels, 2-1/2" thick	S.F.	44.00	5.34	49.34
Continuous vaults, 8' wide				
Single glazed	S.F.	59.00	6.68	65.68
Double glazed	"	95.00	7.63	103

07820.10 Solar Skylight

Skylights	UNIT	MAT.	INST.	TOTAL
Tubular solar skylight, basic kit				
Min.	EA.	350	190	540
Ave.	"	390	280	670
Max.	"	440	560	1,000
Tubular solar skylight dome, 10" Diameter				
Min.	EA.	72.00	56.00	128
Ave.	"	77.00	70.00	147
Max.	"	83.00	93.00	176
14" Diameter				
Min.	EA.	83.00	56.00	139
Ave.	"	88.00	70.00	158
Max.	"	94.00	93.00	187
Straight extension tube, 10" Diameter X 12" long				
Min.	EA.	47.25	46.75	94.00
Ave.	"	55.00	56.00	111
Max.	"	61.00	70.00	131
24" long				
Min.	EA.	55.00	46.75	102
Ave.	"	61.00	56.00	117
Max.	"	66.00	70.00	136
36" long				
Min.	EA.	81.00	46.75	128

Skylights

07820.10 Solar Skylight *(Cont.)*

Skylights	UNIT	MAT.	INST.	TOTAL
Ave.	EA.	88.00	56.00	144
Max.	"	94.00	70.00	164
48" long				
Min.	EA.	96.00	56.00	152
Ave.	"	100	70.00	170
Max.	"	110	93.00	203
14" Diameter X 12" long				
Min.	EA.	61.00	46.75	108
Ave.	"	66.00	56.00	122
Max.	"	72.00	70.00	142
24" long				
Min.	EA.	77.00	46.75	124
Ave.	"	88.00	56.00	144
Max.	"	99.00	70.00	169
36" long				
Min.	EA.	100	46.75	147
Ave.	"	110	56.00	166
Max.	"	120	70.00	190
90 Degree extension tubes, 10" Diameter				
Min.	EA.	55.00	31.00	86.00
Ave.	"	64.00	35.00	99.00
Max.	"	72.00	40.00	112
14" Diameter				
Min.	EA.	68.00	31.00	99.00
Ave.	"	76.00	35.00	111
Max.	"	83.00	40.00	123
Bottom tube adaptor, 10" Diameter				
Min.	EA.	58.00	31.00	89.00
Ave.	"	64.00	35.00	99.00
Max.	"	68.00	40.00	108
14" Diameter				
Min.	EA.	72.00	31.00	103
Ave.	"	77.00	35.00	112
Max.	"	83.00	40.00	123
Top tube adaptor, 10" Diameter				
Min.	EA.	58.00	31.00	89.00
Ave.	"	64.00	35.00	99.00
Max.	"	68.00	40.00	108
14" Diameter				
Min.	EA.	72.00	31.00	103
Ave.	"	77.00	35.00	112
Max.	"	83.00	40.00	123
Tube flashing				
Min.	EA.	77.00	46.75	124
Ave.	"	88.00	56.00	144
Max.	"	99.00	70.00	169
Daylight dimmer switch				
Min.	EA.	57.00	31.00	88.00
Ave.	"	62.00	35.00	97.00
Max.	"	66.00	40.00	106
Dimmer				
Min.	EA.	230	31.00	261
Ave.	"	250	35.00	285

Skylights

Skylights	UNIT	MAT.	INST.	TOTAL
07820.10 Solar Skylight *(Cont.)*				
Max.	EA.	280	40.00	320

Joint Sealers

Joint Sealers	UNIT	MAT.	INST.	TOTAL
07920.10 Caulking				
Caulk exterior, two component				
1/4 x 1/2	L.F.	0.39	2.79	3.18
3/8 x 1/2	"	0.60	3.10	3.70
1/2 x 1/2	"	0.82	3.49	4.31
Caulk interior, single component				
1/4 x 1/2	L.F.	0.26	2.66	2.92
3/8 x 1/2	"	0.37	2.94	3.31
1/2 x 1/2	"	0.49	3.29	3.78

Metal

08110.10 Metal Doors

Metal	UNIT	MAT.	INST.	TOTAL
Flush hollow metal, std. duty, 20 ga., 1-3/8" thick				
2-6 x 6-8	EA.	300	62.00	362
2-8 x 6-8	"	340	62.00	402
3-0 x 6-8	"	360	62.00	422
1-3/4" thick				
2-6 x 6-8	EA.	360	62.00	422
2-8 x 6-8	"	380	62.00	442
3-0 x 6-8	"	410	62.00	472
2-6 x 7-0	"	390	62.00	452
2-8 x 7-0	"	410	62.00	472
3-0 x 7-0	"	430	62.00	492
Heavy duty, 20 ga., unrated, 1-3/4"				
2-8 x 6-8	EA.	390	62.00	452
3-0 x 6-8	"	420	62.00	482
2-8 x 7-0	"	450	62.00	512
3-0 x 7-0	"	430	62.00	492
3-4 x 7-0	"	450	62.00	512
18 ga., 1-3/4", unrated door				
2-0 x 7-0	EA.	420	62.00	482
2-4 x 7-0	"	420	62.00	482
2-6 x 7-0	"	420	62.00	482
2-8 x 7-0	"	460	62.00	522
3-0 x 7-0	"	470	62.00	532
3-4 x 7-0	"	480	62.00	542
2", unrated door				
2-0 x 7-0	EA.	460	70.00	530
2-4 x 7-0	"	460	70.00	530
2-6 x 7-0	"	460	70.00	530
2-8 x 7-0	"	500	70.00	570
3-0 x 7-0	"	520	70.00	590
3-4 x 7-0	"	530	70.00	600
Galvanized metal door				
3-0 x 7-0	EA.	530	70.00	600
For lead lining in doors	"			970
For sound attenuation	"			88.00
Vision glass				
8" x 8"	EA.	120	70.00	190
8" x 48"	"	170	70.00	240
Fixed metal louver	"	250	56.00	306
For fire rating, add				
3 hr door	EA.			410
1-1/2 hr door	"			180
3/4 hr door	"			90.00
1' extra height, add to material, 20%				
1'6" extra height, add to material, 60%				
For dutch doors with shelf, add to material, 100%				

08110.40 Metal Door Frames

Metal	UNIT	MAT.	INST.	TOTAL
Hollow metal, stock, 18 ga., 4-3/4" x 1-3/4"				
2-0 x 7-0	EA.	140	70.00	210
2-4 x 7-0	"	160	70.00	230
2-6 x 7-0	"	160	70.00	230
2-8 x 7-0	"	160	70.00	230

Metal

08110.40 Metal Door Frames *(Cont.)*

Metal	UNIT	MAT.	INST.	TOTAL
3-0 x 7-0	EA.	160	70.00	230
4-0 x 7-0	"	180	93.00	273
5-0 x 7-0	"	190	93.00	283
6-0 x 7-0	"	230	93.00	323
16 ga., 6-3/4" x 1-3/4"				
2-0 x 7-0	EA.	160	77.00	237
2-4 x 7-0	"	150	77.00	227
2-6 x 7-0	"	160	77.00	237
2-8 x 7-0	"	160	77.00	237
3-0 x 7-0	"	170	77.00	247
4-0 x 7-0	"	200	100	300
6-0 x 7-0	"	230	100	330

Wood And Plastic

08210.10 Wood Doors

Wood And Plastic	UNIT	MAT.	INST.	TOTAL
Solid core, 1-3/8" thick				
Birch faced				
2-4 x 7-0	EA.	150	70.00	220
2-8 x 7-0	"	160	70.00	230
3-0 x 7-0	"	160	70.00	230
3-4 x 7-0	"	310	70.00	380
2-4 x 6-8	"	150	70.00	220
2-6 x 6-8	"	150	70.00	220
2-8 x 6-8	"	160	70.00	230
3-0 x 6-8	"	160	70.00	230
Lauan faced				
2-4 x 6-8	EA.	140	70.00	210
2-8 x 6-8	"	150	70.00	220
3-0 x 6-8	"	150	70.00	220
3-4 x 6-8	"	160	70.00	230
Tempered hardboard faced				
2-4 x 7-0	EA.	170	70.00	240
2-8 x 7-0	"	180	70.00	250
3-0 x 7-0	"	210	70.00	280
3-4 x 7-0	"	210	70.00	280
Hollow core, 1-3/8" thick				
Birch faced				
2-4 x 7-0	EA.	140	70.00	210
2-8 x 7-0	"	140	70.00	210
3-0 x 7-0	"	150	70.00	220
3-4 x 7-0	"	160	70.00	230
Lauan faced				
2-4 x 6-8	EA.	58.00	70.00	128
2-6 x 6-8	"	63.00	70.00	133
2-8 x 6-8	"	79.00	70.00	149

Wood And Plastic

08210.10 Wood Doors (Cont.)

Wood And Plastic	UNIT	MAT.	INST.	TOTAL
3-0 x 6-8	EA.	82.00	70.00	152
3-4 x 6-8	"	91.00	70.00	161
Tempered hardboard faced				
2-4 x 7-0	EA.	71.00	70.00	141
2-6 x 7-0	"	77.00	70.00	147
2-8 x 7-0	"	84.00	70.00	154
3-0 x 7-0	"	90.00	70.00	160
3-4 x 7-0	"	98.00	70.00	168
Solid core, 1-3/4" thick				
Birch faced				
2-4 x 7-0	EA.	230	70.00	300
2-6 x 7-0	"	240	70.00	310
2-8 x 7-0	"	250	70.00	320
3-0 x 7-0	"	230	70.00	300
3-4 x 7-0	"	240	70.00	310
Lauan faced				
2-4 x 7-0	EA.	160	70.00	230
2-6 x 7-0	"	180	70.00	250
2-8 x 7-0	"	200	70.00	270
3-4 x 7-0	"	200	70.00	270
3-0 x 7-0	"	220	70.00	290
Tempered hardboard faced				
2-4 x 7-0	EA.	210	70.00	280
2-6 x 7-0	"	230	70.00	300
2-8 x 7-0	"	260	70.00	330
3-0 x 7-0	"	270	70.00	340
3-4 x 7-0	"	290	70.00	360
Hollow core, 1-3/4" thick				
Birch faced				
2-4 x 7-0	EA.	160	70.00	230
2-6 x 7-0	"	160	70.00	230
2-8 x 7-0	"	170	70.00	240
3-0 x 7-0	"	170	70.00	240
3-4 x 7-0	"	180	70.00	250
Lauan faced				
2-4 x 6-8	EA.	97.00	70.00	167
2-6 x 6-8	"	110	70.00	180
2-8 x 6-8	"	97.00	70.00	167
3-0 x 6-8	"	100	70.00	170
3-4 x 6-8	"	110	70.00	180
Tempered hardboard				
2-4 x 7-0	EA.	89.00	70.00	159
2-6 x 7-0	"	93.00	70.00	163
2-8 x 7-0	"	97.00	70.00	167
3-0 x 7-0	"	100	70.00	170
3-4 x 7-0	"	110	70.00	180
Add-on, louver	"	35.00	56.00	91.00
Glass	"	110	56.00	166
Exterior doors, 3-0 x 7-0 x 2-1/2", solid core				
Carved				
One face	EA.	1,460	140	1,600
Two faces	"	2,020	140	2,160
Closet doors, 1-3/4" thick				

Wood And Plastic

	UNIT	MAT.	INST.	TOTAL
08210.10 Wood Doors (Cont.)				
Bi-fold or bi-passing, includes frame and trim				
Paneled				
4-0 x 6-8	EA.	510	93.00	603
6-0 x 6-8	"	580	93.00	673
Louvered				
4-0 x 6-8	EA.	350	93.00	443
6-0 x 6-8	"	420	93.00	513
Flush				
4-0 x 6-8	EA.	260	93.00	353
6-0 x 6-8	"	330	93.00	423
Primed				
4-0 x 6-8	EA.	280	93.00	373
6-0 x 6-8	"	310	93.00	403
08210.90 Wood Frames				
Frame, interior, pine				
2-6 x 6-8	EA.	86.00	80.00	166
2-8 x 6-8	"	93.00	80.00	173
3-0 x 6-8	"	96.00	80.00	176
5-0 x 6-8	"	100	80.00	180
6-0 x 6-8	"	110	80.00	190
2-6 x 7-0	"	99.00	80.00	179
2-8 x 7-0	"	110	80.00	190
3-0 x 7-0	"	120	80.00	200
5-0 x 7-0	"	130	110	240
6-0 x 7-0	"	130	110	240
Exterior, custom, with threshold, including trim				
Walnut				
3-0 x 7-0	EA.	340	140	480
6-0 x 7-0	"	390	140	530
Oak				
3-0 x 7-0	EA.	310	140	450
6-0 x 7-0	"	360	140	500
Pine				
2-4 x 7-0	EA.	130	110	240
2-6 x 7-0	"	130	110	240
2-8 x 7-0	"	140	110	250
3-0 x 7-0	"	140	110	250
3-4 x 7-0	"	160	110	270
6-0 x 7-0	"	170	190	360
08300.10 Special Doors				
Metal clad doors, including electric motor				
Light duty				
Minimum	S.F.	43.00	9.32	52.32
Maximum	"	69.00	22.50	91.50
Accordion folding, tracks and fittings included				
Vinyl covered, 2 layers	S.F.	13.75	22.50	36.25
Woven mahogany and vinyl	"	17.25	22.50	39.75
Economy vinyl	"	11.50	22.50	34.00
Rigid polyvinyl chloride	"	19.00	22.50	41.50
Sectional wood overhead, frames not incl.				
Commercial grade, HD, 1-3/4" thick, manual				

Wood And Plastic

08300.10 Special Doors (Cont.)

Wood And Plastic	UNIT	MAT.	INST.	TOTAL
8' x 8'	EA.	1,080	470	1,550
10' x 10'	"	1,560	510	2,070
Sectional metal overhead doors, complete				
Residential grade, manual				
9' x 7'	EA.	740	220	960
16' x 7'	"	1,330	280	1,610
Commercial grade				
8' x 8'	EA.	860	470	1,330
10' x 10'	"	1,150	510	1,660
12' x 12'	"	1,910	560	2,470
Sliding glass doors				
Tempered plate glass, 1/4" thick				
6' wide				
Economy grade	EA.	1,070	190	1,260
Premium grade	"	1,220	190	1,410
12' wide				
Economy grade	EA.	1,500	280	1,780
Premium grade	"	2,250	280	2,530
Insulating glass, 5/8" thick				
6' wide				
Economy grade	EA.	1,320	190	1,510
Premium grade	"	1,690	190	1,880
12' wide				
Economy grade	EA.	1,640	280	1,920
Premium grade	"	2,630	280	2,910
1" thick				
6' wide				
Economy grade	EA.	1,660	190	1,850
Premium grade	"	1,910	190	2,100
12' wide				
Economy grade	EA.	2,570	280	2,850
Premium grade	"	3,760	280	4,040
Added costs				
Custom quality, add to material, 30%				
Tempered glass, 6' wide, add	S.F.			4.62
Residential storm door				
Minimum	EA.	170	93.00	263
Average	"	230	93.00	323
Maximum	"	510	140	650

08 DOORS AND WINDOWS

Storefronts

Storefronts	UNIT	MAT.	INST.	TOTAL
08410.10 Storefronts				
Storefront, aluminum and glass				
Minimum	S.F.	24.00	7.73	31.73
Average	"	35.75	8.83	44.58
Maximum	"	72.00	10.25	82.25
Entrance doors, premium, closers, panic dev.,etc.				
1/2" thick glass				
3' x 7'	EA.	3,190	520	3,710
3/4" thick glass				
3' x 7'	EA.	3,300	520	3,820

Metal Windows

Metal Windows	UNIT	MAT.	INST.	TOTAL
08510.10 Steel Windows				
Steel windows, primed				
Casements				
Operable				
Minimum	S.F.	38.75	3.63	42.38
Maximum	"	58.00	4.12	62.12
Fixed sash	"	31.00	3.09	34.09
Double hung	"	58.00	3.43	61.43
Picture window	"	29.00	3.43	32.43
Projecting sash				
Minimum	S.F.	50.00	3.86	53.86
Maximum	"	62.00	3.86	65.86
Mullions	L.F.	13.25	3.09	16.34
08520.10 Aluminum Windows				
Jalousie				
3-0 x 4-0	EA.	330	77.00	407
3-0 x 5-0	"	390	77.00	467
Fixed window				
6 sf to 8 sf	S.F.	16.75	8.83	25.58
12 sf to 16 sf	"	14.75	6.87	21.62
Projecting window				
6 sf to 8 sf	S.F.	36.75	15.50	52.25
12 sf to 16 sf	"	33.25	10.25	43.50
Horizontal sliding				
6 sf to 8 sf	S.F.	24.00	7.73	31.73
12 sf to 16 sf	"	22.00	6.18	28.18
Double hung				
6 sf to 8 sf	S.F.	33.25	12.25	45.50
10 sf to 12 sf	"	29.50	10.25	39.75
Storm window, 0.5 cfm, up to				
60 u.i. (united inches)	EA.	77.00	31.00	108
70 u.i.	"	79.00	31.00	110
80 u.i.	"	88.00	31.00	119

Metal Windows

08520.10 Aluminum Windows *(Cont.)*

	UNIT	MAT.	INST.	TOTAL
90 u.i.	EA.	90.00	34.25	124
100 u.i.	"	92.00	34.25	126
2.0 cfm, up to				
60 u.i.	EA.	100	31.00	131
70 u.i.	"	100	31.00	131
80 u.i.	"	100	31.00	131
90 u.i.	"	110	34.25	144
100 u.i.	"	110	34.25	144

Wood And Plastic

08600.10 Wood Windows

	UNIT	MAT.	INST.	TOTAL
Double hung				
24" x 36"				
Minimum	EA.	230	56.00	286
Average	"	340	70.00	410
Maximum	"	450	93.00	543
24" x 48"				
Minimum	EA.	270	56.00	326
Average	"	390	70.00	460
Maximum	"	550	93.00	643
30" x 48"				
Minimum	EA.	280	62.00	342
Average	"	390	80.00	470
Maximum	"	570	110	680
30" x 60"				
Minimum	EA.	300	62.00	362
Average	"	490	80.00	570
Maximum	"	600	110	710
Casement				
1 leaf, 22" x 38" high				
Minimum	EA.	340	56.00	396
Average	"	410	70.00	480
Maximum	"	480	93.00	573
2 leaf, 50" x 50" high				
Minimum	EA.	900	70.00	970
Average	"	1,170	93.00	1,263
Maximum	"	1,340	140	1,480
3 leaf, 71" x 62" high				
Minimum	EA.	1,480	70.00	1,550
Average	"	1,510	93.00	1,603
Maximum	"	1,800	140	1,940
4 leaf, 95" x 75" high				
Minimum	EA.	1,960	80.00	2,040
Average	"	2,240	110	2,350
Maximum	"	2,860	190	3,050

Wood And Plastic

08600.10 Wood Windows *(Cont.)*

Description	UNIT	MAT.	INST.	TOTAL
5 leaf, 119" x 75" high				
Minimum	EA.	2,540	80.00	2,620
Average	"	2,740	110	2,850
Maximum	"	3,510	190	3,700
Picture window, fixed glass, 54" x 54" high				
Minimum	EA.	530	70.00	600
Average	"	590	80.00	670
Maximum	"	1,050	93.00	1,143
68" x 55" high				
Minimum	EA.	950	70.00	1,020
Average	"	1,090	80.00	1,170
Maximum	"	1,420	93.00	1,513
Sliding, 40" x 31" high				
Minimum	EA.	310	56.00	366
Average	"	480	70.00	550
Maximum	"	570	93.00	663
52" x 39" high				
Minimum	EA.	390	70.00	460
Average	"	580	80.00	660
Maximum	"	620	93.00	713
64" x 72" high				
Minimum	EA.	600	70.00	670
Average	"	960	93.00	1,053
Maximum	"	1,060	110	1,170
Awning windows				
34" x 21" high				
Minimum	EA.	310	56.00	366
Average	"	360	70.00	430
Maximum	"	420	93.00	513
40" x 21" high				
Minimum	EA.	370	62.00	432
Average	"	410	80.00	490
Maximum	"	460	110	570
48" x 27" high				
Minimum	EA.	390	62.00	452
Average	"	470	80.00	550
Maximum	"	540	110	650
60" x 36" high				
Minimum	EA.	410	70.00	480
Average	"	730	93.00	823
Maximum	"	820	110	930
Window frame, milled				
Minimum	L.F.	5.80	11.25	17.05
Average	"	6.47	14.00	20.47
Maximum	"	9.74	18.75	28.49

08 DOORS AND WINDOWS

Hardware	UNIT	MAT.	INST.	TOTAL
08710.10 Hinges				
Hinges				
3 x 3 butts, steel, interior, plain bearing	PAIR			20.75
4 x 4 butts, steel, standard	"			30.50
5 x 4-1/2 butts, bronze/s. steel, heavy duty	"			79.00
Pivot hinges				
Top pivot	EA.			88.00
Intermediate pivot	"			94.00
Bottom pivot	"			180
08710.20 Locksets				
Latchset, heavy duty				
Cylindrical	EA.	190	35.00	225
Mortise	"	200	56.00	256
Lockset, heavy duty				
Cylindrical	EA.	310	35.00	345
Mortise	"	350	56.00	406
Preassembled locks and latches, brass				
Latchset, passage or closet latch	EA.	280	46.75	327
Lockset				
Privacy (bath or bathroom)	EA.	340	46.75	387
Entry lock	"	490	46.75	537
Bored locks and latches, satin chrome plated				
Latchset passage or closet latch	EA.	160	46.75	207
Lockset				
Privacy (bath or bedroom)	EA.	220	46.75	267
Entry lock	"	240	46.75	287
08710.30 Closers				
Door closers				
Surface mounted, traditional type, parallel arm				
Standard	EA.	240	70.00	310
Heavy duty	"	280	70.00	350
Modern type, parallel arm, standard duty	"	290	70.00	360
Overhead, concealed, pivot hung, single acting				
Interior	EA.	450	70.00	520
Exterior	"	670	70.00	740
Floor concealed, single acting, offset, pivoted				
Interior	EA.	730	190	920
Exterior	"	930	190	1,120
08710.40 Door Trim				
Panic device				
Mortise	EA.	780	140	920
Vertical rod	"	1,170	140	1,310
Labeled, rim type	"	810	140	950
Mortise	"	1,060	140	1,200
Vertical rod	"	1,130	140	1,270
Door plates				
Kick plate, aluminum, 3 beveled edges				
10" x 28"	EA.	29.00	28.00	57.00
10" x 30"	"	32.00	28.00	60.00
10" x 34"	"	34.75	28.00	62.75
10" x 38"	"	37.75	28.00	65.75

Hardware

08710.40 Door Trim *(Cont.)*

Hardware	UNIT	MAT.	INST.	TOTAL
Push plate, 4" x 16"				
Aluminum	EA.	26.50	11.25	37.75
Bronze	"	84.00	11.25	95.25
Stainless steel	"	67.00	11.25	78.25
Armor plate, 40" x 34"	"	77.00	22.50	99.50
Pull handle, 4" x 16"				
Aluminum	EA.	94.00	11.25	105
Bronze	"	180	11.25	191
Stainless steel	"	140	11.25	151
Hasp assembly				
3"	EA.	4.40	9.32	13.72
4-1/2"	"	5.50	12.50	18.00
6"	"	8.74	16.00	24.74

08710.60 Weatherstripping

Weatherstripping	UNIT	MAT.	INST.	TOTAL
Weatherstrip, head and jamb, metal strip, neoprene bulb				
Standard duty	L.F.	4.95	3.10	8.05
Heavy duty	"	5.50	3.49	8.99
Spring type				
Metal doors	EA.	55.00	140	195
Wood doors	"	55.00	190	245
Sponge type with adhesive backing	"	51.00	56.00	107
Astragal				
1-3/4" x 13 ga., aluminum	L.F.	6.76	4.66	11.42
1-3/8" x 5/8", oak	"	5.50	3.73	9.23
Thresholds				
Bronze	L.F.	53.00	14.00	67.00
Aluminum				10.75
Plain	L.F.	36.50	14.00	50.50
Vinyl insert	"	39.25	14.00	53.25
Aluminum with grit	"	37.50	14.00	51.50
Steel				
Plain	L.F.	29.75	14.00	43.75
Interlocking	"	39.50	46.75	86.25

Glazing

08810.10 Glazing

Glazing	UNIT	MAT.	INST.	TOTAL
Sheet glass, 1/8" thick	S.F.	7.70	3.43	11.13
Plate glass, bronze or grey, 1/4" thick	"	11.25	5.62	16.87
Clear	"	8.80	5.62	14.42
Polished	"	10.50	5.62	16.12
Plexiglass				
1/8" thick	S.F.	4.95	5.62	10.57
1/4" thick	"	8.91	3.43	12.34
Float glass, clear				

Glazing

08810.10 Glazing (Cont.)

Glazing	UNIT	MAT.	INST.	TOTAL
3/16" thick	S.F.	5.99	5.15	11.14
1/4" thick	"	6.10	5.62	11.72
3/8" thick	"	12.25	7.73	19.98
Tinted glass, polished plate, twin ground				
3/16" thick	S.F.	8.25	5.15	13.40
1/4" thick	"	8.25	5.62	13.87
3/8" thick	"	13.25	7.73	20.98
Insulated glass, bronze or gray				
1/2" thick	S.F.	16.75	10.25	27.00
1" thick	"	20.00	15.50	35.50
Spandrel, polished, 1 side, 1/4" thick	"	13.50	5.62	19.12
Tempered glass (safety)				
Clear sheet glass				
1/8" thick	S.F.	9.25	3.43	12.68
3/16" thick	"	11.25	4.75	16.00
Clear float glass				
1/4" thick	S.F.	9.68	5.15	14.83
5/16" thick	"	17.25	6.18	23.43
3/8" thick	"	21.25	7.73	28.98
1/2" thick	"	29.00	10.25	39.25
Tinted float glass				
3/16" thick	S.F.	11.50	4.75	16.25
1/4" thick	"	12.75	5.15	17.90
3/8" thick	"	23.25	7.73	30.98
1/2" thick	"	30.75	10.25	41.00
Laminated glass				
Float safety glass with polyvinyl plastic layer				
1/4", sheet or float				
Two lites, 1/8" thick, clear glass	S.F.	12.25	5.15	17.40
1/2" thick, float glass				
Two lites, 1/4" thick, clear glass	S.F.	18.75	10.25	29.00
Tinted glass	"	22.00	10.25	32.25
Insulating glass, two lites, clear float glass				
1/2" thick	S.F.	12.00	10.25	22.25
5/8" thick	"	13.75	12.25	26.00
3/4" thick	"	15.25	15.50	30.75
Glass seal edge				
3/8" thick	S.F.	10.00	10.25	20.25
Tinted glass				
1/2" thick	S.F.	20.50	10.25	30.75
1" thick	"	22.00	20.50	42.50
Tempered, clear				
1" thick	S.F.	40.25	20.50	60.75
Wire reinforced	"	51.00	20.50	71.50
Plate mirror glass				
1/4" thick				
15 sf	S.F.	10.50	6.18	16.68
Over 15 sf	"	9.68	5.62	15.30
Door type, 1/4" thick	"	11.00	6.18	17.18
Transparent, one way vision, 1/4" thick	"	23.25	6.18	29.43
Sheet mirror glass				
3/16" thick	S.F.	10.25	6.18	16.43
1/4" thick	"	10.75	5.15	15.90

Glazing

08810.10 Glazing *(Cont.)*

Glazing	UNIT	MAT.	INST.	TOTAL
Wall tiles, 12" x 12"				
Clear glass	S.F.	3.46	3.43	6.89
Veined glass	"	4.40	3.43	7.83
Wire glass, 1/4" thick				
Clear	S.F.	20.50	20.50	41.00
Hammered	"	20.75	20.50	41.25
Obscure	"	24.25	20.50	44.75
Glazing accessories				
Neoprene glazing gaskets				
1/4" glass	L.F.	2.17	2.47	4.64
3/8" glass	"	2.42	2.57	4.99
1/2" glass	"	2.54	2.68	5.22

Glazed Curtain Walls

08910.10 Glazed Curtain Walls

Glazed Curtain Walls	UNIT	MAT.	INST.	TOTAL
Curtain wall, aluminum system, framing sections				
2" x 3"				
Jamb	L.F.	11.75	5.15	16.90
Horizontal	"	12.00	5.15	17.15
Mullion	"	16.00	5.15	21.15
2" x 4"				
Jamb	L.F.	16.00	7.73	23.73
Horizontal	"	16.50	7.73	24.23
Mullion	"	16.00	7.73	23.73
3" x 5-1/2"				
Jamb	L.F.	21.25	7.73	28.98
Horizontal	"	23.50	7.73	31.23
Mullion	"	21.25	7.73	28.98
4" corner mullion	"	28.25	10.25	38.50
Coping sections				
1/8" x 8"	L.F.	29.50	10.25	39.75
1/8" x 9"	"	29.75	10.25	40.00
1/8" x 12-1/2"	"	30.50	12.25	42.75
Sill section				
1/8" x 6"	L.F.	29.00	6.18	35.18
1/8" x 7"	"	29.50	6.18	35.68
1/8" x 8-1/2"	"	30.00	6.18	36.18
Column covers, aluminum				
1/8" x 26"	L.F.	29.00	15.50	44.50
1/8" x 34"	"	29.50	16.25	45.75
1/8" x 38"	"	29.75	16.25	46.00
Doors				
Aluminum framed, standard hardware				
Narrow stile				
2-6 x 7-0	EA.	660	310	970

Glazed Curtain Walls	UNIT	MAT.	INST.	TOTAL
08910.10 Glazed Curtain Walls *(Cont.)*				
3-0 x 7-0	EA.	670	310	980
3-6 x 7-0	"	690	310	1,000
Wide stile				
2-6 x 7-0	EA.	1,130	310	1,440
3-0 x 7-0	"	1,220	310	1,530
3-6 x 7-0	"	1,310	310	1,620
Flush panel doors, to match adjacent wall panels				
2-6 x 7-0	EA.	950	390	1,340
3-0 x 7-0	"	1,010	390	1,400
3-6 x 7-0	"	1,040	390	1,430
Window wall system, complete				
Minimum	S.F.	31.75	6.18	37.93
Average	"	51.00	6.87	57.87
Maximum	"	120	8.83	129
Added costs				
For bronze, add 20% to material				
For stainless steel, add 50% to material				

Support Systems

09110.10 Metal Studs

Support Systems	UNIT	MAT.	INST.	TOTAL
Studs, non load bearing, galvanized				
2-1/2", 20 ga.				
12" o.c.	S.F.	0.71	1.16	1.87
16" o.c.	"	0.55	0.93	1.48
25 ga.				
12" o.c.	S.F.	0.48	1.16	1.64
16" o.c.	"	0.38	0.93	1.31
24" o.c.	"	0.29	0.77	1.06
3-5/8", 20 ga.				
12" o.c.	S.F.	0.85	1.39	2.24
16" o.c.	"	0.66	1.11	1.77
24" o.c.	"	0.49	0.93	1.42
25 ga.				
12" o.c.	S.F.	0.57	1.39	1.96
16" o.c.	"	0.46	1.11	1.57
24" o.c.	"	0.35	0.93	1.28
4", 20 ga.				
12" o.c.	S.F.	0.93	1.39	2.32
16" o.c.	"	0.71	1.11	1.82
24" o.c.	"	0.55	0.93	1.48
25 ga.				
12" o.c.	S.F.	0.63	1.39	2.02
16" o.c.	"	0.49	1.11	1.60
24" o.c.	"	0.37	0.93	1.30
6", 20 ga.				
12" o.c.	S.F.	1.19	1.74	2.93
16" o.c.	"	0.88	1.39	2.27
24" o.c.	"	0.71	1.16	1.87
25 ga.				
12" o.c.	S.F.	0.78	1.74	2.52
16" o.c.	"	0.61	1.39	2.00
24" o.c.	"	0.46	1.16	1.62
Load bearing studs, galvanized				
3-5/8", 16 ga.				
12" o.c.	S.F.	1.55	1.39	2.94
16" o.c.	"	1.43	1.11	2.54
18 ga.				
12" o.c.	S.F.	1.11	0.93	2.04
16" o.c.	"	1.21	1.11	2.32
4", 16 ga.				
12" o.c.	S.F.	1.62	1.39	3.01
16" o.c.	"	1.48	1.11	2.59
6", 16 ga.				
12" o.c.	S.F.	2.07	1.74	3.81
16" o.c.	"	1.87	1.39	3.26
Furring				
On beams and columns				
7/8" channel	L.F.	0.55	3.73	4.28
1-1/2" channel	"	0.66	4.30	4.96
On ceilings				
3/4" furring channels				
12" o.c.	S.F.	0.39	2.33	2.72
16" o.c.	"	0.30	2.23	2.53

Support Systems

09110.10 Metal Studs (Cont.)

	UNIT	MAT.	INST.	TOTAL
24" o.c.	S.F.	0.22	1.99	2.21
1-1/2" furring channels				
12" o.c.	S.F.	0.66	2.54	3.20
16" o.c.	"	0.49	2.33	2.82
24" o.c.	"	0.34	2.15	2.49
On walls				
3/4" furring channels				
12" o.c.	S.F.	0.39	1.86	2.25
16" o.c.	"	0.30	1.74	2.04
24" o.c.	"	0.22	1.64	1.86
1-1/2" furring channels				
12" o.c.	S.F.	0.66	1.99	2.65
16" o.c.	"	0.49	1.86	2.35
24" o.c.	"	0.34	1.74	2.08

Lath And Plaster

09205.10 Gypsum Lath

	UNIT	MAT.	INST.	TOTAL
Gypsum lath, 1/2" thick				
Clipped	S.Y.	7.20	3.10	10.30
Nailed	"	7.20	3.49	10.69

09205.20 Metal Lath

	UNIT	MAT.	INST.	TOTAL
Diamond expanded, galvanized				
2.5 lb., on walls				
Nailed	S.Y.	4.22	6.99	11.21
Wired	"	4.22	7.99	12.21
On ceilings				
Nailed	S.Y.	4.22	7.99	12.21
Wired	"	4.22	9.32	13.54
3.4 lb., on walls				
Nailed	S.Y.	5.73	6.99	12.72
Wired	"	5.73	7.99	13.72
On ceilings				
Nailed	S.Y.	5.73	7.99	13.72
Wired	"	5.73	9.32	15.05
Flat rib				
2.75 lb., on walls				
Nailed	S.Y.	3.99	6.99	10.98
Wired	"	3.99	7.99	11.98
On ceilings				
Nailed	S.Y.	3.99	7.99	11.98
Wired	"	3.99	9.32	13.31
3.4 lb., on walls				
Nailed	S.Y.	4.79	6.99	11.78
Wired	"	4.79	7.99	12.78

09 FINISHES

Lath And Plaster	UNIT	MAT.	INST.	TOTAL
09205.20 Metal Lath (Cont.)				
On ceilings				
Nailed	S.Y.	4.79	7.99	12.78
Wired	"	4.79	9.32	14.11
Stucco lath				
1.8 lb.	S.Y.	4.96	6.99	11.95
3.6 lb.	"	5.56	6.99	12.55
Paper backed				
Minimum	S.Y.	3.85	5.59	9.44
Maximum	"	6.21	7.99	14.20
09205.60 Plaster Accessories				
Expansion joint, 3/4", 26 ga., galv.	L.F.	1.48	1.39	2.87
Plaster corner beads, 3/4", galvanized	"	0.41	1.59	2.00
Casing bead, expanded flange, galvanized	"	0.56	1.39	1.95
Expanded wing, 1-1/4" wide, galvanized	"	0.66	1.39	2.05
Joint clips for lath	EA.	0.17	0.27	0.44
Metal base, galvanized, 2-1/2" high	L.F.	0.75	1.86	2.61
Stud clips for gypsum lath	EA.	0.17	0.27	0.44
Tie wire galvanized, 18 ga., 25 lb. hank	"			47.00
Sound deadening board, 1/4"	S.F.	0.31	0.93	1.24
09210.10 Plaster				
Gypsum plaster, trowel finish, 2 coats				
Ceilings	S.Y.	6.20	16.25	22.45
Walls	"	6.20	15.25	21.45
3 coats				
Ceilings	S.Y.	8.60	22.75	31.35
Walls	"	8.60	20.00	28.60
Vermiculite plaster				
2 coats				
Ceilings	S.Y.	7.06	24.75	31.81
Walls	"	7.06	22.75	29.81
3 coats				
Ceilings	S.Y.	11.00	30.75	41.75
Walls	"	11.00	27.50	38.50
Keenes cement plaster				
2 coats				
Ceilings	S.Y.	2.47	20.00	22.47
Walls	"	2.47	17.50	19.97
3 coats				
Ceilings	S.Y.	2.47	22.75	25.22
Walls	"	2.47	20.00	22.47
On columns, add to installation, 50%	"			
Chases, fascia, and soffits, add to installation, 50%	"			
Beams, add to installation, 50%	"			
Patch holes, average size holes				
1 sf to 5 sf				
Minimum	S.F.	2.47	8.70	11.17
Average	"	2.47	10.50	12.97
Maximum	"	2.47	13.00	15.47
Over 5 sf				
Minimum	S.F.	2.47	5.22	7.69

09 FINISHES

Lath And Plaster

	UNIT	MAT.	INST.	TOTAL
09210.10 Plaster *(Cont.)*				
Average	S.F.	2.47	7.45	9.92
Maximum	"	2.47	8.70	11.17
Patch cracks				
Minimum	S.F.	2.47	1.74	4.21
Average	"	2.47	2.61	5.08
Maximum	"	2.47	5.22	7.69
09220.10 Portland Cement Plaster				
Stucco, portland, gray, 3 coat, 1" thick				
Sand finish	S.Y.	7.75	22.75	30.50
Trowel finish	"	7.75	23.75	31.50
White cement				
Sand finish	S.Y.	8.85	23.75	32.60
Trowel finish	"	8.85	26.00	34.85
Scratch coat				
For ceramic tile	S.Y.	2.81	5.22	8.03
For quarry tile	"	2.81	5.22	8.03
Portland cement plaster				
2 coats, 1/2"	S.Y.	5.58	10.50	16.08
3 coats, 7/8"	"	6.66	13.00	19.66
09250.10 Gypsum Board				
Drywall, plasterboard, 3/8" clipped to				
Metal furred ceiling	S.F.	0.35	0.62	0.97
Columns and beams	"	0.35	1.39	1.74
Walls	"	0.35	0.55	0.90
Nailed or screwed to				
Wood framed ceiling	S.F.	0.35	0.55	0.90
Columns and beams	"	0.35	1.24	1.59
Walls	"	0.35	0.50	0.85
1/2", clipped to				
Metal furred ceiling	S.F.	0.36	0.62	0.98
Columns and beams	"	0.33	1.39	1.72
Walls	"	0.33	0.55	0.88
Nailed or screwed to				
Wood framed ceiling	S.F.	0.33	0.55	0.88
Columns and beams	"	0.33	1.24	1.57
Walls	"	0.33	0.50	0.83
5/8", clipped to				
Metal furred ceiling	S.F.	0.36	0.69	1.05
Columns and beams	"	0.36	1.55	1.91
Walls	"	0.36	0.62	0.98
Nailed or screwed to				
Wood framed ceiling	S.F.	0.36	0.69	1.05
Columns and beams	"	0.36	1.55	1.91
Walls	"	0.36	0.62	0.98
Vinyl faced, clipped to metal studs				
1/2"	S.F.	0.99	0.69	1.68
5/8"	"	0.93	0.69	1.62
Add for				
Fire resistant	S.F.			0.11
Water resistant	"			0.17
Water and fire resistant	"			0.22

Lath And Plaster

09250.10 Gypsum Board *(Cont.)*

	UNIT	MAT.	INST.	TOTAL
Taping and finishing joints				
Minimum	S.F.	0.04	0.37	0.41
Average	"	0.06	0.46	0.52
Maximum	"	0.09	0.55	0.64
Casing bead				
Minimum	L.F.	0.15	1.59	1.74
Average	"	0.16	1.86	2.02
Maximum	"	0.20	2.79	2.99
Corner bead				
Minimum	L.F.	0.16	1.59	1.75
Average	"	0.20	1.86	2.06
Maximum	"	0.25	2.79	3.04

Tile

09310.10 Ceramic Tile

	UNIT	MAT.	INST.	TOTAL
Glazed wall tile, 4-1/4" x 4-1/4"				
Minimum	S.F.	2.32	3.81	6.13
Average	"	3.68	4.45	8.13
Maximum	"	13.25	5.34	18.59
Base, 4-1/4" high				
Minimum	L.F.	3.83	6.67	10.50
Average	"	4.46	6.67	11.13
Maximum	"	5.89	6.67	12.56
Unglazed floor tile				
Portland cem., cushion edge, face mtd				
1" x 1"	S.F.	7.72	4.85	12.57
2" x 2"	"	8.16	4.45	12.61
4" x 4"	"	7.59	4.45	12.04
6" x 6"	"	2.71	3.81	6.52
12" x 12"	"	2.38	3.33	5.71
16" x 16"	"	2.07	2.96	5.03
18" x 18"	"	2.01	2.67	4.68
Adhesive bed, with white grout				
1" x 1"	S.F.	6.71	4.85	11.56
2" x 2"	"	7.09	4.45	11.54
4" x 4"	"	7.09	4.45	11.54
6" x 6"	"	2.36	3.81	6.17
12" x 12"	"	2.07	3.33	5.40
16" x 16"	"	1.80	2.96	4.76
18" x 18"	"	1.74	2.67	4.41
Organic adhesive bed, thin set, back mounted				
1" x 1"	S.F.	6.71	4.85	11.56
2" x 2"	"	7.81	4.45	12.26
For group 2 colors, add to material, 10%				
For group 3 colors, add to material, 20%				

Tile

09310.10 Ceramic Tile *(Cont.)*

Tile	UNIT	MAT.	INST.	TOTAL
For abrasive surface, add to material, 25%				
Porcelain floor tile				
1" x 1"	S.F.	8.99	4.85	13.84
2" x 2"	"	8.22	4.64	12.86
4" x 4"	"	7.64	4.45	12.09
6" x 6"	"	2.75	3.81	6.56
12" x 12"	"	2.47	3.33	5.80
16" x 16"	"	1.96	2.96	4.92
18" x 18"	"	1.85	2.67	4.52
Unglazed wall tile				
Organic adhesive, face mounted cushion edge				
1" x 1"				
Minimum	S.F.	3.57	4.45	8.02
Average	"	4.67	4.85	9.52
Maximum	"	6.98	5.34	12.32
2" x 2"				
Minimum	S.F.	4.12	4.10	8.22
Average	"	4.67	4.45	9.12
Maximum	"	7.64	4.85	12.49
Back mounted				
1" x 1"				
Minimum	S.F.	3.57	4.45	8.02
Average	"	4.67	4.85	9.52
Maximum	"	6.98	5.34	12.32
2" x 2"				
Minimum	S.F.	4.12	4.10	8.22
Average	"	4.67	4.45	9.12
Maximum	"	7.64	4.85	12.49
For glazed finish, add to material, 25%				
For glazed mosaic, add to material, 100%				
For metallic colors, add to material, 125%				
For exterior wall use, add to total, 25%				
For exterior soffit, add to total, 25%				
For portland cement bed, add to total, 25%				
For dry set portland cement bed, add to total, 10%				
Conductive floor tile, unglazed square edged				
Portland cement bed				
1 x 1	S.F.	6.93	6.67	13.60
1-9/16 x 1-9/16	"	6.38	6.67	13.05
Dry set				
1 x 1	S.F.	6.93	6.67	13.60
1-9/16 x 1-9/16	"	6.38	6.67	13.05
Epoxy bed with epoxy joints				
1 x 1	S.F.	6.93	6.67	13.60
1-9/16 x 1-9/16	"	6.38	6.67	13.05
For WWF in bed add to total, 15%				
For abrasive surface, add to material, 40%				
Ceramic accessories				
Towel bar, 24" long				
Minimum	EA.	16.25	21.25	37.50
Average	"	20.25	26.75	47.00
Maximum	"	54.00	35.50	89.50
Soap dish				

Tile	UNIT	MAT.	INST.	TOTAL
09310.10 Ceramic Tile (Cont.)				
Minimum	EA.	7.70	35.50	43.20
Average	"	10.50	44.50	55.00
Maximum	"	27.50	53.00	80.50
09330.10 Quarry Tile				
Floor				
4 x 4 x 1/2"	S.F.	6.27	7.12	13.39
6 x 6 x 1/2"	"	6.13	6.67	12.80
6 x 6 x 3/4"	"	7.63	6.67	14.30
12 x 12x 3/4"	"	10.75	5.93	16.68
16x1 6 x 3/4"	"	7.48	5.34	12.82
18 x 18 x 3/4"	"	5.26	4.45	9.71
Medallion				
36" dia.	EA.	320	130	450
48" dia.	"	380	130	510
Wall, applied to 3/4" portland cement bed				
4 x 4 x 1/2"	S.F.	4.62	10.75	15.37
6 x 6 x 3/4"	"	5.16	8.90	14.06
Cove base				
5 x 6 x 1/2" straight top	L.F.	6.05	8.90	14.95
6 x 6 x 3/4" round top	"	5.62	8.90	14.52
Moldings				
2 x 12	L.F.	9.63	5.34	14.97
4 x 12	"	15.00	5.34	20.34
Stair treads 6 x 6 x 3/4"	"	8.28	13.25	21.53
Window sill 6 x 8 x 3/4"	"	7.56	10.75	18.31
For abrasive surface, add to material, 25%				
09410.10 Terrazzo				
Floors on concrete, 1-3/4" thick, 5/8" topping				
Gray cement	S.F.	4.23	7.45	11.68
White cement	"	4.62	7.45	12.07
Sand cushion, 3" thick, 5/8" top, 1/4"				
Gray cement	S.F.	5.00	8.70	13.70
White cement	"	5.55	8.70	14.25
Monolithic terrazzo, 3-1/2" base slab, 5/8" topping	"	3.97	6.52	10.49
Terrazzo wainscot, cast-in-place, 1/2" thick	"	7.48	13.00	20.48
Base, cast in place, terrazzo cove type, 6" high	L.F.	8.86	7.45	16.31
Curb, cast in place, 6" wide x 6" high, polished top	"	9.84	26.00	35.84
For venetian type terrazzo, add to material, 10%				
For abrasive heavy duty terrazzo, add to material, 15%				
Divider strips				
Zinc	L.F.			1.50
Brass	"			2.80
Stairs, cast-in-place, topping on concrete or metal				
1-1/2" thick treads, 12" wide	L.F.	5.90	26.00	31.90
Combined tread and riser	"	8.86	65.00	73.86
Precast terrazzo, thin set				
Terrazzo tiles, non-slip surface				
9" x 9" x 1" thick	S.F.	18.75	7.45	26.20
12" x 12"				
1" thick	S.F.	20.25	6.96	27.21
1-1/2" thick	"	21.00	7.45	28.45

Tile	UNIT	MAT.	INST.	TOTAL
09410.10 Terrazzo (Cont.)				
18" x 18" x 1-1/2" thick	S.F.	27.50	7.45	34.95
24" x 24" x 1-1/2" thick	"	35.50	6.14	41.64
For white cement, add to material, 10%				
For venetian type terrazzo, add to material, 25%				
Terrazzo wainscot				
12" x 12" x 1" thick	S.F.	9.29	13.00	22.29
18" x 18" x 1-1/2" thick	"	15.25	15.00	30.25
Base				
6" high				
Straight	L.F.	13.25	4.01	17.26
Coved	"	15.75	4.01	19.76
8" high				
Straight	L.F.	15.00	4.35	19.35
Coved	"	17.50	4.35	21.85
Terrazzo curbs				
8" wide x 8" high	L.F.	34.75	20.75	55.50
6" wide x 6" high	"	31.25	17.50	48.75
Precast terrazzo stair treads, 12" wide				
1-1/2" thick				
Diamond pattern	L.F.	41.75	9.49	51.24
Non-slip surface	"	44.00	9.49	53.49
2" thick				
Diamond pattern	L.F.	44.00	9.49	53.49
Non-slip surface	"	46.25	10.50	56.75
Stair risers, 1" thick to 6" high				
Straight sections	L.F.	14.00	5.22	19.22
Cove sections	"	16.50	5.22	21.72
Combined tread and riser				
Straight sections				
1-1/2" tread, 3/4" riser	L.F.	61.00	15.00	76.00
3" tread, 1" riser	"	72.00	15.00	87.00
Curved sections				
2" tread, 1" riser	L.F.	77.00	17.50	94.50
3" tread, 1" riser	"	80.00	17.50	97.50
Stair stringers, notched for treads and risers				
1" thick	L.F.	36.50	13.00	49.50
2" thick	"	37.75	17.50	55.25
Landings, structural, nonslip				
1-1/2" thick	S.F.	34.25	8.70	42.95
3" thick	"	48.00	10.50	58.50
Conductive terrazzo, spark proof industrial floor				
Epoxy terrazzo				
Floor	S.F.	6.60	3.26	9.86
Base	"	6.60	4.35	10.95
Polyacrylate				
Floor	S.F.	10.75	3.26	14.01
Base	"	10.75	4.35	15.10
Polyester				
Floor	S.F.	3.79	2.08	5.87
Base	"	3.79	2.61	6.40
Synthetic latex mastic				
Floor	S.F.	6.16	3.26	9.42
Base	"	6.16	4.35	10.51

Acoustical Treatment

09510.10 Ceilings And Walls

Description	UNIT	MAT.	INST.	TOTAL
Acoustical panels, suspension system not included				
Fiberglass panels				
5/8" thick				
2' x 2'	S.F.	1.32	0.79	2.11
2' x 4'	"	1.10	0.62	1.72
3/4" thick				
2' x 2'	S.F.	1.76	0.79	2.55
2' x 4'	"	1.70	0.62	2.32
Glass cloth faced fiberglass panels				
3/4" thick	S.F.	2.50	0.93	3.43
1" thick	"	2.79	0.93	3.72
Mineral fiber panels				
5/8" thick				
2' x 2'	S.F.	1.12	0.79	1.91
2' x 4'	"	1.12	0.62	1.74
3/4" thick				
2' x 2'	S.F.	1.76	0.79	2.55
2' x 4'	"	1.70	0.62	2.32
For aluminum faced panels, add to material, 80%				
For vinyl faced panels, add to total, 125%				
For fire rated panels, add to material, 75%				
Wood fiber panels				
1/2" thick				
2' x 2'	S.F.	1.68	0.79	2.47
2' x 4'	"	1.68	0.62	2.30
5/8" thick				
2' x 2'	S.F.	1.84	0.79	2.63
2' x 4'	"	1.84	0.62	2.46
3/4" thick				
2' x 2'	S.F.	2.26	0.79	3.05
2' x 4'	"	2.26	0.62	2.88
2" thick				
2' x 2'	S.F.	2.64	0.93	3.57
2' x 4'	"	2.64	0.69	3.33
For flameproofing, add to material, 10%				
For sculptured finish, add to material, 15%				
Air distributing panels				
3/4" thick	S.F.	2.64	1.39	4.03
5/8" thick	"	2.26	1.11	3.37
Acoustical tiles, suspension system not included				
Fiberglass tile, 12" x 12"				
5/8" thick	S.F.	1.39	1.01	2.40
3/4" thick	"	1.61	1.24	2.85
Glass cloth faced fiberglass tile				
3/4" thick	S.F.	3.24	1.24	4.48
3" thick	"	3.63	1.39	5.02
Mineral fiber tile, 12" x 12"				
5/8" thick				
Standard	S.F.	0.82	1.11	1.93
Vinyl faced	"	1.65	1.11	2.76
3/4" thick				
Standard	S.F.	1.21	1.11	2.32
Vinyl faced	"	2.11	1.11	3.22

Acoustical Treatment

09510.10 Ceilings And Walls *(Cont.)*

	UNIT	MAT.	INST.	TOTAL
Fire rated	S.F.	2.68	1.11	3.79
Aluminum or mylar faced	"	5.11	1.11	6.22
Wood fiber tile, 12" x 12"				
1/2" thick	S.F.	1.33	1.11	2.44
3/4" thick	"	1.92	1.11	3.03
For flameproofing, add to material, 10%				
For sculptured 3 dimensional, add to material, 50%				
Metal pan units, 24 ga. steel				
12" x 12"	S.F.	5.22	2.23	7.45
12" x 24"	"	5.40	1.86	7.26
Aluminum, .025" thick				
12" x 12"	S.F.	5.55	2.23	7.78
12" x 24"	"	5.73	1.86	7.59
Anodized aluminum, 0.25" thick				
12" x 12"	S.F.	6.40	2.23	8.63
12" x 24"	"	7.59	1.86	9.45
Stainless steel, 24 ga.				
12" x 12"	S.F.	12.75	2.23	14.98
12" x 24"	"	13.25	1.86	15.11
For flameproof sound absorbing pads, add to material	"			1.99
Metal ceiling systems				
.020" thick panels				
10', 12', and 16' lengths	S.F.	4.99	1.59	6.58
Custom lengths, 3' to 20'	"	5.04	1.59	6.63
.025" thick panels				
32 sf, 38 sf, and 52 sf pieces	S.F.	5.01	1.86	6.87
Custom lengths, 10 sf to 65 sf	"	5.85	1.86	7.71
Carriers, black, add	"			3.01
Recess filler strip, add	"			1.01
Custom lengths, add	"			1.50
Sound absorption walls, with fabric cover				
2-6" x 9' x 3/4"	S.F.	9.83	1.86	11.69
2' x 9' x 1"	"	10.75	1.86	12.61
Starter spline	L.F.	1.57	1.39	2.96
Internal spline	"	1.37	1.39	2.76
Acoustical treatment				
Barriers for plenums				
Leaded vinyl				
0.48 lb per sf	S.F.	2.98	2.66	5.64
0.87 lb per sf	"	2.98	2.79	5.77
Aluminum foil, fiberglass reinforcement				
Minimum	S.F.	1.11	1.86	2.97
Maximum	"	1.26	2.79	4.05
Aluminum mesh, paper backed	"	1.04	1.86	2.90
Fibered cement sheet, 3/16" thick	"	2.14	1.99	4.13
Sheet lead, 1/64" thick	"	3.06	1.39	4.45
Sound attenuation blanket				
1" thick	S.F.	0.39	5.59	5.98
1-1/2" thick	"	0.55	5.59	6.14
2" thick	"	0.68	5.59	6.27
3" thick	"	0.82	6.21	7.03
Ceiling suspension systems				

Acoustical Treatment

09510.10 Ceilings And Walls *(Cont.)*

Acoustical Treatment	UNIT	MAT.	INST.	TOTAL
T bar system				
2' x 4'	S.F.	1.15	0.55	1.70
2' x 2'	"	1.25	0.62	1.87
Concealed Z bar suspension system, 12" module	"	1.07	0.93	2.00
For 1-1/2" carrier channels, 4' o.c., add	"			0.38
Carrier channel for recessed light fixtures	"			0.69

Flooring

09550.10 Wood Flooring

Flooring	UNIT	MAT.	INST.	TOTAL
Wood strip flooring, unfinished				
Fir floor				
C and better				
Vertical grain	S.F.	3.52	1.86	5.38
Flat grain	"	3.32	1.86	5.18
Oak floor				
Minimum	S.F.	3.71	2.66	6.37
Average	"	5.12	2.66	7.78
Maximum	"	7.42	2.66	10.08
Maple floor				
25/32" x 2-1/4"				
Minimum	S.F.	3.47	2.66	6.13
Maximum	"	5.07	2.66	7.73
33/32" x 3-1/4"				
Minimum	S.F.	5.56	2.66	8.22
Maximum	"	6.29	2.66	8.95
Added costs				
For factory finish, add to material, 10%				
For random width floor, add to total, 20%				
For simulated pegs, add to total, 10%				
Wood block industrial flooring				
Creosoted				
2" thick	S.F.	4.18	1.47	5.65
2-1/2" thick	"	4.34	1.74	6.08
3" thick	"	4.51	1.86	6.37
Parquet, 5/16", white oak				
Finished	S.F.	9.13	2.79	11.92
Unfinished	"	4.40	2.79	7.19
Gym floor, 2 ply felt, 25/32" maple, finished, in mastic	"	7.75	3.10	10.85
Over wood sleepers	"	8.69	3.49	12.18
Finishing, sand, fill, finish, and wax	"	0.66	1.39	2.05
Refinish sand, seal, and 2 coats of polyurethane	"	1.15	1.86	3.01
Clean and wax floors	"	0.19	0.27	0.46

Flooring

Flooring	UNIT	MAT.	INST.	TOTAL
09630.10 Unit Masonry Flooring				
Clay brick				
9 x 4-1/2 x 3" thick				
Glazed	S.F.	8.03	4.66	12.69
Unglazed	"	7.70	4.66	12.36
8 x 4 x 3/4" thick				
Glazed	S.F.	7.26	4.86	12.12
Unglazed	"	7.15	4.86	12.01
For herringbone pattern, add to labor, 15%				
09660.10 Resilient Tile Flooring				
Solid vinyl tile, 1/8" thick, 12" x 12"				
Marble patterns	S.F.	4.45	1.39	5.84
Solid colors	"	5.77	1.39	7.16
Travertine patterns	"	6.49	1.39	7.88
Conductive resilient flooring, vinyl tile				
1/8" thick, 12" x 12"	S.F.	6.71	1.59	8.30
09665.10 Resilient Sheet Flooring				
Vinyl sheet flooring				
Minimum	S.F.	3.83	0.55	4.38
Average	"	6.19	0.66	6.85
Maximum	"	10.50	0.93	11.43
Cove, to 6"	L.F.	2.28	1.11	3.39
Fluid applied resilient flooring				
Polyurethane, poured in place, 3/8" thick	S.F.	10.50	4.66	15.16
Vinyl sheet goods, backed				
0.070" thick	S.F.	3.90	0.69	4.59
0.093" thick	"	6.05	0.69	6.74
0.125" thick	"	6.98	0.69	7.67
0.250" thick	"	8.03	0.69	8.72
09678.10 Resilient Base And Accessories				
Wall base, vinyl				
Group 1				
4" high	L.F.	1.11	1.86	2.97
6" high	"	1.51	1.86	3.37
Group 2				
4" high	L.F.	0.97	1.86	2.83
6" high	"	1.55	1.86	3.41
Group 3				
4" high	L.F.	2.21	1.86	4.07
6" high	"	2.48	1.86	4.34
Stair accessories				
Treads, 1/4" x 12", rubber diamond surface				
Marbled	L.F.	12.75	4.66	17.41
Plain	"	13.25	4.66	17.91
Grit strip safety tread, 12" wide, colors				
3/16" thick	L.F.	12.50	4.66	17.16
5/16" thick	"	16.75	4.66	21.41
Risers, 7" high, 1/8" thick, colors				
Flat	L.F.	5.17	2.79	7.96
Coved	"	3.57	2.79	6.36
Nosing, rubber				

Flooring

09678.10 Resilient Base And Accessories *(Cont.)*

Flooring	UNIT	MAT.	INST.	TOTAL
3/16" thick, 3" wide				
Black	L.F.	4.01	2.79	6.80
Colors	"	4.18	2.79	6.97
6" wide				
Black	L.F.	5.00	4.66	9.66
Colors	"	5.22	4.66	9.88

Carpet

09680.10 Floor Leveling

Carpet	UNIT	MAT.	INST.	TOTAL
Repair and level floors to receive new flooring				
Minimum	S.Y.	0.88	1.86	2.74
Average	"	3.41	4.66	8.07
Maximum	"	5.06	5.59	10.65

09682.10 Carpet Padding

Carpet	UNIT	MAT.	INST.	TOTAL
Carpet padding				
Foam rubber, waffle type, 0.3" thick	S.Y.	6.16	2.79	8.95
Jute padding				
Minimum	S.Y.	4.18	2.54	6.72
Average	"	5.44	2.79	8.23
Maximum	"	8.19	3.10	11.29
Sponge rubber cushion				
Minimum	S.Y.	4.95	2.54	7.49
Average	"	6.60	2.79	9.39
Maximum	"	9.35	3.10	12.45
Urethane cushion, 3/8" thick				
Minimum	S.Y.	4.95	2.54	7.49
Average	"	5.77	2.79	8.56
Maximum	"	7.53	3.10	10.63

09685.10 Carpet

Carpet	UNIT	MAT.	INST.	TOTAL
Carpet, acrylic				
24 oz., light traffic	S.Y.	16.00	6.21	22.21
28 oz., medium traffic	"	19.25	6.21	25.46
Residential				
Nylon				
15 oz., light traffic	S.Y.	22.50	6.21	28.71
28 oz., medium traffic	"	29.00	6.21	35.21
Commercial				
Nylon				
28 oz., medium traffic	S.Y.	27.75	6.21	33.96
35 oz., heavy traffic	"	34.00	6.21	40.21
Wool				
30 oz., medium traffic	S.Y.	46.00	6.21	52.21
36 oz., medium traffic	"	48.50	6.21	54.71

Carpet

09685.10 Carpet *(Cont.)*

Carpet	UNIT	MAT.	INST.	TOTAL
42 oz., heavy traffic	S.Y.	64.00	6.21	70.21
Carpet tile				
Foam backed				
Minimum	S.F.	3.53	1.11	4.64
Average	"	4.08	1.24	5.32
Maximum	"	6.47	1.39	7.86
Tufted loop or shag				
Minimum	S.F.	3.82	1.11	4.93
Average	"	4.61	1.24	5.85
Maximum	"	7.41	1.39	8.80
Clean and vacuum carpet				
Minimum	S.Y.	0.32	0.21	0.53
Average	"	0.50	0.37	0.87
Maximum	"	0.70	0.55	1.25

09700.10 Special Flooring

Special Flooring	UNIT	MAT.	INST.	TOTAL
Epoxy flooring, marble chips				
Epoxy with colored quartz chips in 1/4" base	S.F.	4.26	3.10	7.36
Heavy duty epoxy topping, 3/16" thick	"	3.45	3.10	6.55
Epoxy terrazzo				
1/4" thick chemical resistant	S.F.	6.27	3.49	9.76

Painting

09905.10 Painting Preparation

Painting	UNIT	MAT.	INST.	TOTAL
Dropcloths				
Minimum	S.F.	0.03	0.02	0.05
Average	"	0.05	0.03	0.08
Maximum	"	0.06	0.05	0.11
Masking				
Paper and tape				
Minimum	L.F.	0.02	0.46	0.48
Average	"	0.03	0.58	0.61
Maximum	"	0.04	0.77	0.81
Doors				
Minimum	EA.	0.04	5.84	5.88
Average	"	0.05	7.79	7.84
Maximum	"	0.06	10.50	10.56
Windows				
Minimum	EA.	0.04	5.84	5.88
Average	"	0.05	7.79	7.84
Maximum	"	0.06	10.50	10.56
Sanding				
Walls and flat surfaces				
Minimum	S.F.		0.31	0.31
Average	"		0.38	0.38

Painting

09905.10 Painting Preparation *(Cont.)*	UNIT	MAT.	INST.	TOTAL
Maximum	S.F.		0.46	0.46
Doors and windows				
Minimum	EA.		7.79	7.79
Average	"		11.75	11.75
Maximum	"		15.50	15.50
Trim				
Minimum	L.F.		0.58	0.58
Average	"		0.77	0.77
Maximum	"		1.03	1.03
Puttying				
Minimum	S.F.	0.01	0.71	0.72
Average	"	0.02	0.93	0.95
Maximum	"	0.03	1.16	1.19
Water cleaning/preparation				
Washing (General)				
Minimum	S.F.		0.02	0.02
Average	"		0.03	0.03
Maximum	"		0.05	0.05
Mildew eradication				
Minimum	S.F.	0.03	0.05	0.08
Average	"	0.04	0.09	0.13
Maximum	"	0.05	0.15	0.20
Remove loose paint				
Minimum	S.F.		0.09	0.09
Average	"		0.15	0.15
Maximum	"		0.23	0.23
Steam clean				
Minimum	S.F.		0.11	0.11
Average	"		0.15	0.15
Maximum	"		0.23	0.23
09910.05 Ext. Painting, Sitework				
Benches				
Brush				
First Coat				
Minimum	S.F.	0.16	0.46	0.62
Average	"	0.16	0.58	0.74
Maximum	"	0.16	0.77	0.93
Brickwork				
Brush				
First Coat				
Minimum	S.F.	0.16	0.29	0.45
Average	"	0.16	0.38	0.54
Maximum	"	0.16	0.58	0.74
Second Coat				
Minimum	S.F.	0.16	0.25	0.41
Average	"	0.16	0.31	0.47
Maximum	"	0.16	0.38	0.54
Roller				
First Coat				
Minimum	S.F.	0.16	0.23	0.39
Average	"	0.16	0.29	0.45
Maximum	"	0.16	0.38	0.54

Painting	UNIT	MAT.	INST.	TOTAL
09910.05 Ext. Painting, Sitework *(Cont.)*				
Second Coat				
Minimum	S.F.	0.16	0.19	0.35
Average	"	0.16	0.23	0.39
Maximum	"	0.16	0.29	0.45
Spray				
First Coat				
Minimum	S.F.	0.13	0.12	0.25
Average	"	0.13	0.16	0.29
Maximum	"	0.13	0.21	0.34
Second Coat				
Minimum	S.F.	0.13	0.12	0.25
Average	"	0.13	0.15	0.28
Maximum	"	0.13	0.19	0.32
Concrete Block				
Roller				
First Coat				
Minimum	S.F.	0.16	0.23	0.39
Average	"	0.16	0.31	0.47
Maximum	"	0.16	0.46	0.62
Second Coat				
Minimum	S.F.	0.16	0.19	0.35
Average	"	0.16	0.25	0.41
Maximum	"	0.16	0.38	0.54
Spray				
First Coat				
Minimum	S.F.	0.13	0.12	0.25
Average	"	0.13	0.15	0.28
Maximum	"	0.13	0.17	0.30
Second Coat				
Minimum	S.F.	0.13	0.08	0.21
Average	"	0.13	0.10	0.23
Maximum	"	0.13	0.14	0.27
Fences, Chain Link				
Brush				
First Coat				
Minimum	S.F.	0.11	0.46	0.57
Average	"	0.11	0.51	0.62
Maximum	"	0.11	0.58	0.69
Second Coat				
Minimum	S.F.	0.11	0.31	0.42
Average	"	0.11	0.35	0.46
Maximum	"	0.11	0.42	0.53
Roller				
First Coat				
Minimum	S.F.	0.11	0.33	0.44
Average	"	0.11	0.38	0.49
Maximum	"	0.11	0.44	0.55
Second Coat				
Minimum	S.F.	0.11	0.19	0.30
Average	"	0.11	0.23	0.34
Maximum	"	0.11	0.29	0.40
Spray				
First Coat				

Painting	UNIT	MAT.	INST.	TOTAL
09910.05 Ext. Painting, Sitework *(Cont.)*				
Minimum	S.F.	0.08	0.14	0.22
Average	"	0.08	0.16	0.24
Maximum	"	0.08	0.19	0.27
Second Coat				
Minimum	S.F.	0.08	0.11	0.19
Average	"	0.08	0.12	0.20
Maximum	"	0.08	0.14	0.22
Fences, Wood or Masonry				
Brush				
First Coat				
Minimum	S.F.	0.16	0.49	0.65
Average	"	0.16	0.58	0.74
Maximum	"	0.16	0.77	0.93
Second Coat				
Minimum	S.F.	0.16	0.29	0.45
Average	"	0.16	0.35	0.51
Maximum	"	0.16	0.46	0.62
Roller				
First Coat				
Minimum	S.F.	0.16	0.25	0.41
Average	"	0.16	0.31	0.47
Maximum	"	0.16	0.35	0.51
Second Coat				
Minimum	S.F.	0.16	0.17	0.33
Average	"	0.16	0.22	0.38
Maximum	"	0.16	0.29	0.45
Spray				
First Coat				
Minimum	S.F.	0.13	0.16	0.29
Average	"	0.13	0.21	0.34
Maximum	"	0.13	0.29	0.42
Second Coat				
Minimum	S.F.	0.13	0.11	0.24
Average	"	0.13	0.14	0.27
Maximum	"	0.13	0.19	0.32
09910.15 Ext. Painting, Buildings				
Decks, Wood, Stained				
Brush				
First Coat				
Minimum	S.F.	0.13	0.23	0.36
Average	"	0.13	0.25	0.38
Maximum	"	0.13	0.29	0.42
Second Coat				
Minimum	S.F.	0.13	0.16	0.29
Average	"	0.13	0.17	0.30
Maximum	"	0.13	0.19	0.32
Roller				
First Coat				
Minimum	S.F.	0.13	0.16	0.29
Average	"	0.13	0.17	0.30
Maximum	"	0.13	0.19	0.32
Second Coat				

Painting	UNIT	MAT.	INST.	TOTAL
09910.15 Ext. Painting, Buildings *(Cont.)*				
Minimum	S.F.	0.13	0.14	0.27
Average	"	0.13	0.15	0.28
Maximum	"	0.13	0.17	0.30
Spray				
First Coat				
Minimum	S.F.	0.11	0.14	0.25
Average	"	0.11	0.15	0.26
Maximum	"	0.11	0.17	0.28
Second Coat				
Minimum	S.F.	0.11	0.12	0.23
Average	"	0.11	0.14	0.25
Maximum	"	0.11	0.15	0.26
Doors, Wood				
Brush				
First Coat				
Minimum	S.F.	0.13	0.71	0.84
Average	"	0.13	0.93	1.06
Maximum	"	0.13	1.16	1.29
Second Coat				
Minimum	S.F.	0.13	0.58	0.71
Average	"	0.13	0.66	0.79
Maximum	"	0.13	0.77	0.90
Roller				
First Coat				
Minimum	S.F.	0.13	0.31	0.44
Average	"	0.13	0.38	0.51
Maximum	"	0.13	0.58	0.71
Second Coat				
Minimum	S.F.	0.13	0.23	0.36
Average	"	0.13	0.25	0.38
Maximum	"	0.13	0.38	0.51
Spray				
First Coat				
Minimum	S.F.	0.11	0.14	0.25
Average	"	0.11	0.17	0.28
Maximum	"	0.11	0.23	0.34
Second Coat				
Minimum	S.F.	0.11	0.11	0.22
Average	"	0.11	0.13	0.24
Maximum	"	0.11	0.15	0.26
Gutters and Downspouts				
Brush				
First Coat				
Minimum	L.F.	0.16	0.58	0.74
Average	"	0.16	0.66	0.82
Maximum	"	0.16	0.77	0.93
Second Coat				
Minimum	L.F.	0.16	0.38	0.54
Average	"	0.16	0.46	0.62
Maximum	"	0.16	0.58	0.74
Siding, Wood				
Roller				
First Coat				

Painting	UNIT	MAT.	INST.	TOTAL
09910.15 Ext. Painting, Buildings *(Cont.)*				
Minimum	S.F.	0.11	0.16	0.27
Average	"	0.11	0.19	0.30
Maximum	"	0.11	0.21	0.32
Second Coat				
Minimum	S.F.	0.11	0.19	0.30
Average	"	0.11	0.21	0.32
Maximum	"	0.11	0.23	0.34
Spray				
First Coat				
Minimum	S.F.	0.11	0.15	0.26
Average	"	0.11	0.16	0.27
Maximum	"	0.11	0.17	0.28
Second Coat				
Minimum	S.F.	0.11	0.11	0.22
Average	"	0.11	0.15	0.26
Maximum	"	0.11	0.23	0.34
Stucco				
Roller				
First Coat				
Minimum	S.F.	0.16	0.21	0.37
Average	"	0.16	0.24	0.40
Maximum	"	0.16	0.29	0.45
Second Coat				
Minimum	S.F.	0.16	0.17	0.33
Average	"	0.16	0.19	0.35
Maximum	"	0.16	0.23	0.39
Spray				
First Coat				
Minimum	S.F.	0.13	0.14	0.27
Average	"	0.13	0.16	0.29
Maximum	"	0.13	0.19	0.32
Second Coat				
Minimum	S.F.	0.13	0.11	0.24
Average	"	0.13	0.13	0.26
Maximum	"	0.13	0.15	0.28
Trim				
Brush				
First Coat				
Minimum	L.F.	0.16	0.19	0.35
Average	"	0.16	0.23	0.39
Maximum	"	0.16	0.29	0.45
Second Coat				
Minimum	L.F.	0.16	0.14	0.30
Average	"	0.16	0.19	0.35
Maximum	"	0.16	0.29	0.45
Walls				
Roller				
First Coat				
Minimum	S.F.	0.13	0.16	0.29
Average	"	0.13	0.17	0.30
Maximum	"	0.13	0.18	0.31
Second Coat				
Minimum	S.F.	0.13	0.14	0.27

Painting

	UNIT	MAT.	INST.	TOTAL
09910.15 Ext. Painting, Buildings *(Cont.)*				
Average	S.F.	0.13	0.15	0.28
Maximum	"	0.13	0.17	0.30
Spray				
First Coat				
Minimum	S.F.	0.09	0.07	0.16
Average	"	0.09	0.09	0.18
Maximum	"	0.09	0.11	0.20
Second Coat				
Minimum	S.F.	0.09	0.06	0.15
Average	"	0.09	0.07	0.16
Maximum	"	0.09	0.10	0.19
Windows				
Brush				
First Coat				
Minimum	S.F.	0.11	0.77	0.88
Average	"	0.11	0.93	1.04
Maximum	"	0.11	1.16	1.27
Second Coat				
Minimum	S.F.	0.11	0.66	0.77
Average	"	0.11	0.77	0.88
Maximum	"	0.11	0.93	1.04
09910.25 Ext. Painting, Misc.				
Shakes				
Spray				
First Coat				
Minimum	S.F.	0.12	0.19	0.31
Average	"	0.12	0.21	0.33
Maximum	"	0.12	0.23	0.35
Second Coat				
Minimum	S.F.	0.12	0.17	0.29
Average	"	0.12	0.19	0.31
Maximum	"	0.12	0.21	0.33
Shingles, Wood				
Roller				
First Coat				
Minimum	S.F.	0.13	0.25	0.38
Average	"	0.13	0.29	0.42
Maximum	"	0.13	0.33	0.46
Second Coat				
Minimum	S.F.	0.13	0.17	0.30
Average	"	0.13	0.19	0.32
Maximum	"	0.13	0.21	0.34
Spray				
First Coat				
Minimum	L.F.	0.11	0.17	0.28
Average	"	0.11	0.19	0.30
Maximum	"	0.11	0.21	0.32
Second Coat				
Minimum	L.F.	0.11	0.13	0.24
Average	"	0.11	0.14	0.25
Maximum	"	0.11	0.15	0.26
Shutters and Louvres				

Painting

09910.25 Ext. Painting, Misc. *(Cont.)*

	UNIT	MAT.	INST.	TOTAL
Brush				
First Coat				
Minimum	EA.	0.16	9.35	9.51
Average	"	0.16	11.75	11.91
Maximum	"	0.16	15.50	15.66
Second Coat				
Minimum	EA.	0.16	5.84	6.00
Average	"	0.16	7.19	7.35
Maximum	"	0.16	9.35	9.51
Spray				
First Coat				
Minimum	EA.	0.12	3.11	3.23
Average	"	0.12	3.74	3.86
Maximum	"	0.12	4.67	4.79
Second Coat				
Minimum	EA.	0.12	2.33	2.45
Average	"	0.12	3.11	3.23
Maximum	"	0.12	3.74	3.86
Stairs, metal				
Brush				
First Coat				
Minimum	S.F.	0.16	0.51	0.67
Average	"	0.16	0.58	0.74
Maximum	"	0.16	0.66	0.82
Second Coat				
Minimum	S.F.	0.16	0.29	0.45
Average	"	0.16	0.33	0.49
Maximum	"	0.16	0.38	0.54
Spray				
First Coat				
Minimum	S.F.	0.12	0.25	0.37
Average	"	0.12	0.33	0.45
Maximum	"	0.12	0.35	0.47
Second Coat				
Minimum	S.F.	0.12	0.19	0.31
Average	"	0.12	0.23	0.35
Maximum	"	0.12	0.29	0.41

09910.35 Int. Painting, Buildings

	UNIT	MAT.	INST.	TOTAL
Acoustical Ceiling				
Roller				
First Coat				
Minimum	S.F.	0.16	0.29	0.45
Average	"	0.16	0.38	0.54
Maximum	"	0.16	0.58	0.74
Second Coat				
Minimum	S.F.	0.16	0.23	0.39
Average	"	0.16	0.29	0.45
Maximum	"	0.16	0.38	0.54
Spray				
First Coat				
Minimum	S.F.	0.13	0.12	0.25
Average	"	0.13	0.15	0.28

Painting	UNIT	MAT.	INST.	TOTAL
09910.35 Int. Painting, Buildings *(Cont.)*				
Maximum	S.F.	0.13	0.19	0.32
Second Coat				
Minimum	S.F.	0.13	0.10	0.23
Average	"	0.13	0.11	0.24
Maximum	"	0.13	0.13	0.26
Cabinets and Casework				
Brush				
First Coat				
Minimum	S.F.	0.16	0.46	0.62
Average	"	0.16	0.51	0.67
Maximum	"	0.16	0.58	0.74
Second Coat				
Minimum	S.F.	0.16	0.38	0.54
Average	"	0.16	0.42	0.58
Maximum	"	0.16	0.46	0.62
Spray				
First Coat				
Minimum	S.F.	0.13	0.23	0.36
Average	"	0.13	0.27	0.40
Maximum	"	0.13	0.33	0.46
Second Coat				
Minimum	S.F.	0.13	0.18	0.31
Average	"	0.13	0.20	0.33
Maximum	"	0.13	0.25	0.38
Ceilings				
Roller				
First Coat				
Minimum	S.F.	0.13	0.19	0.32
Average	"	0.13	0.21	0.34
Maximum	"	0.13	0.23	0.36
Second Coat				
Minimum	S.F.	0.13	0.15	0.28
Average	"	0.13	0.17	0.30
Maximum	"	0.13	0.19	0.32
Spray				
First Coat				
Minimum	S.F.	0.11	0.11	0.22
Average	"	0.11	0.12	0.23
Maximum	"	0.11	0.14	0.25
Second Coat				
Minimum	S.F.	0.11	0.08	0.19
Average	"	0.11	0.10	0.21
Maximum	"	0.11	0.11	0.22
Doors, Wood				
Brush				
First Coat				
Minimum	S.F.	0.16	0.66	0.82
Average	"	0.16	0.85	1.01
Maximum	"	0.16	1.03	1.19
Second Coat				
Minimum	S.F.	0.12	0.51	0.63
Average	"	0.12	0.58	0.70
Maximum	"	0.12	0.66	0.78

Painting

09910.35 Int. Painting, Buildings *(Cont.)*

	UNIT	MAT.	INST.	TOTAL
Spray				
First Coat				
Minimum	S.F.	0.12	0.13	0.25
Average	"	0.12	0.16	0.28
Maximum	"	0.12	0.21	0.33
Second Coat				
Minimum	S.F.	0.12	0.11	0.23
Average	"	0.12	0.12	0.24
Maximum	"	0.12	0.14	0.26
Walls				
Roller				
First Coat				
Minimum	S.F.	0.13	0.16	0.29
Average	"	0.13	0.17	0.30
Maximum	"	0.13	0.19	0.32
Second Coat				
Minimum	S.F.	0.13	0.14	0.27
Average	"	0.13	0.15	0.28
Maximum	"	0.13	0.17	0.30
Spray				
First Coat				
Minimum	S.F.	0.11	0.07	0.18
Average	"	0.11	0.08	0.19
Maximum	"	0.11	0.11	0.22
Second Coat				
Minimum	S.F.	0.11	0.06	0.17
Average	"	0.11	0.08	0.19
Maximum	"	0.11	0.10	0.21

09955.10 Wall Covering

	UNIT	MAT.	INST.	TOTAL
Vinyl wall covering				
Medium duty	S.F.	0.82	0.66	1.48
Heavy duty	"	1.70	0.77	2.47
Over pipes and irregular shapes				
Lightweight, 13 oz.	S.F.	1.43	0.93	2.36
Medium weight, 25 oz.	"	1.70	1.03	2.73
Heavy weight, 34 oz.	"	2.09	1.16	3.25
Cork wall covering				
1' x 1' squares				
1/4" thick	S.F.	4.45	1.16	5.61
1/2" thick	"	5.66	1.16	6.82
3/4" thick	"	6.38	1.16	7.54
Wall fabrics				
Natural fabrics, grass cloths				
Minimum	S.F.	1.43	0.71	2.14
Average	"	1.59	0.77	2.36
Maximum	"	5.33	0.93	6.26
Flexible gypsum coated wall fabric, fire resistant	"	1.60	0.46	2.06
Vinyl corner guards				
3/4" x 3/4" x 8'	EA.	7.53	5.84	13.37
2-3/4" x 2-3/4" x 4'	"	4.45	5.84	10.29

09 FINISHES

Painting

09980.15 Paint	UNIT	MAT.	INST.	TOTAL
Paint, enamel				
600 sf per gal.	GAL			49.50
550 sf per gal.	"			46.25
500 sf per gal.	"			33.00
450 sf per gal.	"			30.75
350 sf per gal.	"			29.75
Filler, 60 sf per gal.	"			35.25
Latex, 400 sf per gal.	"			33.00
Aluminum				
400 sf per gal.	GAL			44.00
500 sf per gal.	"			70.00
Red lead, 350 sf per gal.	"			62.00
Primer				
400 sf per gal.	GAL			29.75
300 sf per gal.	"			29.75
Latex base, interior, white	"			33.00
Sealer and varnish				
400 sf per gal.	GAL			30.75
425 sf per gal.	"			44.00
600 sf per gal.	"			57.00

Specialties

Specialties	UNIT	MAT.	INST.	TOTAL
10110.10 Chalkboards				
Chalkboard, metal frame, 1/4" thick				
48"x60"	EA.	460	56.00	516
48"x96"	"	630	62.00	692
48"x144"	"	840	70.00	910
48"x192"	"	1,130	80.00	1,210
Liquid chalkboard				
48"x60"	EA.	610	56.00	666
48"x96"	"	780	62.00	842
48"x144"	"	1,160	70.00	1,230
48"x192"	"	1,330	80.00	1,410
Map rail, deluxe	L.F.	7.60	2.79	10.39
10165.10 Toilet Partitions				
Toilet partition, plastic laminate				
Ceiling mounted	EA.	1,030	190	1,220
Floor mounted	"	680	140	820
Metal				
Ceiling mounted	EA.	700	190	890
Floor mounted	"	670	140	810
Front door and side divider, floor mounted				
Porcelain enameled steel	EA.	1,130	140	1,270
Painted steel	"	670	140	810
Stainless steel	"	1,650	140	1,790
10185.10 Shower Stalls				
Shower receptors				
Precast, terrazzo				
32" x 32"	EA.	610	51.00	661
32" x 48"	"	650	61.00	711
Concrete				
32" x 32"	EA.	250	51.00	301
48" x 48"	"	280	68.00	348
Shower door, trim and hardware				
Economy, 24" wide, chrome, tempered glass	EA.	280	61.00	341
Porcelain enameled steel, flush	"	500	61.00	561
Baked enameled steel, flush	"	300	61.00	361
Aluminum, tempered glass, 48" wide, sliding	"	620	77.00	697
Folding	"	590	77.00	667
Aluminum and tempered glass, molded plastic				
Complete with receptor and door				
32" x 32"	EA.	750	150	900
36" x 36"	"	850	150	1,000
40" x 40"	"	880	180	1,060
10210.10 Vents And Wall Louvers				
Block vent, 8"x16"x4" alum., w/screen, mill finish	EA.	150	20.50	171
Standard	"	85.00	19.25	104
Vents w/screen, 4" deep, 8" wide, 5" high				
Modular	EA.	98.00	19.25	117
Aluminum gable louvers	S.F.	18.00	10.25	28.25
Vent screen aluminum, 4" wide, continuous	L.F.	5.22	2.06	7.28

10 SPECIALTIES

Specialties	UNIT	MAT.	INST.	TOTAL
10225.10 Door Louvers				
Fixed, 1" thick, enameled steel				
8"x8"	EA.	54.00	6.99	60.99
12"x8"	"	62.00	6.99	68.99
12"x12"	"	69.00	7.99	76.99
16"x12"	"	99.00	8.60	108
20"x8"	"	110	7.99	118
10290.10 Pest Control				
Termite control				
Under slab spraying				
Minimum	S.F.	1.14	0.10	1.24
Average	"	1.14	0.21	1.35
Maximum	"	1.66	0.43	2.09
10350.10 Flagpoles				
Installed in concrete base				
Fiberglass				
25' high	EA.	1,420	370	1,790
Aluminum				
25' high	EA.	1,380	370	1,750
Bonderized steel				
25' high	EA.	1,550	430	1,980
Freestanding tapered, fiberglass				
30' high	EA.	1,690	400	2,090
Wall mounted, with collar, brushed aluminum finish				
15' long	EA.	1,310	280	1,590
Outrigger, wall, including base				
10' long	EA.	1,330	370	1,700
10450.10 Control				
Access control, 7' high, indoor or outdoor impenetrability				
Remote or card control, type B	EA.	1,630	760	2,390
Free passage, type B	"	1,320	760	2,080
Remote or card control, type AA	"	2,590	760	3,350
Free passage, type AA	"	2,350	760	3,110
10550.10 Postal Specialties				
Single mail chute				
Finished aluminum	L.F.	740	140	880
Bronze	"	1,030	140	1,170
Single mail chute receiving box				
Finished aluminum	EA.	1,100	280	1,380
Bronze	"	1,320	280	1,600
Receiving box, 36" x 20" x 12"				
Finished aluminum	EA.	2,530	470	3,000
Bronze	"	3,410	470	3,880
Locked receiving mail box				
Finished aluminum	EA.	1,100	280	1,380
Bronze	"	2,140	280	2,420
Residential postal accessories				
Letter slot	EA.	83.00	28.00	111
Rural letter box	"	160	70.00	230
Apartment house, keyed, 3.5" x 4.5" x 16"	"	150	18.75	169

Specialties	UNIT	MAT.	INST.	TOTAL
10550.10 Postal Specialties *(Cont.)*				
Ranch style	EA.	160	28.00	188
10800.10 Bath Accessories				
Grab bar, 1-1/2" dia., stainless steel, wall mounted				
24" long	EA.	46.00	28.00	74.00
36" long	"	52.00	29.50	81.50
42" long	"	57.00	31.00	88.00
48" long	"	63.00	33.00	96.00
52" long	"	69.00	35.00	104
1" dia., stainless steel				
12" long	EA.	28.75	24.25	53.00
18" long	"	34.50	25.50	60.00
24" long	"	39.00	28.00	67.00
30" long	"	46.00	29.50	75.50
36" long	"	52.00	31.00	83.00
48" long	"	57.00	33.00	90.00
Medicine cabinet, 16 x 22, baked enamel, lighted	"	130	22.50	153
With mirror, lighted	"	190	37.25	227
Mirror, 1/4" plate glass, up to 10 sf	S.F.	10.00	5.59	15.59
Mirror, stainless steel frame				
18"x24"	EA.	77.00	18.75	95.75
18"x32"	"	88.00	22.50	111
18"x36"	"	91.00	28.00	119
24"x30"	"	94.00	28.00	122
24"x36"	"	100	31.00	131
24"x48"	"	140	46.75	187
24"x60"	"	350	56.00	406
30"x30"	"	310	56.00	366
30"x72"	"	550	70.00	620
48"x72"	"	600	93.00	693
Shower rod, 1" diameter				
Chrome finish over brass	EA.	210	28.00	238
Stainless steel	"	140	28.00	168
Soap dish, stainless steel, wall mounted	"	130	37.25	167
Toilet tissue dispenser, stainless, wall mounted				
Single roll	EA.	64.00	14.00	78.00
Towel bar, stainless steel				
18" long	EA.	79.00	22.50	102
24" long	"	110	25.50	136
30" long	"	110	28.00	138
36" long	"	120	31.00	151
Toothbrush and tumbler holder	"	50.00	18.75	68.75

11 EQUIPMENT

Architectural Equipment	UNIT	MAT.	INST.	TOTAL
11010.10 Maintenance Equipment				
Vacuum cleaning system				
3 valves				
1.5 hp	EA.	940	620	1,560
2.5 hp	"	1,130	800	1,930
5 valves	"	1,770	1,120	2,890
7 valves	"	2,360	1,400	3,760
11400.10 Food Service Equipment				
Unit kitchens				
30" compact kitchen				
Refrigerator, with range, sink	EA.	1,320	290	1,610
Sink only	"	1,680	190	1,870
Range only	"	1,370	140	1,510
Cabinet for upper wall section	"	340	81.00	421
Stainless shield, for rear wall	"	140	22.75	163
Side wall	"	100	22.75	123
42" compact kitchen				
Refrigerator with range, sink	EA.	1,610	320	1,930
Sink only	"	880	290	1,170
Cabinet for upper wall section	"	680	95.00	775
Stainless shield, for rear wall	"	550	23.75	574
Side wall	"	150	23.75	174
54" compact kitchen				
Refrigerator, oven, range, sink	EA.	2,160	410	2,570
Cabinet for upper wall section	"	880	110	990
Stainless shield, for				
Rear wall	EA.	550	26.00	576
Side wall	"	150	26.00	176
60" compact kitchen				
Refrigerator, oven, range, sink	EA.	2,930	410	3,340
Cabinet for upper wall section	"	200	110	310
Stainless shield, for				
Rear wall	EA.	650	26.00	676
Side wall	"	160	26.00	186
72" compact kitchen				
Refrigerator, oven, range, sink	EA.	3,100	480	3,580
Cabinet for upper wall section	"	210	110	320
Stainless shield for				
Rear wall	EA.	700	28.50	729
Side wall	"	160	28.50	189
Bake oven				
Single deck				
Minimum	EA.	3,120	71.00	3,191
Maximum	"	5,960	140	6,100
Double deck				
Minimum	EA.	5,550	95.00	5,645
Maximum	"	17,350	140	17,490
Triple deck				
Minimum	EA.	19,690	95.00	19,785
Maximum	"	35,130	190	35,320
Convection type oven, electric, 40" x 45" x 57"				

Architectual Equipment

Architectural Equipment	UNIT	MAT.	INST.	TOTAL
11400.10 Food Service Equipment (Cont.)				
Minimum	EA.	3,070	71.00	3,141
Maximum	"	5,410	140	5,550
Range				
Heavy duty, single oven, open top				
Minimum	EA.	6,430	71.00	6,501
Maximum	"	13,540	190	13,730
11450.10 Residential Equipment				
Compactor, 4 to 1 compaction	EA.	1,360	140	1,500
Dishwasher, built-in				
2 cycles	EA.	670	290	960
4 or more cycles	"	1,810	290	2,100
Disposal				
Garbage disposer	EA.	180	190	370
Heaters, electric, built-in				
Ceiling type	EA.	380	190	570
Wall type				
Minimum	EA.	190	140	330
Maximum	"	670	190	860
Hood for range, 2-speed, vented				
30" wide	EA.	530	190	720
42" wide	"	980	190	1,170
Ice maker, automatic				
30 lb per day	EA.	1,790	81.00	1,871
50 lb per day	"	2,270	290	2,560
Folding access stairs, disappearing metal stair				
8' long	EA.	930	81.00	1,011
11' long	"	970	81.00	1,051
12' long	"	1,040	81.00	1,121
Wood frame, wood stair				
22" x 54" x 8'9" long	EA.	180	57.00	237
25" x 54" x 10' long	"	220	57.00	277
Ranges electric				
Built-in, 30", 1 oven	EA.	1,960	190	2,150
2 oven	"	2,270	190	2,460
Counter top, 4 burner, standard	"	1,130	140	1,270
With grill	"	2,830	140	2,970
Free standing, 21", 1 oven	"	1,020	190	1,210
30", 1 oven	"	1,980	110	2,090
2 oven	"	3,230	110	3,340
Water softener				
30 grains per gallon	EA.	1,110	190	1,300
70 grains per gallon	"	1,400	290	1,690
11470.10 Darkroom Equipment				
Dryers				
36" x 25" x 68"	EA.	9,860	310	10,170
48" x 25" x 68"	"	10,180	310	10,490
Processors, film				
Black and white	EA.	15,710	310	16,020
Color negatives	"	17,790	310	18,100

11 EQUIPMENT

Architectural Equipment

11470.10 Darkroom Equipment *(Cont.)*

Architectural Equipment	UNIT	MAT.	INST.	TOTAL
Prints	EA.	20,400	310	20,710
Transparencies	"	22,380	310	22,690
Sinks with cabinet and/or stand				
5" sink with stand				
24" x 48"	EA.	880	150	1,030
32" x 64"	"	1,350	200	1,550

12 FURNISHINGS

Interior	UNIT	MAT.	INST.	TOTAL
12302.10 Casework				
Kitchen base cabinet, standard, 24" deep, 35" high				
12"wide	EA.	190	56.00	246
18" wide	"	230	56.00	286
24" wide	"	290	62.00	352
27" wide	"	330	62.00	392
36" wide	"	390	70.00	460
48" wide	"	470	70.00	540
Drawer base, 24" deep, 35" high				
15"wide	EA.	250	56.00	306
18" wide	"	260	56.00	316
24" wide	"	420	62.00	482
27" wide	"	480	62.00	542
30" wide	"	560	62.00	622
Sink-ready, base cabinet				
30" wide	EA.	260	62.00	322
36" wide	"	270	62.00	332
42" wide	"	300	62.00	362
60" wide	"	350	70.00	420
Corner cabinet, 36" wide	"	490	70.00	560
Wall cabinet, 12" deep, 12" high				
30" wide	EA.	250	56.00	306
36" wide	"	260	56.00	316
15" high				
30" wide	EA.	290	62.00	352
36" wide	"	440	62.00	502
24" high				
30" wide	EA.	320	62.00	382
36" wide	"	330	62.00	392
30" high				
12" wide	EA.	180	70.00	250
18" wide	"	210	70.00	280
24" wide	"	230	70.00	300
27" wide	"	270	70.00	340
30" wide	"	300	80.00	380
36" wide	"	310	80.00	390
Corner cabinet, 30" high				
24" wide	EA.	340	93.00	433
30" wide	"	410	93.00	503
36" wide	"	450	93.00	543
Wardrobe	"	910	140	1,050
Vanity with top, laminated plastic				
24" wide	EA.	750	140	890
30" wide	"	840	140	980
36" wide	"	970	190	1,160
48" wide	"	1,080	220	1,300
12390.10 Counter Tops				
Stainless steel, counter top, with backsplash	S.F.	230	14.00	244
Acid-proof, kemrock surface	"	93.00	9.32	102

12 FURNISHINGS

Interior

Interior	UNIT	MAT.	INST.	TOTAL
12500.10 Window Treatment				
Drapery tracks, wall or ceiling mounted				
Basic traverse rod				
50 to 90"	EA.	50.00	28.00	78.00
84 to 156"	"	67.00	31.00	98.00
136 to 250"	"	97.00	31.00	128
165 to 312"	"	150	35.00	185
Traverse rod with stationary curtain rod				
30 to 50"	EA.	76.00	28.00	104
50 to 90"	"	87.00	28.00	115
84 to 156"	"	120	31.00	151
136 to 250"	"	150	35.00	185
Double traverse rod				
30 to 50"	EA.	89.00	28.00	117
50 to 84"	"	110	28.00	138
84 to 156"	"	120	31.00	151
136 to 250"	"	150	35.00	185
12510.10 Blinds				
Venetian blinds				
2" slats	S.F.	35.50	1.39	36.89
1" slats	"	38.00	1.39	39.39
12690.40 Floor Mats				
Recessed entrance mat, 3/8" thick, aluminum link	S.F.	51.00	28.00	79.00
Steel, flexible	"	18.50	28.00	46.50

13 SPECIAL

Construction	UNIT	MAT.	INST.	TOTAL
13056.10 Vaults				
Floor safes				
1.0 cf	EA.	840	46.75	887
1.3 cf	"	920	70.00	990
1.9 cf	"	1,210	93.00	1,303
5.2 cf	"	2,470	93.00	2,563
13121.10 Pre-engineered Buildings				
Pre-engineered metal building, 40'x100'				
14' eave height	S.F.	7.30	4.77	12.07
16' eave height	"	8.28	5.50	13.78
13152.10 Swimming Pool Equipment				
Diving boards				
14' long				
Aluminum	EA.	4,210	240	4,450
Fiberglass	"	3,180	240	3,420
Lights, underwater				
12 volt, with transformer, 100 watt				
Incandescent	EA.	210	110	320
Halogen	"	180	110	290
LED	"	570	110	680
110 volt				
Minimum	EA.	950	110	1,060
Maximum	"	2,310	110	2,420
Ground fault interrupter for 110 volt, each light	"	210	36.50	247
Pool cover				
Reinforced polyethylene	S.F.	2.04	3.37	5.41
Vinyl water tube				
Minimum	S.F.	1.25	3.37	4.62
Maximum	"	1.87	3.37	5.24
Slides with water tube				
Minimum	EA.	1,000	370	1,370
Maximum	"	21,340	370	21,710
13200.10 Storage Tanks				
Oil storage tank, underground, single wall, no excv.				
Steel				
500 gals	EA.	3,510	300	3,810
1,000 gals	"	4,760	400	5,160
Fiberglass, double wall				
550 gals	EA.	9,890	400	10,290
1,000 gals	"	12,720	400	13,120
Above ground				
Steel, single wall				
275 gals	EA.	1,990	240	2,230
500 gals	"	4,970	400	5,370
1,000 gals	"	6,790	480	7,270
Fill cap	"	120	61.00	181
Vent cap	"	120	61.00	181
Level indicator	"	190	61.00	251

13 SPECIAL

Hazardous Waste

	UNIT	MAT.	INST.	TOTAL
13280.10 Asbestos Removal				
Enclosure using wood studs & poly, install & remove	S.F.	510	1.09	511
Trailer (change room)	DAY			110
Disposal suits (4 suits per man day)	"			45.25
Type C respirator mask, includes hose & filters, per man	"			22.75
Respirator mask & filter, light contamination	"			9.06
Air monitoring test, 12 tests per day				
Off job testing	DAY			1,190
On the job testing	"			1,590
Asbestos vacuum with attachments	EA.			690
Hydraspray piston pump	"			920
Negative air pressure system	"			920
Grade D breathing air equipment	"			2,070
Glove bag, 44" x 60" x 6 mil plastic	"			6.66
40 CY asbestos dumpster				
Weekly rental	EA.			770
Pick up/delivery	"			350
Asbestos dump fee	"			220
13280.12 Duct Insulation Removal				
Remove duct insulation, duct size				
6" x 12"	L.F.	240	2.43	242
x 18"	"	170	3.37	173
x 24"	"	120	4.87	125
8" x 12"	"	160	3.65	164
x 18"	"	150	3.98	154
x 24"	"	110	5.48	115
12" x 12"	"	160	3.65	164
x 18"	"	120	4.87	125
x 24"	"	93.00	6.26	99.26
13280.15 Pipe Insulation Removal				
Removal, asbestos insulation				
2" thick, pipe				
1" to 3" dia.	L.F.	160	3.65	164

14 CONVEYING

Elevators	UNIT	MAT.	INST.	TOTAL
14210.10 Elevators				
Hydraulic, based on a shaft of 3 stops, 3 openings				
50 fpm				
2000 lb	EA.	79,960	2,000	81,960
2500 lb	"	85,510	2,000	87,510
3000 lb	"	90,240	2,090	92,330
For each additional; 50 fpm add per stop, $3500				
500 lb, add per stop, $3500				
Opening, add, $4200				
Stop, add per stop, $5300				
Bonderized steel door, add per opening, $400				
Colored aluminum door, add per opening, $1500				
Stainless steel door, add per opening, $650				
Cast bronze door, add per opening, $1200				
Custom cab interior, add per cab, $5000				
Small elevators, 4 to 6 passenger capacity				
Electric, push				
2 stops	EA.	29,200	2,000	31,200
3 stops	"	36,500	2,180	38,680
4 stops	"	41,550	2,400	43,950

Lifts	UNIT	MAT.	INST.	TOTAL
14410.10 Personnel Lifts				
Electrically operated, 1 or 2 person lift				
With attached foot platforms				
3 stops	EA.			10,230
5 stops	"			15,950
7 stops	"			18,590
For each additional stop, add $1250				
Residential stair climber, per story	EA.	4,820	480	5,300
14410.20 Wheelchair Lifts				
600 lb, Residential	EA.	5,720	570	6,290

14 CONVEYING

Material Handling

14560.10 Chutes	UNIT	MAT.	INST.	TOTAL
Linen chutes, stainless steel, with supports				
18" dia.	L.F.	140	4.41	144
Hopper	EA.	2,310	41.25	2,351
Skylight	"	1,400	62.00	1,462

Basic Materials

15100.10 Specialties

Description	UNIT	MAT.	INST.	TOTAL
Wall penetration				
Concrete wall, 6" thick				
2" dia.	EA.		14.50	14.50
4" dia.	"		22.00	22.00
8" dia.	"		31.25	31.25
12" thick				
2" dia.	EA.		20.00	20.00
4" dia.	"		31.25	31.25
8" dia.	"		48.75	48.75

15120.10 Backflow Preventers

Description	UNIT	MAT.	INST.	TOTAL
Backflow preventer, flanged, cast iron, with valves				
3" pipe	EA.	3,330	310	3,640
4" pipe	"	4,270	340	4,610
6" pipe	"	7,300	510	7,810
Threaded				
3/4" pipe	EA.	680	38.50	719
2" pipe	"	1,200	61.00	1,261
Reduced pressure assembly, bronze, threaded				
3/4"	EA.	640	38.50	679
1"	"	660	44.00	704
1-1/4"	"	960	51.00	1,011
1-1/2"	"	980	61.00	1,041

15140.11 Pipe Hangers, Light

Description	UNIT	MAT.	INST.	TOTAL
A band, black iron				
1/2"	EA.	1.00	4.38	5.38
1"	"	1.07	4.55	5.62
1-1/4"	"	1.19	4.72	5.91
1-1/2"	"	1.25	5.12	6.37
2"	"	1.32	5.58	6.90
2-1/2"	"	1.98	6.14	8.12
3"	"	2.40	6.82	9.22
4"	"	3.16	7.68	10.84
5"	"	3.34	8.19	11.53
6"	"	5.77	8.77	14.54
Copper				
1/2"	EA.	1.62	4.38	6.00
3/4"	"	1.89	4.55	6.44
1"	"	1.89	4.55	6.44
1-1/4"	"	2.02	4.72	6.74
1-1/2"	"	2.17	5.12	7.29
2"	"	2.31	5.58	7.89
2-1/2"	"	4.65	6.14	10.79
3"	"	4.85	6.82	11.67
4"	"	5.34	7.68	13.02
2 hole clips, galvanized				
3/4"	EA.	0.26	4.09	4.35
1"	"	0.29	4.23	4.52
1-1/4"	"	0.38	4.38	4.76
1-1/2"	"	0.47	4.55	5.02
2"	"	0.61	4.72	5.33
2-1/2"	"	1.11	4.91	6.02

Basic Materials

15140.11 Pipe Hangers, Light *(Cont.)*

Basic Materials	UNIT	MAT.	INST.	TOTAL
3"	EA.	1.61	5.12	6.73
4"	"	3.46	5.58	9.04
Perforated strap				
3/4"				
Galvanized, 20 ga.	L.F.	0.40	3.07	3.47
Copper, 22 ga.	"	0.63	3.07	3.70
J-Hooks				
1/2"	EA.	0.73	2.79	3.52
3/4"	"	0.78	2.79	3.57
1"	"	0.80	2.92	3.72
1-1/4"	"	0.84	2.99	3.83
1-1/2"	"	0.86	3.07	3.93
2"	"	0.90	3.07	3.97
3"	"	1.03	3.23	4.26
4"	"	1.11	3.23	4.34
PVC coated hangers, galvanized, 28 ga.				
1-1/2" x 12"	EA.	1.26	4.09	5.35
2" x 12"	"	1.37	4.38	5.75
3" x 12"	"	1.54	4.72	6.26
4" x 12"	"	1.70	5.12	6.82
Copper, 30 ga.				
1-1/2" x 12"	EA.	1.93	4.09	6.02
2" x 12"	"	2.29	4.38	6.67
3" x 12"	"	2.54	4.72	7.26
4" x 12"	"	2.78	5.12	7.90
2" x 24"	"	4.47	4.72	9.19
3" x 24"	"	5.14	5.12	10.26
4" x 24"	"	8.04	5.58	13.62
Wire hook hangers				
Black wire, 1/2" x				
4"	EA.	0.42	3.07	3.49
6"	"	0.49	3.23	3.72
3/4" x				
4"	EA.	0.52	3.23	3.75
6"	"	0.56	3.41	3.97
1" x				
4"	EA.	0.52	3.41	3.93
6"	"	0.53	3.61	4.14
1-1/4" x				
4"	EA.	0.56	3.61	4.17
6"	"	0.57	3.84	4.41
1-1/2" x				
6"	EA.	0.63	4.09	4.72
Copper wire hooks				
1/2" x				
4"	EA.	0.59	3.07	3.66
6"	"	0.67	3.23	3.90
3/4" x				
4"	EA.	0.59	3.23	3.82
6"	"	0.73	3.41	4.14
1" x				

Basic Materials

Basic Materials	UNIT	MAT.	INST.	TOTAL
15140.11 Pipe Hangers, Light *(Cont.)*				
4"	EA.	0.63	3.41	4.04
6"	"	0.71	3.61	4.32
1-1/4" x				
6"	EA.	0.83	3.61	4.44
1-1/2" x				
6"	EA.	0.99	4.09	5.08
15240.10 Vibration Control				
Vibration isolator, in-line, stainless connector				
1/2"	EA.	93.00	34.25	127
3/4"	"	110	36.25	146
1"	"	110	38.50	149
1-1/4"	"	150	41.00	191

Insulation

Insulation	UNIT	MAT.	INST.	TOTAL
15260.10 Fiberglass Pipe Insulation				
Fiberglass insulation on 1/2" pipe				
1" thick	L.F.	1.19	2.04	3.23
1-1/2" thick	"	2.52	2.56	5.08
3/4" pipe				
1" thick	L.F.	1.46	2.04	3.50
1-1/2" thick	"	2.65	2.56	5.21
1" pipe				
1" thick	L.F.	1.46	2.04	3.50
1-1/2" thick	"	2.78	2.56	5.34
1-1/4" pipe				
1" thick	L.F.	1.66	2.56	4.22
1-1/2" thick	"	3.05	2.79	5.84
1-1/2" pipe				
1" thick	L.F.	1.79	2.56	4.35
1-1/2" thick	"	3.12	2.79	5.91
2" pipe				
1" thick	L.F.	1.99	2.56	4.55
1-1/2" thick	"	3.45	2.79	6.24
2-1/2" pipe				
1" thick	L.F.	2.12	2.56	4.68
1-1/2" thick	"	3.71	2.79	6.50
3" pipe				
1" thick	L.F.	2.39	2.92	5.31
1-1/2" thick	"	3.85	3.07	6.92
4" pipe				
1" thick	L.F.	3.05	2.92	5.97
1-1/2" thick	"	4.38	3.07	7.45

Insulation

15260.60 Exterior Pipe Insulation	UNIT	MAT.	INST.	TOTAL
Fiberglass insulation, aluminum jacket				
1/2" pipe				
1" thick	L.F.	1.89	4.72	6.61
1-1/2" thick	"	3.54	5.12	8.66
1" pipe				
1" thick	L.F.	2.29	4.72	7.01
1-1/2" thick	"	3.96	5.12	9.08
2" pipe				
1" thick	L.F.	3.13	5.58	8.71
1-1/2" thick	"	4.66	5.85	10.51
3" pipe				
1" thick	L.F.	3.75	6.14	9.89
1-1/2" thick	"	5.56	6.46	12.02
4" pipe				
1" thick	L.F.	4.73	6.14	10.87
1-1/2" thick	"	6.40	6.46	12.86

15290.10 Ductwork Insulation	UNIT	MAT.	INST.	TOTAL
Fiberglass duct insulation, plain blanket				
1-1/2" thick	S.F.	0.20	0.76	0.96
2" thick	"	0.27	1.02	1.29
With vapor barrier				
1-1/2" thick	S.F.	0.24	0.76	1.00
2" thick	"	0.30	1.02	1.32
Rigid with vapor barrier				
2" thick	S.F.	1.32	2.04	3.36
3" thick	"	1.81	2.45	4.26
4" thick	"	2.31	3.07	5.38
6" thick	"	3.63	4.09	7.72
Weatherproof, poly, 3" thick, w/vapor barrier	"	2.80	6.14	8.94
Urethane board with vapor barrier	"	3.96	7.68	11.64

Plumbing

15410.05 C.I. Pipe, Above Ground	UNIT	MAT.	INST.	TOTAL
No hub pipe				
1-1/2" pipe	L.F.	8.26	4.38	12.64
2" pipe	"	7.32	5.12	12.44
3" pipe	"	10.00	6.14	16.14
4" pipe	"	13.25	10.25	23.50
6" pipe	"	23.25	12.25	35.50
No hub fittings, 1-1/2" pipe				
1/4 bend	EA.	8.89	20.50	29.39
1/8 bend	"	7.44	20.50	27.94
Sanitary tee	"	12.25	30.75	43.00
Sanitary cross	"	16.75	30.75	47.50

Plumbing	UNIT	MAT.	INST.	TOTAL
15410.05 C.I. Pipe, Above Ground *(Cont.)*				
Plug	EA.			4.91
Coupling	"			17.50
Wye	"	15.50	30.75	46.25
Tapped tee	"	16.25	20.50	36.75
P-trap	"	14.00	20.50	34.50
Tapped cross	"	18.50	20.50	39.00
2" pipe				
1/4 bend	EA.	10.25	24.50	34.75
1/8 bend	"	8.28	24.50	32.78
Sanitary tee	"	14.00	41.00	55.00
Sanitary cross	"	23.75	41.00	64.75
Plug	"			4.91
Coupling	"			15.50
Wye	"	13.25	51.00	64.25
Double wye	"	20.25	51.00	71.25
2x1-1/2" wye & 1/8 bend	"	24.50	38.50	63.00
Double wye & 1/8 bend	"	20.25	51.00	71.25
Test tee less 2" plug	"	12.75	24.50	37.25
Tapped tee				
2"x2"	EA.	16.50	24.50	41.00
2"x1-1/2"	"	15.50	24.50	40.00
P-trap				
2"x2"	EA.	15.00	24.50	39.50
Tapped cross				
2"x1-1/2"	EA.	21.25	24.50	45.75
3" pipe				
1/4 bend	EA.	14.00	30.75	44.75
1/8 bend	"	11.75	30.75	42.50
Sanitary tee	"	17.25	38.50	55.75
3"x2" sanitary tee	"	15.50	38.50	54.00
3"x1-1/2" sanitary tee	"	16.25	38.50	54.75
Sanitary cross	"	36.50	51.00	87.50
3x2" sanitary cross	"	32.50	51.00	83.50
Plug	"			7.28
Coupling	"			17.75
Wye	"	18.75	51.00	69.75
3x2" wye	"	14.00	51.00	65.00
Double wye	"	37.50	51.00	88.50
3x2" double wye	"	31.75	51.00	82.75
3x2" wye & 1/8 bend	"	17.50	44.00	61.50
3x1-1/2" wye & 1/8 bend	"	17.50	44.00	61.50
Double wye & 1/8 bend	"	37.50	51.00	88.50
3x2" double wye & 1/8 bend	"	31.75	51.00	82.75
3x2" reducer	"	7.11	28.00	35.11
Test tee, less 3" plug	"	19.50	30.75	50.25
Plug	"			7.28
3x3" tapped tee	"	44.75	30.75	75.50
3x2" tapped tee	"	24.25	30.75	55.00
3x1-1/2" tapped tee	"	20.75	30.75	51.50
P-trap	"	32.75	30.75	63.50
3x2" tapped cross	"	30.50	30.75	61.25
3x1-1/2" tapped cross	"	28.50	30.75	59.25
Closet flange, 3-1/2" deep	"	20.75	15.25	36.00

Plumbing

15410.05 C.I. Pipe, Above Ground *(Cont.)*

Plumbing	UNIT	MAT.	INST.	TOTAL
4" pipe				
1/4 bend	EA.	20.25	30.75	51.00
1/8 bend	"	15.00	30.75	45.75
Sanitary tee	"	26.50	51.00	77.50
4x3" sanitary tee	"	24.50	51.00	75.50
4x2" sanitary tee	"	20.25	51.00	71.25
Sanitary cross	"	69.00	61.00	130
4x3" sanitary cross	"	56.00	61.00	117
4x2" sanitary cross	"	46.50	61.00	108
Plug	"			11.25
Coupling	"			17.25
Wye	"	30.50	51.00	81.50
4x3" wye	"	26.50	51.00	77.50
4x2" wye	"	19.50	51.00	70.50
Double wye	"	77.00	61.00	138
4x3" double wye	"	48.00	61.00	109
4x2" double wye	"	42.25	61.00	103
Wye & 1/8 bend	"	41.50	51.00	92.50
4x3" wye & 1/8 bend	"	30.25	51.00	81.25
4x2" wye & 1/8 bend	"	23.50	51.00	74.50
Double wye & 1/8 bend	"	110	61.00	171
4x3" double wye & 1/8 bend	"	70.00	61.00	131
4x2" double wye & 1/8 bend	"	67.00	61.00	128
4x3" reducer	"	11.00	30.75	41.75
4x2" reducer	"	11.00	30.75	41.75
Test tee, less 4" plug	"	33.00	30.75	63.75
Plug	"			11.25
4x2" tapped tee	"	24.25	30.75	55.00
4x1-1/2" tapped tee	"	21.50	30.75	52.25
P-trap	"	57.00	30.75	87.75
4x2" tapped cross	"	43.50	30.75	74.25
4x1-1/2" tapped cross	"	34.00	30.75	64.75
Closet flange				
3" deep	EA.	22.75	30.75	53.50
8" deep	"	58.00	30.75	88.75
6" pipe				
1/4 bend	EA.	51.00	51.00	102
1/8 bend	"	34.50	51.00	85.50
Sanitary tee	"	77.00	61.00	138
6x4" sanitary tee	"	59.00	61.00	120
Coupling	"			43.75
Wye	"	81.00	61.00	142
6x4" wye	"	64.00	61.00	125
6x3" wye	"	62.00	61.00	123
6x2" wye	"	49.00	61.00	110
Double wye	"	140	77.00	217
6x4" double wye	"	110	77.00	187
Wye & 1/8 bend	"	120	61.00	181
6x4" wye & 1/8 bend	"	73.00	61.00	134
6x3" wye & 1/8 bend	"	71.00	61.00	132
6x2" wye & 1/8 bend	"	56.00	61.00	117
6x4" reducer	"	29.50	34.25	63.75

Plumbing	UNIT	MAT.	INST.	TOTAL
15410.05 C.I. Pipe, Above Ground *(Cont.)*				
6x3" reducer	EA.	29.50	34.25	63.75
6x2" reducer	"	30.25	30.75	61.00
Test tee				
Less 6" plug	EA.	79.00	38.50	118
Plug	"			22.00
P-trap	"	140	38.50	179
15410.06 C.I. Pipe, Below Ground				
No hub pipe				
1-1/2" pipe	L.F.	7.82	3.07	10.89
2" pipe	"	8.02	3.41	11.43
3" pipe	"	11.00	3.84	14.84
4" pipe	"	14.50	5.12	19.62
6" pipe	"	24.75	5.58	30.33
Fittings, 1-1/2"				
1/4 bend	EA.	9.31	17.50	26.81
1/8 bend	"	7.79	17.50	25.29
Plug	"			4.91
Wye	"	13.00	24.50	37.50
Wye & 1/8 bend	"	14.00	17.50	31.50
P-trap	"	15.50	17.50	33.00
2"				
1/4 bend	EA.	10.25	20.50	30.75
1/8 bend	"	8.72	20.50	29.22
Plug	"			4.91
Double wye	"	20.25	38.50	58.75
Wye & 1/8 bend	"	14.25	30.75	45.00
Double wye & 1/8 bend	"	35.00	38.50	73.50
P-trap	"	15.00	20.50	35.50
3"				
1/4 bend	EA.	14.00	24.50	38.50
1/8 bend	"	11.75	24.50	36.25
Plug	"			7.28
Wye	"	18.75	38.50	57.25
3x2" wye	"	14.00	38.50	52.50
Wye & 1/8 bend	"	22.50	38.50	61.00
Double wye & 1/8 bend	"	54.00	38.50	92.50
3x2" double wye & 1/8 bend	"	40.75	38.50	79.25
3x2" reducer	"	7.11	24.50	31.61
P-trap	"	32.75	24.50	57.25
4"				
1/4 bend	EA.	20.25	24.50	44.75
1/8 bend	"	15.00	24.50	39.50
Plug	"			11.25
Wye	"	30.50	38.50	69.00
4x3" wye	"	26.50	38.50	65.00
4x2" wye	"	19.50	38.50	58.00
Double wye	"	77.00	51.00	128
4x3" double wye	"	48.00	51.00	99.00
4x2" double wye	"	42.25	51.00	93.25
Wye & 1/8 bend	"	41.50	38.50	80.00
4x3" wye & 1/8 bend	"	30.25	38.50	68.75
4x2" wye & 1/8 bend	"	23.50	38.50	62.00

Plumbing	UNIT	MAT.	INST.	TOTAL
15410.06 C.I. Pipe, Below Ground (Cont.)				
Double wye & 1/8 bend	EA.	110	51.00	161
4x3" double wye & 1/8 bend	"	70.00	51.00	121
4x2" double wye & 1/8 bend	"	67.00	51.00	118
4x3" reducer	"	11.00	24.50	35.50
4x2" reducer	"	11.00	24.50	35.50
6"				
1/4 bend	EA.	32.25	38.50	70.75
1/8 bend	"	21.75	38.50	60.25
Wye & 1/8 bend	"	51.00	51.00	102
6x4" wye & 1/8 bend	"	40.25	51.00	91.25
6x3" wye & 1/8 bend	"	39.00	51.00	90.00
6x2" wye & 1/8 bend	"	31.00	51.00	82.00
6x3" reducer	"	18.50	28.00	46.50
P-trap	"	87.00	30.75	118
15410.09 Service Weight Pipe				
Service weight pipe, single hub				
2" x 5'	EA.	49.00	12.25	61.25
3" x 5'	"	55.00	13.00	68.00
4" x 5'	"	63.00	13.75	76.75
5" x 5'	"	98.00	14.50	113
6" x 5'	"	120	15.25	135
Double hub				
2" x 5'	EA.	55.00	15.25	70.25
3" x 5'	"	61.00	16.50	77.50
4" x 5'	"	69.00	17.50	86.50
5" x 5'	"	110	19.25	129
6" x 5'	"	130	20.50	151
Single hub				
2" x 10'	EA.	52.00	15.25	67.25
3" x 10'	"	72.00	16.50	88.50
4" x 10'	"	94.00	17.50	112
5" x 10'	"	130	19.25	149
6" x 10'	"	160	20.50	181
Shorty				
2" x 42"	EA.	35.75	12.25	48.00
3" x 42"	"	39.75	13.00	52.75
4" x 42"	"	51.00	13.75	64.75
5" x 42"	"	78.00	14.50	92.50
6" x 42"	"	93.00	15.25	108
1/8 bend				
2"	EA.	7.32	20.50	27.82
3"	"	11.50	24.50	36.00
4"	"	16.75	28.00	44.75
5"	"	23.25	29.25	52.50
6"	"	28.50	30.75	59.25
1/4 bend				
2"	EA.	10.25	20.50	30.75
3"	"	13.75	24.50	38.25
4"	"	21.50	28.00	49.50
5"	"	30.00	29.25	59.25
6"	"	37.50	30.75	68.25
Sweep				

Plumbing	UNIT	MAT.	INST.	TOTAL
15410.09 Service Weight Pipe *(Cont.)*				
2"	EA.	15.50	20.50	36.00
3"	"	22.25	24.50	46.75
4"	"	32.75	28.00	60.75
5"	"	54.00	29.25	83.25
6"	"	66.00	30.75	96.75
Sanitary T				
2"	EA.	14.50	38.50	53.00
3" x 2"	"	20.00	41.00	61.00
3"	"	23.25	44.00	67.25
4" x 2"	"	24.25	47.25	71.50
4" x 3"	"	26.00	51.00	77.00
4"	"	28.50	51.00	79.50
5"	"	57.00	56.00	113
6"	"	64.00	56.00	120
Tapped sanitary T				
2" x 1-1/2"	EA.	20.25	44.00	64.25
2" x 2"	"	19.25	44.00	63.25
3" x 1-1/2"	"	23.50	47.25	70.75
3" x 2"	"	21.50	47.25	68.75
4" x 1-1/2"	"	29.50	51.00	80.50
4" x 2"	"	35.25	51.00	86.25
Cleanout, dandy, with brass plug				
2", 1-1/2" plug	EA.	25.25	44.00	69.25
3", 2" plug	"	29.50	47.25	76.75
4", 3" plug	"	56.00	51.00	107
5", 4" plug	"	65.00	56.00	121
6", 4" plug	"	81.00	61.00	142
15410.10 Copper Pipe				
Type "K" copper				
1/2"	L.F.	4.13	1.92	6.05
3/4"	"	7.71	2.04	9.75
1"	"	10.00	2.19	12.19
DWV, copper				
1-1/4"	L.F.	11.25	2.56	13.81
1-1/2"	"	14.25	2.79	17.04
2"	"	18.50	3.07	21.57
3"	"	31.75	3.41	35.16
4"	"	55.00	3.84	58.84
6"	"	210	4.38	214
Refrigeration tubing, copper, sealed				
1/8"	L.F.	0.81	2.45	3.26
3/16"	"	0.94	2.56	3.50
1/4"	"	1.12	2.67	3.79
5/16"	"	1.45	2.79	4.24
3/8"	"	1.65	2.92	4.57
1/2"	"	2.17	3.07	5.24
Type "L" copper				
1/4"	L.F.	1.66	1.80	3.46
3/8"	"	2.56	1.80	4.36
1/2"	"	2.97	1.92	4.89
3/4"	"	4.75	2.04	6.79
1"	"	7.15	2.19	9.34

Plumbing	UNIT	MAT.	INST.	TOTAL
15410.10 Copper Pipe *(Cont.)*				
Type "M" copper				
1/2"	L.F.	2.10	1.92	4.02
3/4"	"	3.42	2.04	5.46
1"	"	5.56	2.19	7.75
Type "K" tube, coil				
1/4" x 60'	EA.			120
1/2" x 60'	"			260
1/2" x 100'	"			430
3/4" x 60'	"			480
3/4" x 100'	"			800
1" x 60'	"			630
1" x 100'	"			1,040
Type "L" tube, coil				
1/4" x 60'	EA.			130
3/8" x 60'	"			210
1/2" x 60'	"			280
1/2" x 100'	"			460
3/4" x 60'	"			440
3/4" x 100'	"			740
1" x 60'	"			640
1" x 100'	"			1,070
15410.11 Copper Fittings				
Coupling, with stop				
1/4"	EA.	0.91	20.50	21.41
3/8"	"	1.18	24.50	25.68
1/2"	"	0.95	26.75	27.70
5/8"	"	2.75	30.75	33.50
3/4"	"	1.89	34.25	36.14
1"	"	3.89	36.25	40.14
Reducing coupling				
1/4" x 1/8"	EA.	2.42	24.50	26.92
3/8" x 1/4"	"	2.67	26.75	29.42
1/2" x				
3/8"	EA.	2.00	30.75	32.75
1/4"	"	2.43	30.75	33.18
1/8"	"	2.68	30.75	33.43
3/4" x				
3/8"	EA.	4.31	34.25	38.56
1/2"	"	3.41	34.25	37.66
1" x				
3/8"	EA.	7.73	38.50	46.23
1" x 1/2"	"	7.49	38.50	45.99
1" x 3/4"	"	6.31	38.50	44.81
Slip coupling				
1/4"	EA.	0.75	20.50	21.25
1/2"	"	1.26	24.50	25.76
3/4"	"	2.63	30.75	33.38
1"	"	5.56	34.25	39.81
Coupling with drain				
1/2"	EA.	9.60	30.75	40.35
3/4"	"	14.00	34.25	48.25
1"	"	17.50	38.50	56.00

Plumbing

15410.11 Copper Fittings (Cont.)

	UNIT	MAT.	INST.	TOTAL
Reducer				
3/8" x 1/4"	EA.	2.72	24.50	27.22
1/2" x 3/8"	"	2.19	24.50	26.69
3/4" x				
1/4"	EA.	4.45	28.00	32.45
3/8"	"	4.65	28.00	32.65
1/2"	"	4.85	28.00	32.85
1" x				
1/2"	EA.	6.68	30.75	37.43
3/4"	"	5.12	30.75	35.87
Female adapters				
1/4"	EA.	7.08	24.50	31.58
3/8"	"	7.25	28.00	35.25
1/2"	"	3.45	30.75	34.20
3/4"	"	4.73	34.25	38.98
1"	"	11.00	34.25	45.25
Increasing female adapters				
1/8" x				
3/8"	EA.	6.96	24.50	31.46
1/2"	"	6.48	24.50	30.98
1/4" x 1/2"	"	6.79	26.75	33.54
3/8" x 1/2"	"	7.29	28.00	35.29
1/2" X				
3/4"	EA.	7.73	30.75	38.48
1"	"	15.50	30.75	46.25
3/4" X				
1"	EA.	16.50	34.25	50.75
Reducing female adapters				
3/8" x 1/4"	EA.	6.27	28.00	34.27
1/2" x				
1/4"	EA.	5.40	30.75	36.15
3/8"	"	5.40	30.75	36.15
3/4" x 1/2"	"	7.52	34.25	41.77
1" x				
1/2"	EA.	20.25	34.25	54.50
3/4"	"	16.25	34.25	50.50
Female fitting adapters				
1/2"	EA.	9.61	30.75	40.36
3/4"	"	12.50	30.75	43.25
3/4" x 1/2"	"	14.75	32.25	47.00
1"	"	16.50	34.25	50.75
Male adapters				
1/4"	EA.	11.00	28.00	39.00
3/8"	"	5.40	28.00	33.40
Increasing male adapters				
3/8" x 1/2"	EA.	7.34	28.00	35.34
1/2" x				
3/4"	EA.	6.37	30.75	37.12
1"	"	14.25	30.75	45.00
3/4" x				
1"	EA.	14.00	32.25	46.25
1-1/4"	"	18.00	32.25	50.25
1" x 1-1/4"	"	18.00	34.25	52.25

Plumbing	UNIT	MAT.	INST.	TOTAL
15410.11 Copper Fittings *(Cont.)*				
Reducing male adapters				
1/2" x				
1/4"	EA.	9.33	30.75	40.08
3/8"	"	7.75	30.75	38.50
3/4" x 1/2"	"	8.85	32.25	41.10
1" x				
1/2"	EA.	24.25	34.25	58.50
3/4"	"	19.50	34.25	53.75
Fitting x male adapters				
1/2"	EA.	13.50	30.75	44.25
3/4"	"	17.25	32.25	49.50
1"	"	17.50	34.25	51.75
90 ells				
1/8"	EA.	2.03	24.50	26.53
1/4"	"	3.23	24.50	27.73
3/8"	"	3.07	28.00	31.07
1/2"	"	1.02	30.75	31.77
3/4"	"	2.29	32.25	34.54
1"	"	5.65	34.25	39.90
Reducing 90 ell				
3/8" x 1/4"	EA.	5.26	28.00	33.26
1/2" x				
1/4"	EA.	7.52	30.75	38.27
3/8"	"	7.52	30.75	38.27
3/4" x 1/2"	"	6.61	32.25	38.86
1" x				
1/2"	EA.	11.50	34.25	45.75
3/4"	"	10.75	34.25	45.00
Street ells, copper				
1/4"	EA.	5.45	24.50	29.95
3/8"	"	3.75	28.00	31.75
1/2"	"	1.51	30.75	32.26
3/4"	"	3.19	32.25	35.44
1"	"	8.26	34.25	42.51
Female, 90 ell				
1/2"	EA.	1.03	30.75	31.78
3/4"	"	2.32	32.25	34.57
1"	"	5.71	34.25	39.96
Female increasing, 90 ell				
3/8" x 1/2"	EA.	11.50	28.00	39.50
1/2" x				
3/4"	EA.	7.97	30.75	38.72
1"	"	16.25	30.75	47.00
3/4" x 1"	"	14.50	32.25	46.75
1" x 1-1/4"	"	37.25	34.25	71.50
Female reducing, 90 ell				
1/2" x 3/8"	EA.	12.75	30.75	43.50
3/4" x 1/2"	"	14.00	32.25	46.25
1" x				
1/2"	EA.	19.75	34.25	54.00
3/4"	"	21.50	34.25	55.75
Male, 90 ell				
1/4"	EA.	9.75	24.50	34.25

Plumbing	UNIT	MAT.	INST.	TOTAL
15410.11 Copper Fittings *(Cont.)*				
3/8"	EA.	10.50	28.00	38.50
1/2"	"	5.42	30.75	36.17
3/4"	"	12.25	32.25	44.50
1"	"	14.25	34.25	48.50
Male, increasing 90 ell				
1/2" x				
3/4"	EA.	19.25	30.75	50.00
1"	"	38.25	30.75	69.00
3/4" x 1"	"	36.25	32.25	68.50
1" x 1-1/4"	"	33.50	34.25	67.75
Male, reducing 90 ell				
1/2" x 3/8"	EA.	11.00	30.75	41.75
3/4" x 1/2"	"	19.25	32.25	51.50
1" x				
1/2"	EA.	36.50	34.25	70.75
3/4"	"	34.50	34.25	68.75
Drop ear ells				
1/2"	EA.	6.99	30.75	37.74
Female drop ear ells				
1/2"	EA.	6.99	30.75	37.74
1/2" x 3/8"	"	12.25	30.75	43.00
3/4"	"	20.50	32.25	52.75
Female flanged sink ell				
1/2"	EA.	12.75	30.75	43.50
45 ells				
1/4"	EA.	5.83	24.50	30.33
3/8"	"	4.74	28.00	32.74
45 street ell				
1/4"	EA.	6.60	24.50	31.10
3/8"	"	7.13	28.00	35.13
1/2"	"	2.12	30.75	32.87
3/4"	"	3.19	32.25	35.44
1"	"	8.41	34.25	42.66
Tee				
1/8"	EA.	4.96	24.50	29.46
1/4"	"	5.22	24.50	29.72
3/8"	"	3.99	28.00	31.99
Caps				
1/4"	EA.	0.90	24.50	25.40
3/8"	"	1.44	28.00	29.44
Test caps				
1/2"	EA.	0.84	30.75	31.59
3/4"	"	0.95	32.25	33.20
1"	"	1.70	34.25	35.95
Flush bushing				
1/4" x 1/8"	EA.	1.71	24.50	26.21
1/2" x				
1/4"	EA.	2.32	30.75	33.07
3/8"	"	2.05	30.75	32.80
3/4" x				
3/8"	EA.	4.32	32.25	36.57
1/2"	"	3.81	32.25	36.06
1" x				

Plumbing

15410.11 Copper Fittings (Cont.)

Plumbing	UNIT	MAT.	INST.	TOTAL
1/2"	EA.	6.58	34.25	40.83
3/4"	"	5.84	34.25	40.09
Female flush bushing				
1/2" x				
1/2" x 1/8"	EA.	5.19	30.75	35.94
1/4"	"	5.46	30.75	36.21
Union				
1/4"	EA.	32.50	24.50	57.00
3/8"	"	44.50	28.00	72.50
Female				
1/2"	EA.	15.50	30.75	46.25
3/4"	"	15.50	32.25	47.75
Male				
1/2"	EA.	16.75	30.75	47.50
3/4"	"	22.25	32.25	54.50
1"	"	48.25	34.25	82.50
45 degree wye				
1/2"	EA.	20.75	30.75	51.50
3/4"	"	29.75	32.25	62.00
1"	"	40.25	34.25	74.50
1" x 3/4" x 3/4"	"	56.00	34.25	90.25
Twin ells				
1" x 3/4" x 3/4"	EA.	14.50	34.25	48.75
1" x 1" x 1"	"	14.50	34.25	48.75
90 union ells, male				
1/2"	EA.	22.50	30.75	53.25
3/4"	"	37.50	32.25	69.75
1"	"	56.00	34.25	90.25
DWV fittings, coupling with stop				
1-1/4"	EA.	5.13	36.25	41.38
1-1/2"	"	6.38	38.50	44.88
1-1/2" x 1-1/4"	"	10.25	38.50	48.75
2"	"	8.84	41.00	49.84
2" x 1-1/4"	"	12.00	41.00	53.00
2" x 1-1/2"	"	12.00	41.00	53.00
3"	"	17.00	51.00	68.00
3" x 1-1/2"	"	41.00	51.00	92.00
3" x 2"	"	39.00	51.00	90.00
4"	"	55.00	61.00	116
Slip coupling				
1-1/2"	EA.	9.92	38.50	48.42
2"	"	11.75	41.00	52.75
3"	"	21.50	51.00	72.50
90 ells				
1-1/2"	EA.	12.00	38.50	50.50
1-1/2" x 1-1/4"	"	33.25	38.50	71.75
2"	"	22.00	41.00	63.00
2" x 1-1/2"	"	44.75	41.00	85.75
3"	"	59.00	51.00	110
4"	"	200	61.00	261
Street, 90 elbows				
1-1/2"	EA.	15.50	38.50	54.00
2"	"	33.75	41.00	74.75

Plumbing	UNIT	MAT.	INST.	TOTAL
15410.11 Copper Fittings (Cont.)				
3"	EA.	86.00	51.00	137
4"	"	220	61.00	281
Female, 90 elbows				
1-1/2"	EA.	15.25	38.50	53.75
2"	"	29.50	41.00	70.50
Male, 90 elbows				
1-1/2"	EA.	27.00	38.50	65.50
2"	"	56.00	41.00	97.00
90 with side inlet				
3" x 3" x 1"	EA.	81.00	51.00	132
3" x 3" x 1-1/2"	"	84.00	51.00	135
3" x 3" x 2"	"	84.00	51.00	135
45 ells				
1-1/4"	EA.	10.00	36.25	46.25
1-1/2"	"	8.33	38.50	46.83
2"	"	19.25	41.00	60.25
3"	"	40.75	51.00	91.75
4"	"	180	61.00	241
Street, 45 ell				
1-1/2"	EA.	13.50	38.50	52.00
2"	"	23.75	41.00	64.75
3"	"	69.00	51.00	120
60 ell				
1-1/2"	EA.	21.00	38.50	59.50
2"	"	38.50	41.00	79.50
3"	"	86.00	51.00	137
22-1/2 ell				
1-1/2"	EA.	25.50	38.50	64.00
2"	"	32.75	41.00	73.75
3"	"	57.00	51.00	108
11-1/4 ell				
1-1/2"	EA.	28.00	38.50	66.50
2"	"	39.75	41.00	80.75
3"	"	80.00	51.00	131
Wye				
1-1/4"	EA.	42.75	36.25	79.00
1-1/2"	"	46.50	38.50	85.00
2"	"	61.00	41.00	102
2" x 1-1/2" x 1-1/2"	"	67.00	41.00	108
2" x 1-1/2" x 2"	"	75.00	41.00	116
2" x 1-1/2" x 2"	"	75.00	41.00	116
3"	"	150	51.00	201
3" x 3" x 1-1/2"	"	130	51.00	181
3" x 3" x 2"	"	130	51.00	181
4"	"	300	61.00	361
4" x 4" x 2"	"	210	61.00	271
4" x 4" x 3"	"	210	61.00	271
Sanitary tee				
1-1/4"	EA.	21.75	36.25	58.00
1-1/2"	"	27.00	38.50	65.50
2"	"	31.50	41.00	72.50
2" x 1-1/2" x 1-1/2"	"	50.00	41.00	91.00
2" x 1-1/2" x 2"	"	51.00	41.00	92.00

Plumbing	UNIT	MAT.	INST.	TOTAL
15410.11 Copper Fittings *(Cont.)*				
2" x 2" x 1-1/2"	EA.	29.75	41.00	70.75
3"	"	110	51.00	161
3" x 3" x 1-1/2"	"	89.00	51.00	140
3" x 3" x 2"	"	89.00	51.00	140
4"	"	300	61.00	361
4" x 4" x 3"	"	250	61.00	311
Female sanitary tee				
1-1/2"	EA.	52.00	38.50	90.50
Long turn tee				
1-1/2"	EA.	52.00	38.50	90.50
2"	"	120	41.00	161
3" x 1-1/2"	"	150	51.00	201
Double wye				
1-1/2"	EA.	77.00	38.50	116
2"	"	140	41.00	181
2" x 2" x 1-1/2" x 1-1/2"	"	110	41.00	151
3"	"	220	51.00	271
3" x 3" x 1-1/2" x 1-1/2"	"	220	51.00	271
3" x 3" x 2" x 2"	"	220	51.00	271
4" x 4" x 1-1/2" x 1-1/2"	"	230	61.00	291
Double sanitary tee				
1-1/2"	EA.	53.00	38.50	91.50
2"	"	120	41.00	161
2" x 2" x 1-1/2"	"	110	41.00	151
3"	"	140	51.00	191
3" x 3" x 1-1/2" x 1-1/2"	"	180	51.00	231
3" x 3" x 2" x 2"	"	150	51.00	201
4" x 4" x 1-1/2" x 1-1/2"	"	330	61.00	391
Long				
2" x 1-1/2"	EA.	140	41.00	181
Twin elbow				
1-1/2"	EA.	69.00	38.50	108
2"	"	110	41.00	151
2" x 1-1/2" x 1-1/2"	"	96.00	41.00	137
Spigot adapter, manoff				
1-1/2" x 2"	EA.	45.50	38.50	84.00
1-1/2" x 3"	"	55.00	38.50	93.50
2"	"	22.50	41.00	63.50
2" x 3"	"	53.00	41.00	94.00
2" x 4"	"	75.00	41.00	116
3"	"	77.00	51.00	128
3" x 4"	"	140	51.00	191
4"	"	120	61.00	181
No-hub adapters				
1-1/2" x 2"	EA.	27.50	38.50	66.00
2"	"	26.00	41.00	67.00
2" x 3"	"	60.00	41.00	101
3"	"	52.00	51.00	103
3" x 4"	"	110	51.00	161
4"	"	110	61.00	171
Fitting reducers				
1-1/2" x 1-1/4"	EA.	9.67	38.50	48.17
2" x 1-1/2"	"	15.25	41.00	56.25

Plumbing

15410.11 Copper Fittings (Cont.)

	UNIT	MAT.	INST.	TOTAL
3" x 1-1/2"	EA.	42.75	51.00	93.75
3" x 2"	"	38.25	51.00	89.25
Slip joint (Desanco)				
1-1/4"	EA.	16.25	36.25	52.50
1-1/2"	"	17.00	38.50	55.50
1-1/2" x 1-1/4"	"	17.75	38.50	56.25
Street x slip joint (Desanco)				
1-1/2"	EA.	21.00	38.50	59.50
1-1/2" x 1-1/4"	"	22.50	38.50	61.00
Flush bushing				
1-1/2" x 1-1/4"	EA.	12.00	38.50	50.50
2" x 1-1/2"	"	20.75	41.00	61.75
3" x 1-1/2"	"	37.25	51.00	88.25
3" x 2"	"	37.25	51.00	88.25
Male hex trap bushing				
1-1/4" x 1-1/2"	EA.	17.75	36.25	54.00
1-1/2"	"	13.00	38.50	51.50
1-1/2" x 2"	"	19.50	38.50	58.00
2"	"	15.25	41.00	56.25
Round trap bushing				
1-1/2"	EA.	15.00	38.50	53.50
2"	"	16.00	41.00	57.00
Female adapter				
1-1/4"	EA.	17.50	36.25	53.75
1-1/2"	"	27.25	38.50	65.75
1-1/2" x 2"	"	69.00	38.50	108
2"	"	37.00	41.00	78.00
2" x 1-1/2"	"	60.00	41.00	101
3"	"	150	51.00	201
Fitting x female adapter				
1-1/2"	EA.	36.75	38.50	75.25
2"	"	48.75	41.00	89.75
Male adapters				
1-1/4"	EA.	15.25	36.25	51.50
1-1/4" x 1-1/2"	"	35.50	36.25	71.75
1-1/2"	"	17.50	38.50	56.00
1-1/2" x 2"	"	66.00	38.50	105
2"	"	29.50	41.00	70.50
2" x 1-1/2"	"	68.00	41.00	109
3"	"	150	51.00	201
Male x slip joint adapters				
1-1/2" x 1-1/4"	EA.	27.75	38.50	66.25
Dandy cleanout				
1-1/2"	EA.	49.25	38.50	87.75
2"	"	58.00	41.00	99.00
3"	"	210	51.00	261
End cleanout, flush pattern				
1-1/2" x 1"	EA.	30.00	38.50	68.50
2" x 1-1/2"	"	35.75	41.00	76.75
3" x 2-1/2"	"	76.00	51.00	127
Copper caps				
1-1/2"	EA.	10.25	38.50	48.75
2"	"	18.75	41.00	59.75

Plumbing	UNIT	MAT.	INST.	TOTAL
15410.11 Copper Fittings *(Cont.)*				
Closet flanges				
3"	EA.	44.50	51.00	95.50
4"	"	78.00	61.00	139
Drum traps, with cleanout				
1-1/2" x 3" x 6"	EA.	170	38.50	209
P-trap, swivel, with cleanout				
1-1/2"	EA.	110	38.50	149
P-trap, solder union				
1-1/2"	EA.	44.00	38.50	82.50
2"	"	77.00	41.00	118
With cleanout				
1-1/2"	EA.	48.50	38.50	87.00
2"	"	86.00	41.00	127
2" x 1-1/2"	"	86.00	41.00	127
Swivel joint, with cleanout				
1-1/2" x 1-1/4"	EA.	62.00	38.50	101
1-1/2"	"	79.00	38.50	118
2" x 1-1/2"	"	97.00	41.00	138
Estabrook TY, with inlets				
3", with 1-1/2" inlet	EA.	130	51.00	181
Fine thread adapters				
1/2"	EA.	3.55	30.75	34.30
1/2" x 1/2" IPS	"	4.01	30.75	34.76
1/2" x 3/4" IPS	"	6.53	30.75	37.28
1/2" x male	"	2.46	30.75	33.21
1/2" x female	"	5.03	30.75	35.78
Copper pipe fittings				
1/2"				
90 deg ell	EA.	1.56	13.75	15.31
45 deg ell	"	1.98	13.75	15.73
Tee	"	2.62	17.50	20.12
Cap	"	1.06	6.82	7.88
Coupling	"	1.15	13.75	14.90
Union	"	7.94	15.25	23.19
3/4"				
90 deg ell	EA.	3.41	15.25	18.66
45 deg ell	"	3.99	15.25	19.24
Tee	"	5.71	20.50	26.21
Cap	"	2.08	7.22	9.30
Coupling	"	2.32	15.25	17.57
Union	"	11.75	17.50	29.25
1"				
90 deg ell	EA.	7.94	20.50	28.44
45 deg ell	"	10.25	20.50	30.75
Tee	"	13.00	24.50	37.50
Cap	"	3.86	10.25	14.11
Coupling	"	5.71	20.50	26.21
Union	"	15.25	20.50	35.75

Plumbing

15410.14 Brass I.p.s. Fittings

Plumbing	UNIT	MAT.	INST.	TOTAL
Fittings, iron pipe size, 45 deg ell				
1/8"	EA.	9.82	24.50	34.32
1/4"	"	9.82	24.50	34.32
3/8"	"	9.82	28.00	37.82
1/2"	"	9.82	30.75	40.57
3/4"	"	14.00	32.25	46.25
1"	"	23.75	34.25	58.00
90 deg ell				
1/8"	EA.	9.24	24.50	33.74
1/4"	"	9.24	24.50	33.74
3/8"	"	9.24	28.00	37.24
1/2"	"	9.24	30.75	39.99
3/4"	"	11.00	32.25	43.25
1"	"	19.75	34.25	54.00
90 deg ell, reducing				
1/4" x 1/8"	EA.	11.00	24.50	35.50
3/8" x 1/8"	"	11.00	28.00	39.00
3/8" x 1/4"	"	11.00	28.00	39.00
1/2" x 1/4"	"	11.00	30.75	41.75
1/2" x 3/8"	"	11.00	30.75	41.75
3/4" x 1/2"	"	16.00	32.25	48.25
1" x 3/8"	"	22.75	34.25	57.00
1" x 1/2"	"	22.75	34.25	57.00
1" x 3/4"	"	22.75	34.25	57.00
Street ell, 45 deg				
1/2"	EA.	11.00	30.75	41.75
3/4"	"	16.00	32.25	48.25
90 deg				
1/8"	EA.	11.00	24.50	35.50
1/4"	"	11.00	24.50	35.50
3/8"	"	11.00	28.00	39.00
1/2"	"	11.00	30.75	41.75
3/4"	"	13.50	32.25	45.75
1"	"	17.75	34.25	52.00
Tee, 1/8"	"	9.38	24.50	33.88
1/4"	"	9.38	24.50	33.88
3/8"	"	9.38	28.00	37.38
1/2"	"	9.38	30.75	40.13
3/4"	"	12.75	32.25	45.00
1"	"	18.75	34.25	53.00
Tee, reducing, 3/8" x				
1/4"	EA.	13.00	28.00	41.00
1/2"	"	13.00	28.00	41.00
1/2" x				
1/4"	EA.	13.00	30.75	43.75
3/8"	"	13.00	30.75	43.75
3/4"	"	15.25	30.75	46.00
3/4" x				
1/4"	EA.	15.25	32.25	47.50
1/2"	"	15.25	32.25	47.50
1"	"	30.75	32.25	63.00
1" x				
1/2"	EA.	30.50	34.25	64.75

Plumbing	UNIT	MAT.	INST.	TOTAL
15410.14 Brass I.p.s. Fittings *(Cont.)*				
3/4"	EA.	30.50	34.25	64.75
Tee, reducing				
1/2" x 3/8" x 1/2"	EA.	11.75	30.75	42.50
3/4" x 1/2" x 1/2"	"	17.00	32.25	49.25
3/4" x 1/2" x 3/4"	"	16.00	32.25	48.25
1" x 1/2" x 1/2"	"	29.50	34.25	63.75
1" x 1/2" x 3/4"	"	29.50	34.25	63.75
1" x 3/4" x 1/2"	"	36.00	34.25	70.25
1" x 3/4" x 3/4"	"	29.50	34.25	63.75
Union				
1/8"	EA.	24.50	24.50	49.00
1/4"	"	24.50	24.50	49.00
3/8"	"	24.50	28.00	52.50
1/2"	"	24.50	30.75	55.25
3/4"	"	33.75	32.25	66.00
1"	"	45.00	34.25	79.25
Brass face bushing				
3/8" x 1/4"	EA.	8.48	28.00	36.48
1/2" x 3/8"	"	8.48	30.75	39.23
3/4" x 1/2"	"	10.50	32.25	42.75
1" x 3/4"	"	18.00	34.25	52.25
Hex bushing, 1/4" x 1/8"	"	5.99	24.50	30.49
1/2" x				
1/4"	EA.	5.47	30.75	36.22
3/8"	"	5.47	30.75	36.22
5/8" x				
1/8"	EA.	5.48	30.75	36.23
1/4"	"	5.48	30.75	36.23
3/4" x				
1/8"	EA.	7.95	32.25	40.20
1/4"	"	7.95	32.25	40.20
3/8"	"	6.90	32.25	39.15
1/2"	"	6.90	32.25	39.15
1" x				
1/4"	EA.	10.50	34.25	44.75
3/8"	"	10.50	34.25	44.75
1/2"	"	9.74	34.25	43.99
3/4"	"	9.74	34.25	43.99
Caps				
1/8"	EA.	5.46	24.50	29.96
1/4"	"	5.89	24.50	30.39
3/8"	"	5.89	28.00	33.89
1/2"	"	5.89	30.75	36.64
3/4"	"	6.21	32.25	38.46
1"	"	11.25	34.25	45.50
Couplings				
1/8"	EA.	6.36	24.50	30.86
1/4"	"	6.36	24.50	30.86
3/8"	"	6.36	28.00	34.36
1/2"	"	6.36	30.75	37.11
3/4"	"	8.73	32.25	40.98
1"	"	13.75	34.25	48.00
Couplings, reducing, 1/4" x 1/8"	"	7.43	24.50	31.93

Plumbing	UNIT	MAT.	INST.	TOTAL
15410.14 Brass I.p.s. Fittings *(Cont.)*				
3/8" x				
1/8"	EA.	7.43	28.00	35.43
1/4"	"	7.43	28.00	35.43
1/2" x				
1/8"	EA.	9.16	30.75	39.91
1/4"	"	8.11	30.75	38.86
3/8"	"	8.11	30.75	38.86
3/4" x				
1/4"	EA.	12.75	32.25	45.00
3/8"	"	10.50	32.25	42.75
1/2"	"	10.50	32.25	42.75
1" x				
1/2"	EA.	14.75	32.25	47.00
3/4"	"	14.75	32.25	47.00
Square head plug, solid				
1/8"	EA.	6.21	24.50	30.71
1/4"	"	6.21	24.50	30.71
3/8"	"	6.21	28.00	34.21
1/2"	"	6.21	30.75	36.96
3/4"	"	7.43	32.25	39.68
Cored				
1/2"	EA.	4.94	30.75	35.69
3/4"	"	6.21	32.25	38.46
1"	"	9.94	34.25	44.19
Countersunk				
1/2"	EA.	6.89	30.75	37.64
3/4"	"	7.35	32.25	39.60
Locknut				
3/4"	EA.	6.21	32.25	38.46
1"	"	7.75	34.25	42.00
Close standard red nipple, 1/8"	"	2.02	24.50	26.52
1/8" x				
1-1/2"	EA.	3.74	24.50	28.24
2"	"	4.12	24.50	28.62
2-1/2"	"	4.67	24.50	29.17
3"	"	5.28	24.50	29.78
3-1/2"	"	6.32	24.50	30.82
4"	"	6.76	24.50	31.26
4-1/2"	"	7.09	24.50	31.59
5"	"	7.48	24.50	31.98
5-1/2"	"	8.69	24.50	33.19
6"	"	9.24	24.50	33.74
1/4" x close	"	4.12	24.50	28.62
1/4" x				
1-1/2"	EA.	6.10	24.50	30.60
2"	"	6.49	24.50	30.99
2-1/2"	"	6.76	24.50	31.26
3"	"	7.09	24.50	31.59
3-1/2"	"	7.92	24.50	32.42
4"	"	8.25	24.50	32.75
4-1/2"	"	8.80	24.50	33.30
5"	"	9.07	24.50	33.57
5-1/2"	"	9.95	24.50	34.45

Plumbing	UNIT	MAT.	INST.	TOTAL
15410.14 Brass I.p.s. Fittings *(Cont.)*				
6"	EA.	10.25	24.50	34.75
3/8" x close	"	4.12	28.00	32.12
3/8" x				
1-1/2"	EA.	4.84	28.00	32.84
2"	"	5.32	28.00	33.32
2-1/2"	"	6.41	28.00	34.41
3"	"	7.80	28.00	35.80
3-1/2"	"	8.53	28.00	36.53
4"	"	11.00	28.00	39.00
4-1/2"	"	11.25	28.00	39.25
5"	"	12.00	28.00	40.00
5-1/2"	"	12.75	28.00	40.75
6"	"	14.50	28.00	42.50
1/2" x close	"	5.44	30.75	36.19
1/2" x				
1-1/2"	EA.	6.11	30.75	36.86
2"	"	7.44	30.75	38.19
2-1/2"	"	8.53	30.75	39.28
3"	"	9.68	30.75	40.43
3-1/2"	"	10.75	30.75	41.50
4"	"	11.25	30.75	42.00
4-1/2"	"	12.00	30.75	42.75
5"	"	12.25	30.75	43.00
5-1/2"	"	12.75	30.75	43.50
6"	"	14.00	30.75	44.75
7-1/2"	"	42.25	30.75	73.00
8"	"	42.25	30.75	73.00
3/4" x close	"	15.00	32.25	47.25
3/4" x				
1-1/2"	EA.	8.40	32.25	40.65
2"	"	9.68	32.25	41.93
2-1/2"	"	10.75	32.25	43.00
3"	"	11.50	32.25	43.75
3-1/2"	"	12.75	32.25	45.00
4"	"	13.50	32.25	45.75
4-1/2"	"	14.50	32.25	46.75
5"	"	15.00	32.25	47.25
5-1/2"	"	17.00	32.25	49.25
6"	"	17.75	32.25	50.00
1" x close	"	12.00	34.25	46.25
1" x				
2"	EA.	17.50	34.25	51.75
2-1/2"	"	17.75	34.25	52.00
3"	"	18.50	34.25	52.75
3-1/2"	"	19.25	34.25	53.50
4"	"	21.00	34.25	55.25
4-1/2"	"	21.25	34.25	55.50
5"	"	24.75	34.25	59.00
5-1/2"	"	25.50	34.25	59.75
6"	"	28.00	34.25	62.25

Plumbing

15410.15 Brass Fittings

Plumbing	UNIT	MAT.	INST.	TOTAL
Compression fittings, union				
3/8"	EA.	3.54	10.25	13.79
1/2"	"	5.81	10.25	16.06
5/8"	"	6.67	10.25	16.92
Union elbow				
3/8"	EA.	2.45	10.25	12.70
1/2"	"	3.39	10.25	13.64
5/8"	"	4.47	10.25	14.72
Union tee				
3/8"	EA.	3.01	10.25	13.26
1/2"	"	4.16	10.25	14.41
5/8"	"	5.35	10.25	15.60
Male connector				
3/8"	EA.	2.49	10.25	12.74
1/2"	"	1.99	10.25	12.24
5/8"	"	1.75	10.25	12.00
Female connector				
3/8"	EA.	2.26	10.25	12.51
1/2"	"	2.84	10.25	13.09
5/8"	"	3.32	10.25	13.57
Brass flare fittings, union				
3/8"	EA.	1.99	9.90	11.89
1/2"	"	2.73	9.90	12.63
5/8"	"	2.94	9.90	12.84
90 deg elbow union				
3/8"	EA.	3.86	9.90	13.76
1/2"	"	6.30	9.90	16.20
5/8"	"	9.35	9.90	19.25
Three way tee				
3/8"	EA.	4.30	16.50	20.80
1/2"	"	5.70	16.50	22.20
5/8"	"	9.57	16.50	26.07
Cross				
3/8"	EA.	9.12	22.00	31.12
1/2"	"	19.50	22.00	41.50
5/8"	"	41.00	22.00	63.00
Male connector, half union				
3/8"	EA.	1.34	9.90	11.24
1/2"	"	2.36	9.90	12.26
5/8"	"	3.30	9.90	13.20
Female connector, half union				
3/8"	EA.	1.84	9.90	11.74
1/2"	"	1.71	9.90	11.61
5/8"	"	3.30	9.90	13.20
Long forged nut				
3/8"	EA.	1.46	9.90	11.36
1/2"	"	2.12	9.90	12.02
5/8"	"	7.32	9.90	17.22
Short forged nut				
3/8"	EA.	1.17	9.90	11.07
1/2"	"	1.60	9.90	11.50
5/8"	"	2.04	9.90	11.94
Nut				

Plumbing

15410.15 Brass Fittings *(Cont.)*

Plumbing	UNIT	MAT.	INST.	TOTAL
1/8"	EA.			0.26
1/4"	"			0.26
5/16"	"			0.29
3/8"	"			0.39
1/2"	"			0.58
5/8"	"			1.21
Sleeve				
1/8"	EA.	0.18	12.25	12.43
1/4"	"	0.06	12.25	12.31
5/16"	"	0.18	12.25	12.43
3/8"	"	0.27	12.25	12.52
1/2"	"	0.35	12.25	12.60
5/8"	"	0.50	12.25	12.75
Tee				
1/4"	EA.	3.03	17.50	20.53
5/16"	"	4.55	17.50	22.05
Male tee				
5/16" x 1/8"	EA.	6.26	17.50	23.76
Female union				
1/8" x 1/8"	EA.	1.51	15.25	16.76
1/4" x 3/8"	"	2.90	15.25	18.15
3/8" x 1/4"	"	2.40	15.25	17.65
3/8" x 1/2"	"	3.03	15.25	18.28
5/8" x 1/2"	"	4.74	17.50	22.24
Male union, 1/4"				
1/4" x 1/4"	EA.	1.44	15.25	16.69
3/8"	"	1.89	15.25	17.14
1/2"	"	2.84	15.25	18.09
5/16" x				
1/8"	EA.	1.39	15.25	16.64
1/4"	"	1.77	15.25	17.02
3/8"	"	2.70	15.25	17.95
3/8" x				
1/8"	EA.	1.64	15.25	16.89
1/4"	"	1.89	15.25	17.14
1/2"	"	2.60	15.25	17.85
5/8" x				
3/8"	EA.	3.78	17.50	21.28
1/2"	"	3.26	17.50	20.76
Female elbow, 1/4" x 1/4"	"	3.58	17.50	21.08
5/16" x				
1/8"	EA.	3.92	17.50	21.42
1/4"	"	5.12	17.50	22.62
3/8" x				
3/8"	EA.	3.16	17.50	20.66
1/2"	"	2.53	17.50	20.03
Male elbow, 1/8" x 1/8"	"	3.39	17.50	20.89
3/16" x 1/4"	"	3.23	17.50	20.73
1/4" x				
1/8"	EA.	1.99	17.50	19.49
1/4"	"	2.35	17.50	19.85
3/8"	"	2.03	17.50	19.53
5/16" x				

Plumbing	UNIT	MAT.	INST.	TOTAL
15410.15 Brass Fittings *(Cont.)*				
1/8"	EA.	2.09	17.50	19.59
1/4"	"	2.45	17.50	19.95
3/8"	"	3.59	17.50	21.09
3/8" x				
1/8"	EA.	2.01	17.50	19.51
1/4"	"	2.77	17.50	20.27
3/8"	"	2.09	17.50	19.59
1/2"	"	2.79	17.50	20.29
1/2" x				
1/4"	EA.	4.40	20.50	24.90
3/8"	"	3.90	20.50	24.40
1/2"	"	3.40	20.50	23.90
5/8" x				
3/8"	EA.	4.42	20.50	24.92
1/2"	"	4.68	20.50	25.18
3/4"	"	9.53	20.50	30.03
Union				
1/8"	EA.	2.02	17.50	19.52
3/16"	"	2.02	17.50	19.52
1/4"	"	1.72	17.50	19.22
5/16"	"	1.96	17.50	19.46
3/8"	"	2.21	17.50	19.71
Reducing union				
3/8" x 1/4"	EA.	2.41	20.50	22.91
5/8" x				
3/8"	EA.	3.83	20.50	24.33
1/2"	"	4.19	20.50	24.69
15410.17 Chrome Plated Fittings				
Fittings				
90 ell				
3/8"	EA.	27.50	15.25	42.75
1/2"	"	35.50	15.25	50.75
45 ell				
3/8"	EA.	35.50	15.25	50.75
1/2"	"	46.75	15.25	62.00
Tee				
3/8"	EA.	34.00	20.50	54.50
1/2"	"	40.50	20.50	61.00
Coupling				
3/8"	EA.	21.50	15.25	36.75
1/2"	"	21.50	15.25	36.75
Union				
3/8"	EA.	35.50	15.25	50.75
1/2"	"	36.75	15.25	52.00
Tee				
1/2" x 3/8" x 3/8"	EA.	40.50	20.50	61.00
1/2" x 3/8" x 1/2"	"	41.25	20.50	61.75

Plumbing

15410.30 PVC/CPVC Pipe

Plumbing	UNIT	MAT.	INST.	TOTAL
PVC schedule 40				
1/2" pipe	L.F.	0.50	2.56	3.06
3/4" pipe	"	0.69	2.79	3.48
1" pipe	"	0.88	3.07	3.95
1-1/4" pipe	"	1.13	3.41	4.54
1-1/2" pipe	"	1.70	3.84	5.54
2" pipe	"	2.15	4.38	6.53
2-1/2" pipe	"	3.47	5.12	8.59
3" pipe	"	4.42	6.14	10.56
4" pipe	"	6.32	7.68	14.00
6" pipe	"	11.50	15.25	26.75
Fittings, 1/2"				
90 deg ell	EA.	0.48	7.68	8.16
45 deg ell	"	0.66	7.68	8.34
Tee	"	0.49	8.77	9.26
Reducing insert	"	0.50	10.25	10.75
Threaded	"	1.18	7.68	8.86
Male adapter	"	0.47	10.25	10.72
Female adapter	"	0.50	7.68	8.18
Coupling	"	0.37	7.68	8.05
Union	"	3.96	12.25	16.21
Cap	"	0.46	10.25	10.71
Flange	"	8.36	12.25	20.61
3/4"				
90 deg elbow	EA.	0.47	10.25	10.72
45 deg elbow	"	1.12	10.25	11.37
Tee	"	0.64	12.25	12.89
Reducing insert	"	0.47	8.77	9.24
Threaded	"	0.71	10.25	10.96
1"				
90 deg elbow	EA.	0.82	12.25	13.07
45 deg elbow	"	1.21	12.25	13.46
Tee	"	1.10	13.75	14.85
Reducing insert	"	0.82	12.25	13.07
Threaded	"	1.10	13.75	14.85
Male adapter	"	0.77	15.25	16.02
Female adapter	"	0.66	15.25	15.91
Coupling	"	0.60	15.25	15.85
Union	"	5.94	20.50	26.44
Cap	"	0.66	12.25	12.91
Flange	"	8.41	20.50	28.91
1-1/4"				
90 deg elbow	EA.	1.43	17.50	18.93
45 deg elbow	"	1.70	17.50	19.20
Tee	"	1.65	20.50	22.15
Reducing insert	"	0.99	20.50	21.49
Threaded	"	1.65	20.50	22.15
Male adapter	"	0.93	20.50	21.43
Female adapter	"	1.04	20.50	21.54
Coupling	"	0.88	20.50	21.38
Union	"	13.50	24.50	38.00
Cap	"	0.93	20.50	21.43
Flange	"	8.52	24.50	33.02

Plumbing

15410.30 PVC/CPVC Pipe *(Cont.)*

	UNIT	MAT.	INST.	TOTAL
1-1/2"				
90 deg elbow	EA.	1.59	17.50	19.09
45 deg elbow	"	2.36	17.50	19.86
Tee	"	2.20	20.50	22.70
Reducing insert	"	1.10	20.50	21.60
Threaded	"	1.98	20.50	22.48
Male adapter	"	1.32	20.50	21.82
Female adapter	"	1.32	20.50	21.82
Coupling	"	0.99	20.50	21.49
Union	"	18.75	30.75	49.50
Cap	"	1.04	20.50	21.54
Flange	"	14.50	30.75	45.25
2"				
90 deg elbow	EA.	2.53	20.50	23.03
45 deg elbow	"	3.19	20.50	23.69
Tee	"	3.35	24.50	27.85
Reducing insert	"	2.14	24.50	26.64
Threaded	"	2.86	24.50	27.36
Male adapter	"	1.76	24.50	26.26
Female adapter	"	1.81	24.50	26.31
Coupling	"	1.48	24.50	25.98
Union	"	25.50	38.50	64.00
Cap	"	1.37	24.50	25.87
Flange	"	15.25	38.50	53.75
2-1/2"				
90 deg elbow	EA.	7.64	38.50	46.14
45 deg elbow	"	11.00	38.50	49.50
Tee	"	9.84	41.00	50.84
Reducing insert	"	3.08	41.00	44.08
Threaded	"	4.40	41.00	45.40
Male adapter	"	5.11	41.00	46.11
Female adapter	"	4.23	41.00	45.23
Coupling	"	3.05	41.00	44.05
Union	"	34.00	51.00	85.00
Cap	"	4.07	38.50	42.57
Flange	"	20.50	51.00	71.50
3"				
90 deg elbow	EA.	8.25	51.00	59.25
45 deg elbow	"	10.75	51.00	61.75
Tee	"	13.25	56.00	69.25
Reducing insert	"	3.96	51.00	54.96
Threaded	"	5.11	51.00	56.11
Male adapter	"	6.27	51.00	57.27
Female adapter	"	5.06	51.00	56.06
Coupling	"	4.67	51.00	55.67
Union	"	35.75	61.00	96.75
Cap	"	4.07	51.00	55.07
Flange	"	18.75	61.00	79.75
4"				
90 deg elbow	EA.	14.75	61.00	75.75
45 deg elbow	"	19.25	61.00	80.25
Tee	"	22.00	68.00	90.00
Reducing insert	"	8.96	61.00	69.96

Plumbing

15410.30 PVC/CPVC Pipe *(Cont.)*

Plumbing	UNIT	MAT.	INST.	TOTAL
Threaded	EA.	11.50	61.00	72.50
Male adapter	"	7.97	61.00	68.97
Female adapter	"	8.58	61.00	69.58
Coupling	"	6.82	61.00	67.82
Union	"	43.50	77.00	121
Cap	"	9.24	61.00	70.24
Flange	"	25.25	77.00	102
PVC schedule 80 pipe				
1-1/2" pipe	L.F.	2.15	3.84	5.99
2" pipe	"	2.90	4.38	7.28
3" pipe	"	6.00	6.14	12.14
4" pipe	"	7.84	7.68	15.52
Fittings, 1-1/2"				
90 deg elbow	EA.	6.69	20.50	27.19
45 deg elbow	"	14.75	20.50	35.25
Tee	"	23.00	30.75	53.75
Reducing insert	"	4.24	20.50	24.74
Threaded	"	5.06	20.50	25.56
Male adapter	"	8.05	20.50	28.55
Female adapter	"	8.71	20.50	29.21
Coupling	"	9.31	20.50	29.81
Union	"	16.75	30.75	47.50
Cap	"	4.73	20.50	25.23
Flange	"	11.00	30.75	41.75
2"				
90 deg elbow	EA.	8.10	24.50	32.60
45 deg elbow	"	19.00	24.50	43.50
Tee	"	28.75	38.50	67.25
Reducing insert	"	6.04	24.50	30.54
Threaded	"	6.09	24.50	30.59
Male adapter	"	11.00	24.50	35.50
Female adapter	"	15.25	24.50	39.75
2-1/2"				
90 deg elbow	EA.	19.00	38.50	57.50
45 deg elbow	"	40.00	38.50	78.50
Tee	"	31.25	51.00	82.25
Reducing insert	"	10.50	38.50	49.00
Threaded	"	13.00	38.50	51.50
Male adapter	"	13.25	38.50	51.75
Female adapter	"	24.00	38.50	62.50
Coupling	"	13.00	38.50	51.50
Union	"	36.00	51.00	87.00
Cap	"	15.25	38.50	53.75
Flange	"	19.25	51.00	70.25
3"				
90 deg elbow	EA.	17.00	51.00	68.00
45 deg elbow	"	48.75	51.00	99.75
Tee	"	39.25	61.00	100
Reducing insert	"	16.75	51.00	67.75
Threaded	"	24.25	51.00	75.25
Male adapter	"	14.75	51.00	65.75
Female adapter	"	27.00	51.00	78.00
Coupling	"	14.75	51.00	65.75

Plumbing

15410.30 PVC/CPVC Pipe *(Cont.)*

	UNIT	MAT.	INST.	TOTAL
Union	EA.	46.00	61.00	107
Cap	"	19.25	51.00	70.25
Flange	"	22.00	61.00	83.00
4"				
90 deg elbow	EA.	43.50	61.00	105
45 deg elbow	"	88.00	61.00	149
Tee	"	45.25	77.00	122
Reducing insert	"	23.00	61.00	84.00
Threaded	"	37.25	61.00	98.25
Male adapter	"	26.00	61.00	87.00
Coupling	"	18.50	61.00	79.50
Union	"	43.50	77.00	121
Cap	"	23.50	61.00	84.50
Flange	"		77.00	77.00
CPVC schedule 40				
1/2" pipe	L.F.	0.62	2.56	3.18
3/4" pipe	"	0.82	2.79	3.61
1" pipe	"	1.20	3.07	4.27
1-1/4" pipe	"	1.57	3.41	4.98
1-1/2" pipe	"	1.90	3.84	5.74
2" pipe	"	2.52	4.38	6.90
Fittings, CPVC, schedule 80				
1/2", 90 deg ell	EA.	3.43	6.14	9.57
Tee	"	10.50	10.25	20.75
3/4", 90 deg ell	"	4.48	6.14	10.62
Tee	"	15.50	10.25	25.75
1", 90 deg ell	"	7.09	6.82	13.91
Tee	"	16.50	11.25	27.75
1-1/4", 90 deg ell	"	13.00	6.82	19.82
Tee	"	15.50	11.25	26.75
1-1/2", 90 deg ell	"	14.25	12.25	26.50
Tee	"	17.50	15.25	32.75
2", 90 deg ell	"	15.50	12.25	27.75
Tee	"	19.75	15.25	35.00

15410.33 ABS DWV Pipe

	UNIT	MAT.	INST.	TOTAL
Schedule 40 ABS				
1-1/2" pipe	L.F.	1.25	3.07	4.32
2" pipe	"	1.66	3.41	5.07
3" pipe	"	3.42	4.38	7.80
4" pipe	"	4.84	6.14	10.98
6" pipe	"	9.93	7.68	17.61
Fittings				
1/8 bend				
1-1/2"	EA.	1.81	12.25	14.06
2"	"	2.68	15.25	17.93
3"	"	6.43	20.50	26.93
4"	"	11.50	24.50	36.00
6"	"	47.00	30.75	77.75
Tee, sanitary				
1-1/2"	EA.	2.64	20.50	23.14
2"	"	4.07	24.50	28.57
3"	"	11.00	30.75	41.75

Plumbing

15410.33 ABS DWV Pipe *(Cont.)*

	UNIT	MAT.	INST.	TOTAL
4"	EA.	20.25	38.50	58.75
6"	"	87.00	51.00	138
Tee, sanitary reducing				
2 x 1-1/2 x 1-1/2	EA.	3.74	24.50	28.24
2 x 1-1/2 x 2	"	3.85	25.50	29.35
2 x 2 x 1-1/2	"	3.57	28.00	31.57
3 x 3 x 1-1/2	"	6.49	30.75	37.24
3 x 3 x 2	"	8.08	34.25	42.33
4 x 4 x 1-1/2	"	20.25	38.50	58.75
4 x 4 x 2	"	18.75	44.00	62.75
4 x 4 x 3	"	16.50	47.25	63.75
6 x 6 x 4	"	85.00	51.00	136
Wye				
1-1/2"	EA.	3.85	17.50	21.35
2"	"	5.39	24.50	29.89
3"	"	12.25	30.75	43.00
4"	"	26.50	38.50	65.00
6"	"	81.00	51.00	132
Reducer				
2 x 1-1/2	EA.	2.58	15.25	17.83
3 x 1-1/2	"	6.65	20.50	27.15
3 x 2	"	5.66	20.50	26.16
4 x 2	"	11.50	24.50	36.00
4 x 3	"	11.75	24.50	36.25
6 x 4	"	24.25	30.75	55.00
P-trap				
1-1/2"	EA.	5.99	20.50	26.49
2"	"	8.08	22.75	30.83
3"	"	31.00	26.75	57.75
4"	"	63.00	30.75	93.75
6"	"	100	38.50	139
Double sanitary, tee				
1-1/2"	EA.	5.83	24.50	30.33
2"	"	8.47	30.75	39.22
3"	"	23.25	38.50	61.75
4"	"	37.00	51.00	88.00
Long sweep, 1/4 bend				
1-1/2"	EA.	3.02	12.25	15.27
2"	"	3.85	15.25	19.10
3"	"	9.29	20.50	29.79
4"	"	17.00	30.75	47.75
Wye, standard				
1-1/2"	EA.	3.89	20.50	24.39
2"	"	5.39	24.50	29.89
3"	"	12.25	30.75	43.00
4"	"	26.50	38.50	65.00
Wye, reducing				
2 x 1-1/2 x 1-1/2	EA.	7.20	20.50	27.70
2 x 2 x 1-1/2	"	6.87	24.50	31.37
4 x 4 x 2	"	14.75	38.50	53.25
4 x 4 x 3	"	20.00	41.00	61.00
Double wye				
1-1/2"	EA.	8.80	24.50	33.30

Plumbing

	UNIT	MAT.	INST.	TOTAL
15410.33 ABS DWV Pipe *(Cont.)*				
2"	EA.	10.50	30.75	41.25
3"	"	27.00	38.50	65.50
4"	"	55.00	51.00	106
2 x 2 x 1-1/2 x 1-1/2	"	10.50	30.75	41.25
3 x 3 x 2 x 2	"	22.00	38.50	60.50
4 x 4 x 3 x 3	"	51.00	51.00	102
Combination wye and 1/8 bend				
1-1/2"	EA.	6.16	20.50	26.66
2"	"	7.42	24.50	31.92
3"	"	16.00	30.75	46.75
4"	"	32.75	38.50	71.25
2 x 2 x 1-1/2	"	6.38	24.50	30.88
3 x 3 x 1-1/2	"	15.25	30.75	46.00
3 x 3 x 2	"	10.75	30.75	41.50
4 x 4 x 2	"	21.00	38.50	59.50
4 x 4 x 3	"	25.75	38.50	64.25
15410.70 Stainless Steel Pipe				
Stainless steel, schedule 40, threaded				
1/2" pipe	L.F.	10.00	8.77	18.77
3/4" pipe	"	14.00	9.03	23.03
1" pipe	"	16.50	9.45	25.95
Fittings, 1/2"				
90 deg ell	EA.	16.50	77.00	93.50
45 deg ell	"	23.75	77.00	101
Tee	"	24.75	100	125
Cap	"	8.56	38.50	47.06
Reducer	"	13.25	51.00	64.25
Union	"	46.00	77.00	123
Flange	"	24.50	77.00	102
3/4"				
90 deg ell	EA.	18.75	77.00	95.75
45 deg ell	"	22.00	77.00	99.00
Tee	"	26.50	100	127
Cap	"	12.50	38.50	51.00
Reducer	"	17.50	51.00	68.50
Union	"	63.00	77.00	140
Flange	"	27.25	77.00	104
1"				
90 deg ell	EA.	21.00	77.00	98.00
45 deg ell	"	24.00	77.00	101
Tee	"	26.50	100	127
Cap	"	17.00	38.50	55.50
Reducer	"	25.00	77.00	102
Union	"	78.00	77.00	155
Flange	"	25.50	77.00	103
Type 304 tubing				
.035 wall				
1/4"	L.F.	3.11	3.41	6.52
3/8"	"	3.90	3.84	7.74
1/2"	"	4.80	4.38	9.18
5/8"	"	6.38	5.12	11.50
3/4"	"	7.52	6.14	13.66

Plumbing	UNIT	MAT.	INST.	TOTAL
15410.70 Stainless Steel Pipe *(Cont.)*				
7/8"	L.F.	8.36	6.82	15.18
1"	"	8.95	7.68	16.63
.049 wall				
1/4"	L.F.	4.10	3.61	7.71
3/8"	"	4.80	4.09	8.89
1/2"	"	5.19	4.72	9.91
5/8"	"	6.97	5.58	12.55
3/4"	"	8.16	6.82	14.98
7/8"	"	9.75	7.68	17.43
1"	"	10.50	8.77	19.27
.065 wall				
1/4"	L.F.	4.45	4.09	8.54
3/8"	"	5.74	5.12	10.86
1/2"	"	7.92	5.58	13.50
5/8"	"	8.56	6.82	15.38
3/4"	"	10.50	8.77	19.27
7/8"	"	11.75	10.25	22.00
1"	"	12.75	12.25	25.00
Type 316 tubing				
.035 wall				
1/4"	L.F.	3.86	3.41	7.27
3/8"	"	4.95	3.84	8.79
1/2"	"	7.92	4.38	12.30
5/8"	"	10.00	5.12	15.12
3/4"	"	12.75	6.14	18.89
7/8"	"	17.25	6.82	24.07
1"	"	19.00	7.68	26.68
.049 wall				
1/4"	L.F.	5.19	4.09	9.28
3/8"	"	5.69	5.12	10.81
1/2"	"	7.17	5.58	12.75
5/8"	"	9.05	6.82	15.87
3/4"	"	9.55	8.77	18.32
7/8"	"	12.00	10.25	22.25
1"	"	13.25	12.25	25.50
.065 wall				
1/4"	L.F.	6.43	4.09	10.52
3/8"	"	8.91	5.12	14.03
1/2"	"	10.00	5.58	15.58
5/8"	"	10.50	6.82	17.32
3/4"	"	12.00	8.77	20.77
7/8"	"	15.50	10.25	25.75
1"	"	18.25	12.25	30.50
Fittings, 1/4"				
90 deg elbow	EA.	12.75	12.25	25.00
Union tee	"	22.25	20.50	42.75
Union	"	8.66	20.50	29.16
Male connector	"	6.03	15.25	21.28
3/8"				
90 deg elbow	EA.	15.00	15.25	30.25
Union tee	"	30.25	23.75	54.00
Union	"	10.50	23.75	34.25
Male connector	"	8.61	15.25	23.86

Plumbing

15410.70 Stainless Steel Pipe *(Cont.)*

Plumbing	UNIT	MAT.	INST.	TOTAL
1/2"				
90 deg elbow	EA.	21.25	16.25	37.50
Union tee	"	40.50	25.50	66.00
Union	"	18.25	25.50	43.75
Male connector	"	11.50	15.25	26.75
5/8"				
90 deg elbow	EA.	26.25	20.50	46.75
Union tee	"	43.00	30.75	73.75
Union	"	24.75	30.75	55.50
Male connector	"	13.25	20.50	33.75
3/4"				
90 deg elbow	EA.	47.50	20.50	68.00
Union tee	"	55.00	30.75	85.75
Union	"	33.75	30.75	64.50
Male connector	"	19.25	20.50	39.75
7/8"				
90 deg elbow	EA.	71.00	22.00	93.00
Union tee	"	91.00	34.25	125
Union	"	52.00	34.25	86.25
Male connector	"	28.00	22.00	50.00
1"				
90 deg elbow	EA.	94.00	28.00	122
Union tee	"	120	38.50	159
Union	"	54.00	38.50	92.50
Male connector	"	36.25	30.75	67.00
Type 316 valves				
Gate valves				
1/4"	EA.	270	20.50	291
3/8"	"	270	24.50	295
1/2"	"	300	26.75	327
3/4"	"	370	30.75	401
1"	"	420	41.00	461
Globe valves				
1/4"	EA.	170	20.50	191
3/8"	"	280	24.50	305
1/2"	"	350	26.75	377
3/4"	"	390	30.75	421
1"	"	430	41.00	471
Check valves				
1/4"	EA.	120	20.50	141
3/8"	"	120	24.50	145
1/2"	"	120	26.75	147
3/4"	"	140	30.75	171
1"	"	150	41.00	191
Test and balance	"	46.25	51.00	97.25
Pipe identification	"	0.25	12.25	12.50
Disinfect	"	46.25	51.00	97.25

Plumbing

15410.80 Steel Pipe

Plumbing	UNIT	MAT.	INST.	TOTAL
Black steel, extra heavy pipe, threaded				
1/2" pipe	L.F.	2.59	2.45	5.04
3/4" pipe	"	3.35	2.45	5.80
1" pipe	"	4.30	3.07	7.37
Fittings, malleable iron, threaded, 1/2" pipe				
90 deg ell	EA.	3.24	20.50	23.74
45 deg ell	"	4.38	20.50	24.88
Tee	"	3.53	30.75	34.28
Reducing tee	"	7.92	30.75	38.67
Cap	"	2.74	12.25	14.99
Coupling	"	3.66	24.50	28.16
Union	"	15.50	20.50	36.00
Nipple, 4" long	"	2.88	20.50	23.38
3/4" pipe				
90 deg ell	EA.	3.79	20.50	24.29
45 deg ell	"	6.02	30.75	36.77
Tee	"	5.10	30.75	35.85
Reducing tee	"	8.83	20.50	29.33
Cap	"	3.66	12.25	15.91
Coupling	"	4.32	20.50	24.82
Union	"	17.50	20.50	38.00
Nipple, 4" long	"	3.33	20.50	23.83
1" pipe				
90 deg ell	EA.	5.89	24.50	30.39
45 deg ell	"	7.79	24.50	32.29
Tee	"	8.83	34.25	43.08
Reducing tee	"	12.00	34.25	46.25
Cap	"	4.97	12.25	17.22
Coupling	"	6.41	24.50	30.91
Union	"	20.75	24.50	45.25
Nipple, 4" long	"	4.71	24.50	29.21
Cast iron fittings				
1/2" pipe				
90 deg. ell	EA.	4.27	20.50	24.77
45 deg. ell	"	8.67	20.50	29.17
Tee	"	5.64	30.75	36.39
Reducing tee	"	10.75	30.75	41.50
3/4" pipe				
90 deg. ell	EA.	4.57	20.50	25.07
45 deg. ell	"	5.64	20.50	26.14
Tee	"	7.06	30.75	37.81
Reducing tee	"	9.14	30.75	39.89
1" pipe				
90 deg. ell	EA.	5.46	24.50	29.96
45 deg. ell	"	7.54	24.50	32.04
Tee	"	10.50	34.25	44.75
Reducing tee	"	9.02	34.25	43.27

Plumbing	UNIT	MAT.	INST.	TOTAL
15410.82 Galvanized Steel Pipe				
Galvanized pipe				
1/2" pipe	L.F.	3.52	6.14	9.66
3/4" pipe	"	4.59	7.68	12.27
1" pipe	"	7.04	8.77	15.81
90 degree ell, 150 lb malleable iron, galvanized				
1/2"	EA.	2.37	12.25	14.62
3/4"	"	3.14	15.25	18.39
1"	"	5.13	16.25	21.38
45 degree ell, 150 lb m.i., galv.				
1/2"	EA.	3.77	12.25	16.02
3/4"	"	5.13	15.25	20.38
1"	"	5.75	16.25	22.00
Tees, straight, 150 lb m.i., galv.				
1/2"	EA.	3.14	15.25	18.39
3/4"	"	5.22	17.50	22.72
1"	"	7.69	20.50	28.19
Tees, reducing, out, 150 lb m.i., galv.				
1/2"	EA.	5.42	15.25	20.67
3/4"	"	6.29	17.50	23.79
1"	"	9.29	20.50	29.79
Couplings, straight, 150 lb m.i., galv.				
1/2"	EA.	2.90	12.25	15.15
3/4"	"	3.48	13.75	17.23
1"	"	5.95	15.25	21.20
Couplings, reducing, 150 lb m.i., galv				
1/2"	EA.	3.38	12.25	15.63
3/4"	"	3.77	13.75	17.52
1"	"	6.92	15.25	22.17
Caps, 150 lb m.i., galv.				
1/2"	EA.	2.42	6.14	8.56
3/4"	"	3.19	6.46	9.65
1"	"	4.35	6.82	11.17
Unions, 150 lb m.i., galv.				
1/2"	EA.	13.50	15.25	28.75
3/4"	"	15.25	17.50	32.75
1"	"	18.25	20.50	38.75
Nipples, galvanized steel, 4" long				
1/2"	EA.	3.48	7.68	11.16
3/4"	"	4.64	8.19	12.83
1"	"	6.38	8.77	15.15
90 degree reducing ell, 150 lb m.i., galv.				
3/4" x 1/2"	EA.	3.77	12.25	16.02
1" x 3/4"	"	5.13	13.75	18.88
Square head plug (C.I.)				
1/2"	EA.	2.39	6.82	9.21
3/4"	"	5.32	7.68	13.00
1"	"	5.59	8.19	13.78

Plumbing	UNIT	MAT.	INST.	TOTAL
15430.23 Cleanouts				
Cleanout, wall				
2"	EA.	210	41.00	251
3"	"	290	41.00	331
4"	"	300	51.00	351
6"	"	490	61.00	551
Floor				
2"	EA.	190	51.00	241
3"	"	250	51.00	301
4"	"	260	61.00	321
6"	"	350	77.00	427
15430.25 Hose Bibbs				
Hose bibb				
1/2"	EA.	9.07	20.50	29.57
3/4"	"	9.62	20.50	30.12
15430.60 Valves				
Gate valve, 125 lb, bronze, soldered				
1/2"	EA.	28.50	15.25	43.75
3/4"	"	34.00	15.25	49.25
1"	"	41.75	20.50	62.25
Threaded				
1/4", 125 lb	EA.	32.75	24.50	57.25
1/2"				
125 lb	EA.	31.25	24.50	55.75
150 lb	"	41.75	24.50	66.25
300 lb	"	79.00	24.50	104
3/4"				
125 lb	EA.	36.75	24.50	61.25
150 lb	"	49.75	24.50	74.25
300 lb	"	94.00	24.50	119
1"				
125 lb	EA.	47.50	24.50	72.00
150 lb	"	65.00	24.50	89.50
300 lb	"	130	30.75	161
Check valve, bronze, soldered, 125 lb				
1/2"	EA.	54.00	15.25	69.25
3/4"	"	67.00	15.25	82.25
1"	"	85.00	20.50	106
Threaded				
1/2"				
125 lb	EA.	62.00	20.50	82.50
150 lb	"	58.00	20.50	78.50
200 lb	"	61.00	20.50	81.50
3/4"				
125 lb	EA.	46.50	24.50	71.00
150 lb	"	72.00	24.50	96.50
200 lb	"	80.00	24.50	105
1"				
125 lb	EA.	64.00	30.75	94.75
150 lb	"	98.00	30.75	129
200 lb	"	98.00	30.75	129
Vertical check valve, bronze, 125 lb, threaded				

Plumbing

15430.60 Valves (Cont.)

Plumbing	UNIT	MAT.	INST.	TOTAL
1/2"	EA.	71.00	24.50	95.50
3/4"	"	100	28.00	128
1"	"	120	30.75	151
Globe valve, bronze, soldered, 125 lb				
1/2"	EA.	66.00	17.50	83.50
3/4"	"	82.00	19.25	101
1"	"	110	20.50	131
Threaded				
1/2"				
125 lb	EA.	61.00	20.50	81.50
150 lb	"	80.00	20.50	101
300 lb	"	150	20.50	171
3/4"				
125 lb	EA.	86.00	24.50	111
150 lb	"	97.00	24.50	122
300 lb	"	180	24.50	205
1"				
125 lb	EA.	110	30.75	141
150 lb	"	190	30.75	221
300 lb	"	230	30.75	261
Ball valve, bronze, 250 lb, threaded				
1/2"	EA.	18.50	24.50	43.00
3/4"	"	27.75	24.50	52.25
1"	"	35.25	30.75	66.00
Angle valve, bronze, 150 lb, threaded				
1/2"	EA.	87.00	22.00	109
3/4"	"	120	24.50	145
1"	"	170	24.50	195
Balancing valve, meter connections, circuit setter				
1/2"	EA.	82.00	24.50	107
3/4"	"	86.00	28.00	114
1"	"	110	30.75	141
Balancing valve, straight type				
1/2"	EA.	22.25	24.50	46.75
3/4"	"	27.00	24.50	51.50
Angle type				
1/2"	EA.	30.00	24.50	54.50
3/4"	"	41.50	24.50	66.00
Square head cock, 125 lb, bronze body				
1/2"	EA.	17.50	20.50	38.00
3/4"	"	21.00	24.50	45.50
1"	"	29.25	28.00	57.25
Radiator temp control valve, with control and sensor				
1/2" valve	EA.	120	38.50	159
1" valve	"	130	38.50	169
Pressure relief valve, 1/2", bronze				
Low pressure	EA.	28.00	24.50	52.50
High pressure	"	32.75	24.50	57.25
Pressure and temperature relief valve				
Bronze, 3/4"	EA.	100	24.50	125
Cast iron, 3/4"				
High pressure	EA.	48.50	24.50	73.00
Temperature relief	"	65.00	24.50	89.50

Plumbing

15430.60 Valves *(Cont.)*

	UNIT	MAT.	INST.	TOTAL
Pressure & temp relief valve	EA.	78.00	24.50	103
Pressure reducing valve, bronze, threaded, 250 lb				
1/2"	EA.	160	38.50	199
3/4"	"	160	38.50	199
1"	"	260	38.50	299
Solar water temperature regulating valve				
3/4"	EA.	640	51.00	691
Tempering valve, threaded				
3/4"	EA.	350	20.50	371
1"	"	440	24.50	465
Thermostatic mixing valve, threaded				
1/2"	EA.	120	22.00	142
3/4"	"	120	24.50	145
1"	"	440	26.75	467
Sweat connection				
1/2"	EA.	130	22.00	152
3/4"	"	160	24.50	185
Mixing valve, sweat connection				
1/2"	EA.	72.00	22.00	94.00
3/4"	"	72.00	24.50	96.50
Liquid level gauge, aluminum body				
3/4"	EA.	360	24.50	385
4125 psi, PVC body				
3/4"	EA.	430	24.50	455
150 psi, crs body				
3/4"	EA.	340	24.50	365
1"	"	370	24.50	395
175 psi, bronze body, 1/2"	"	690	22.00	712

15430.65 Vacuum Breakers

	UNIT	MAT.	INST.	TOTAL
Vacuum breaker, atmospheric, threaded connection				
3/4"	EA.	45.75	24.50	70.25
1"	"	67.00	24.50	91.50
Anti-siphon, brass				
3/4"	EA.	49.50	24.50	74.00
1"	"	77.00	24.50	102

15430.68 Strainers

	UNIT	MAT.	INST.	TOTAL
Strainer, Y pattern, 125 psi, cast iron body, threaded				
3/4"	EA.	11.50	22.00	33.50
1"	"	14.75	24.50	39.25
250 psi, brass body, threaded				
3/4"	EA.	29.75	24.50	54.25
1"	"	41.50	24.50	66.00
Cast iron body, threaded				
3/4"	EA.	17.50	24.50	42.00
1"	"	22.00	24.50	46.50

Plumbing

	UNIT	MAT.	INST.	TOTAL
15430.70 Drains, Roof & Floor				
Floor drain, cast iron, with cast iron top				
2"	EA.	150	51.00	201
3"	"	150	51.00	201
4"	"	320	51.00	371
6"	"	410	61.00	471
Roof drain, cast iron				
2"	EA.	230	51.00	281
3"	"	240	51.00	291
4"	"	300	51.00	351
5"	"	450	61.00	511
6"	"	450	61.00	511
15430.80 Traps				
Bucket trap, threaded				
3/4"	EA.	210	38.50	249
1"	"	590	41.00	631
Inverted bucket steam trap, threaded				
3/4"	EA.	250	38.50	289
1"	"	500	38.50	539
With stainless interior				
1/2"	EA.	160	38.50	199
3/4"	"	180	38.50	219
1"	"	380	38.50	419
Brass interior				
3/4"	EA.	280	38.50	319
1"	"	560	41.00	601
Cast steel body, threaded, high temperature				
3/4"	EA.	730	38.50	769
Float trap, 15 psi				
3/4"	EA.	180	38.50	219
Float and thermostatic trap, 15 psi				
3/4"	EA.	190	38.50	229
Steam trap, cast iron body, threaded, 125 psi				
3/4"	EA.	230	38.50	269
Thermostatic trap, low pressure, angle type, 25 psi				
1/2"	EA.	70.00	38.50	109
3/4"	"	120	38.50	159
1"	"	160	41.00	201
50 psi				
1/2"	EA.	110	38.50	149
3/4"	"	150	38.50	189
1"	"	170	41.00	211
Cast iron body, threaded, 125 psi				
3/4"	EA.	160	38.50	199
1"	"	190	44.00	234

Plumbing Fixtures

15440.10 Baths

Plumbing Fixtures	UNIT	MAT.	INST.	TOTAL
Bath tub, 5' long				
Minimum	EA.	530	200	730
Average	"	1,160	310	1,470
Maximum	"	2,640	610	3,250
6' long				
Minimum	EA.	590	200	790
Average	"	1,210	310	1,520
Maximum	"	3,420	610	4,030
Square tub, whirlpool, 4'x4'				
Minimum	EA.	1,810	310	2,120
Average	"	2,570	610	3,180
Maximum	"	7,850	770	8,620
5'x5'				
Minimum	EA.	1,810	310	2,120
Average	"	2,570	610	3,180
Maximum	"	8,000	770	8,770
6'x6'				
Minimum	EA.	2,210	310	2,520
Average	"	3,230	610	3,840
Maximum	"	9,270	770	10,040
For trim and rough-in				
Minimum	EA.	190	200	390
Average	"	280	310	590
Maximum	"	780	610	1,390

15440.12 Disposals & Accessories

Plumbing Fixtures	UNIT	MAT.	INST.	TOTAL
Continuous feed				
Minimum	EA.	72.00	120	192
Average	"	200	150	350
Maximum	"	390	200	590
Batch feed, 1/2 hp				
Minimum	EA.	280	120	400
Average	"	550	150	700
Maximum	"	950	200	1,150
Hot water dispenser				
Minimum	EA.	200	120	320
Average	"	320	150	470
Maximum	"	510	200	710
Epoxy finish faucet	"	290	120	410
Lock stop assembly	"	61.00	77.00	138
Mounting gasket	"	7.04	51.00	58.04
Tailpipe gasket	"	1.03	51.00	52.03
Stopper assembly	"	24.00	61.00	85.00
Switch assembly, on/off	"	27.50	100	128
Tailpipe gasket washer	"	1.10	30.75	31.85
Stop gasket	"	2.42	34.25	36.67
Tailpipe flange	"	0.27	30.75	31.02
Tailpipe	"	3.13	38.50	41.63

Plumbing Fixtures

Plumbing Fixtures	UNIT	MAT.	INST.	TOTAL
15440.15 Faucets				
Kitchen				
Minimum	EA.	83.00	100	183
Average	"	230	120	350
Maximum	"	290	150	440
Bath				
Minimum	EA.	83.00	100	183
Average	"	240	120	360
Maximum	"	370	150	520
Lavatory, domestic				
Minimum	EA.	88.00	100	188
Average	"	280	120	400
Maximum	"	460	150	610
Hospital, patient rooms				
Minimum	EA.	120	150	270
Average	"	390	200	590
Maximum	"	670	310	980
Washroom				
Minimum	EA.	110	100	210
Average	"	280	120	400
Maximum	"	510	150	660
Handicapped				
Minimum	EA.	120	120	240
Average	"	360	150	510
Maximum	"	560	200	760
Shower				
Minimum	EA.	110	100	210
Average	"	320	120	440
Maximum	"	510	150	660
For trim and rough-in				
Minimum	EA.	77.00	120	197
Average	"	120	150	270
Maximum	"	200	310	510
15440.18 Hydrants				
Wall hydrant				
8" thick	EA.	360	100	460
12" thick	"	430	120	550
18" thick	"	460	140	600
24" thick	"	510	150	660
15440.20 Lavatories				
Lavatory, counter top, porcelain enamel on cast iron				
Minimum	EA.	190	120	310
Average	"	290	150	440
Maximum	"	520	200	720
Wall hung, china				
Minimum	EA.	260	120	380
Average	"	310	150	460
Maximum	"	770	200	970
Handicapped				
Minimum	EA.	430	150	580

Plumbing Fixtures

	UNIT	MAT.	INST.	TOTAL
15440.20 Lavatories *(Cont.)*				
Average	EA.	500	200	700
Maximum	"	830	310	1,140
For trim and rough-in				
Minimum	EA.	220	150	370
Average	"	370	200	570
Maximum	"	460	310	770
15440.30 Showers				
Shower, fiberglass, 36"x34"x84"				
Minimum	EA.	570	440	1,010
Average	"	800	610	1,410
Maximum	"	1,160	610	1,770
Steel, 1 piece, 36"x36"				
Minimum	EA.	530	440	970
Average	"	800	610	1,410
Maximum	"	950	610	1,560
Receptor, molded stone, 36"x36"				
Minimum	EA.	220	200	420
Average	"	370	310	680
Maximum	"	570	510	1,080
For trim and rough-in				
Minimum	EA.	220	280	500
Average	"	370	340	710
Maximum	"	460	610	1,070
15440.40 Sinks				
Service sink, 24"x29"				
Minimum	EA.	640	150	790
Average	"	790	200	990
Maximum	"	1,170	310	1,480
Kitchen sink, single, stainless steel, single bowl				
Minimum	EA.	280	120	400
Average	"	320	150	470
Maximum	"	580	200	780
Double bowl				
Minimum	EA.	320	150	470
Average	"	360	200	560
Maximum	"	620	310	930
Porcelain enamel, cast iron, single bowl				
Minimum	EA.	200	120	320
Average	"	260	150	410
Maximum	"	410	200	610
Double bowl				
Minimum	EA.	280	150	430
Average	"	390	200	590
Maximum	"	550	310	860
Mop sink, 24"x36"x10"				
Minimum	EA.	480	120	600
Average	"	580	150	730
Maximum	"	780	200	980
Washing machine box				
Minimum	EA.	180	150	330

Plumbing Fixtures

15440.40 Sinks *(Cont.)*

	UNIT	MAT.	INST.	TOTAL
Average	EA.	250	200	450
Maximum	"	310	310	620
For trim and rough-in				
Minimum	EA.	290	200	490
Average	"	440	310	750
Maximum	"	560	410	970

15440.50 Urinals

	UNIT	MAT.	INST.	TOTAL
Wall mounted				
Minimum	EA.	410	150	560
Average	"	560	200	760
Maximum	"	730	310	1,040
For trim and rough-in				
Minimum	EA.	180	150	330
Average	"	260	310	570
Maximum	"	360	410	770

15440.60 Water Closets

	UNIT	MAT.	INST.	TOTAL
Water closet flush tank, floor mounted				
Minimum	EA.	330	150	480
Average	"	650	200	850
Maximum	"	1,020	310	1,330
Handicapped				
Minimum	EA.	370	200	570
Average	"	670	310	980
Maximum	"	1,280	610	1,890
Bowl, with flush valve, floor mounted				
Minimum	EA.	460	150	610
Average	"	510	200	710
Maximum	"	990	310	1,300
Wall mounted				
Minimum	EA.	460	150	610
Average	"	540	200	740
Maximum	"	1,030	310	1,340
For trim and rough-in				
Minimum	EA.	210	150	360
Average	"	250	200	450
Maximum	"	330	310	640

15440.70 Water Heaters

	UNIT	MAT.	INST.	TOTAL
Water heater, electric				
6 gal	EA.	400	100	500
10 gal	"	420	100	520
15 gal	"	410	100	510
20 gal	"	580	120	700
30 gal	"	590	120	710
40 gal	"	650	120	770
52 gal	"	730	150	880
66 gal	"	880	150	1,030
80 gal	"	960	150	1,110
100 gal	"	1,180	200	1,380
120 gal	"	1,520	200	1,720
Oil fired				

Plumbing Fixtures

	UNIT	MAT.	INST.	TOTAL
15440.70 Water Heaters *(Cont.)*				
20 gal	EA.	1,300	310	1,610
50 gal	"	2,020	440	2,460
15440.95 Fixture Carriers				
Lavatory, wall carrier				
Minimum	EA.	140	61.00	201
Average	"	210	77.00	287
Maximum	"	260	100	360
Sink, industrial, wall carrier				
Minimum	EA.	190	61.00	251
Average	"	220	77.00	297
Maximum	"	280	100	380
Toilets, water closets, wall carrier				
Minimum	EA.	280	61.00	341
Average	"	330	77.00	407
Maximum	"	430	100	530
Floor support				
Minimum	EA.	140	51.00	191
Average	"	160	61.00	221
Maximum	"	180	77.00	257
Urinals, wall carrier				
Minimum	EA.	150	61.00	211
Average	"	190	77.00	267
Maximum	"	230	100	330
15450.40 Storage Tanks				
Hot water storage tank, cement lined				
10 gallon	EA.	490	200	690
70 gallon	"	1,540	310	1,850

Heating & Ventilating

	UNIT	MAT.	INST.	TOTAL
15555.10 Boilers				
Cast iron, gas fired, hot water				
115 mbh	EA.	2,600	2,000	4,600
175 mbh	"	3,100	2,180	5,280
235 mbh	"	3,970	2,400	6,370
Steam				
115 mbh	EA.	2,860	2,000	4,860
175 mbh	"	3,450	2,180	5,630
235 mbh	"	4,090	2,400	6,490
Electric, hot water				
115 mbh	EA.	4,650	1,200	5,850
175 mbh	"	5,150	1,200	6,350
235 mbh	"	5,870	1,200	7,070
Steam				

Heating & Ventilating

15555.10 Boilers *(Cont.)*

	UNIT	MAT.	INST.	TOTAL
115 mbh	EA.	5,870	1,200	7,070
175 mbh	"	7,190	1,200	8,390
235 mbh	"	7,850	1,200	9,050
Oil fired, hot water				
115 mbh	EA.	3,430	1,600	5,030
175 mbh	"	4,360	1,850	6,210
235 mbh	"	6,020	2,180	8,200
Steam				
115 mbh	EA.	3,430	1,600	5,030
175 mbh	"	4,360	1,850	6,210
235 mbh	"	5,560	2,180	7,740

15610.10 Furnaces

	UNIT	MAT.	INST.	TOTAL
Electric, hot air				
40 mbh	EA.	810	310	1,120
60 mbh	"	880	320	1,200
80 mbh	"	960	340	1,300
100 mbh	"	1,080	360	1,440
125 mbh	"	1,320	370	1,690
160 mbh	"	1,810	380	2,190
200 mbh	"	2,640	400	3,040
Gas fired hot air				
40 mbh	EA.	810	310	1,120
60 mbh	"	870	320	1,190
80 mbh	"	1,000	340	1,340
100 mbh	"	1,040	360	1,400
125 mbh	"	1,140	370	1,510
160 mbh	"	1,360	380	1,740
200 mbh	"	2,430	400	2,830
Oil fired hot air				
40 mbh	EA.	1,090	310	1,400
60 mbh	"	1,800	320	2,120
80 mbh	"	1,820	340	2,160
100 mbh	"	1,850	360	2,210
125 mbh	"	1,910	370	2,280
160 mbh	"	2,200	380	2,580
200 mbh	"	2,590	400	2,990

Refrigeration

15670.10 Condensing Units

	UNIT	MAT.	INST.	TOTAL
Air cooled condenser, single circuit				
3 ton	EA.	1,640	100	1,740
5 ton	"	2,590	100	2,690
With low ambient dampers				
3 ton	EA.	1,880	150	2,030

Refrigeration

Refrigeration	UNIT	MAT.	INST.	TOTAL
15670.10 Condensing Units *(Cont.)*				
5 ton	EA.	2,970	150	3,120
15780.20 Rooftop Units				
Packaged, single zone rooftop unit, with roof curb				
2 ton	EA.	3,740	610	4,350
3 ton	"	3,930	610	4,540
4 ton	"	4,300	770	5,070
5 ton	"	4,660	1,020	5,680
15830.10 Radiation Units				
Baseboard radiation unit				
1.7 mbh/lf	L.F.	85.00	24.50	110
2.1 mbh/lf	"	110	30.75	141
Enclosure only				
Two tier	L.F.	44.25	10.25	54.50
Three tier	"	57.00	10.25	67.25
Copper element only, 3/4" dia.				
Two tier	L.F.	64.00	15.25	79.25
Three tier	"	100	20.50	121
Fin-tube, 16 ga, sloping cover, 1-1/4" steel				
One tier	L.F.	61.00	20.50	81.50
Two tier	"	97.00	24.50	122
2" steel				
Two tier	L.F.	100	24.50	125
Three tier	"	140	30.75	171
1-1/4" copper				
Two tier	L.F.	130	20.50	151
18 ga flat cover, 1-1/4" steel				
One tier	L.F.	40.75	20.50	61.25
Two tier	"	65.00	24.50	89.50
Three tier	"	110	30.75	141
2" steel				
One tier	L.F.	48.50	20.50	69.00
Two tier	"	77.00	24.50	102
Three tier	"	110	30.75	141
1-1/4" copper				
One tier	L.F.	47.00	20.50	67.50
Two tier	"	85.00	24.50	110
Three tier	"	110	30.75	141
15830.20 Fan Coil Units				
Fan coil unit, 2 pipe, complete				
200 cfm ceiling hung	EA.	1,110	200	1,310
15830.70 Unit Heaters				
Steam unit heater, horizontal				
12,500 btuh, 200 cfm	EA.	540	100	640
17,000 btuh, 300 cfm	"	710	100	810
Gas unit heater, horizontal				
27,400 btuh	EA.	820	250	1,070
38,000 btuh	"	860	250	1,110
Hot water unit heater, horizontal				
12,500 btuh, 200 cfm	EA.	430	100	530

Refrigeration

15830.70 Unit Heaters (Cont.)	UNIT	MAT.	INST.	TOTAL
17,000 btuh, 300 cfm	EA.	480	100	580
25,000 btuh, 500 cfm	"	550	100	650
30,000 btuh, 700 cfm	"	650	100	750
Cabinet unit heaters, ceiling, exposed, hot water				
200 cfm	EA.	1,240	200	1,440
300 cfm	"	1,330	250	1,580
400 cfm	"	1,380	290	1,670

Air Handling

15855.10 Air Handling Units	UNIT	MAT.	INST.	TOTAL
Air handling unit, medium pressure, single zone				
1500 cfm	EA.	4,000	380	4,380
3000 cfm	"	5,260	680	5,940
Rooftop air handling units				
4950 cfm	EA.	11,500	680	12,180
7370 cfm	"	14,580	880	15,460

15870.20 Exhaust Fans	UNIT	MAT.	INST.	TOTAL
Belt drive roof exhaust fans				
640 cfm, 2618 fpm	EA.	1,030	77.00	1,107
940 cfm, 2604 fpm	"	1,340	77.00	1,417

Air Distribution

15890.10 Metal Ductwork	UNIT	MAT.	INST.	TOTAL
Rectangular duct				
Galvanized steel				
Minimum	Lb.	0.88	5.58	6.46
Average	"	1.10	6.82	7.92
Maximum	"	1.68	10.25	11.93
Aluminum				
Minimum	Lb.	2.29	12.25	14.54
Average	"	3.05	15.25	18.30
Maximum	"	3.79	20.50	24.29
Fittings				
Minimum	EA.	7.26	20.50	27.76
Average	"	11.00	30.75	41.75
Maximum	"	16.00	61.00	77.00

Air Distribution	UNIT	MAT.	INST.	TOTAL
15890.10 Metal Ductwork (Cont.)				
For work				
10-20' high, add per pound, $.30				
15890.30 Flexible Ductwork				
Flexible duct, 1.25" fiberglass				
5" dia.	L.F.	3.31	3.07	6.38
6" dia.	"	3.68	3.41	7.09
7" dia.	"	4.54	3.61	8.15
8" dia.	"	4.76	3.84	8.60
10" dia.	"	6.34	4.38	10.72
12" dia.	"	6.93	4.72	11.65
Flexible duct connector, 3" wide fabric	"	2.31	10.25	12.56
15910.10 Dampers				
Horizontal parallel aluminum backdraft damper				
12" x 12"	EA.	55.00	15.25	70.25
16" x 16"	"	57.00	17.50	74.50
20" x 20"	"	73.00	22.00	95.00
24" x 24"	"	88.00	30.75	119
15940.10 Diffusers				
Ceiling diffusers, round, baked enamel finish				
6" dia.	EA.	36.50	20.50	57.00
8" dia.	"	44.00	25.50	69.50
10" dia.	"	48.75	25.50	74.25
12" dia.	"	63.00	25.50	88.50
Rectangular				
6x6"	EA.	39.00	20.50	59.50
9x9"	"	47.25	30.75	78.00
12x12"	"	69.00	30.75	99.75
15x15"	"	86.00	30.75	117
18x18"	"	110	30.75	141
15940.40 Registers And Grilles				
Lay in flush mounted, perforated face, return				
6x6/24x24	EA.	48.50	24.50	73.00
8x8/24x24	"	48.50	24.50	73.00
9x9/24x24	"	53.00	24.50	77.50
10x10/24x24	"	57.00	24.50	81.50
12x12/24x24	"	57.00	24.50	81.50
Rectangular, ceiling return, single deflection				
10x10	EA.	29.25	30.75	60.00
12x12	"	34.00	30.75	64.75
14x14	"	41.50	30.75	72.25
16x8	"	34.00	30.75	64.75
16x16	"	34.00	30.75	64.75
18x8	"	39.00	30.75	69.75
20x20	"	63.00	30.75	93.75
24x12	"	92.00	30.75	123
24x18	"	120	30.75	151
36x24	"	220	34.25	254
36x30	"	330	34.25	364
Wall, return air register				

Air Distribution

15940.40 Registers And Grilles (Cont.)

Air Distribution	UNIT	MAT.	INST.	TOTAL
12x12	EA.	48.50	15.25	63.75
16x16	"	71.00	15.25	86.25
18x18	"	85.00	15.25	100
20x20	"	100	15.25	115
24x24	"	140	15.25	155
Ceiling, return air grille				
6x6	EA.	28.00	20.50	48.50
8x8	"	35.00	24.50	59.50
10x10	"	43.25	24.50	67.75
Ceiling, exhaust grille, aluminum egg crate				
6x6	EA.	19.25	20.50	39.75
8x8	"	19.25	24.50	43.75
10x10	"	21.25	24.50	45.75
12x12	"	26.25	30.75	57.00
14x14	"	34.50	30.75	65.25
16x16	"	40.50	30.75	71.25
18x18	"	48.50	30.75	79.25

Basic Materials

	UNIT	MAT.	INST.	TOTAL
16050.30 Bus Duct				
Bus duct, 100a, plug-in				
10', 600v	EA.	310	200	510
With ground	"	410	300	710
10', 277/480v	"	400	200	600
With ground	"	490	300	790
Circuit breakers, with enclosure				
1 pole				
15a-60a	EA.	310	71.00	381
70a-100a	"	350	89.00	439
2 pole				
15a-60a	EA.	450	78.00	528
70a-100a	"	540	93.00	633
Circuit breaker, adapter cubicle				
225a	EA.	5,180	110	5,290
400a	"	6,120	110	6,230
Fusible switches, 240v, 3 phase				
30a	EA.	590	71.00	661
60a	"	730	89.00	819
100a	"	970	110	1,080
200a	"	1,690	150	1,840
16110.20 Conduit Specialties				
Rod beam clamp, 1/2"	EA.	6.29	3.56	9.85
Hanger rod				
3/8"	L.F.	1.33	2.85	4.18
1/2"	"	3.31	3.56	6.87
All thread rod				
1/4"	L.F.	0.42	2.15	2.57
3/8"	"	0.48	2.85	3.33
1/2"	"	0.90	3.56	4.46
5/8"	"	1.60	5.70	7.30
Hanger channel, 1-1/2"				
No holes	EA.	4.18	2.15	6.33
Holes	"	5.17	2.15	7.32
Channel strap				
1/2"	EA.	1.30	3.56	4.86
3/4"	"	1.75	3.56	5.31
Conduit penetrations, roof and wall, 8" thick				
1/2"	EA.		43.75	43.75
3/4"	"		43.75	43.75
1"	"		57.00	57.00
Threaded rod couplings				
1/4"	EA.	1.39	3.56	4.95
3/8"	"	1.46	3.56	5.02
1/2"	"	1.65	3.56	5.21
5/8"	"	2.54	3.56	6.10
3/4"	"	2.78	3.56	6.34
Hex nuts				
1/4"	EA.	0.15	3.56	3.71
3/8"	"	0.24	3.56	3.80
1/2"	"	0.52	3.56	4.08
5/8"	"	1.11	3.56	4.67
3/4"	"	1.47	3.56	5.03

Basic Materials

	UNIT	MAT.	INST.	TOTAL
16110.20 Conduit Specialties (Cont.)				
Square nuts				
1/4"	EA.	0.14	3.56	3.70
3/8"	"	0.27	3.56	3.83
3/8"	"	0.45	3.56	4.01
5/8"	"	0.60	3.56	4.16
3/4"	"	1.06	3.56	4.62
Flat washers				
1/4"	EA.			0.15
3/8"	"			0.21
1/2"	"			0.30
5/8"	"			0.60
3/4"	"			0.84
Lockwashers				
1/4"	EA.			0.09
3/8"	"			0.16
1/2"	"			0.20
5/8"	"			0.36
3/4"	"			0.60
16110.21 Aluminum Conduit				
Aluminum conduit				
1/2"	L.F.	2.00	2.15	4.15
3/4"	"	2.58	2.85	5.43
1"	"	3.61	3.56	7.17
90 deg. elbow				
1/2"	EA.	15.50	13.50	29.00
3/4"	"	21.00	17.75	38.75
1"	"	29.25	22.00	51.25
Coupling				
1/2"	EA.	3.68	3.56	7.24
3/4"	"	5.58	4.22	9.80
1"	"	7.37	5.70	13.07
16110.22 Emt Conduit				
EMT conduit				
1/2"	L.F.	0.61	2.15	2.76
3/4"	"	1.12	2.85	3.97
1"	"	1.87	3.56	5.43
90 deg. elbow				
1/2"	EA.	5.79	6.33	12.12
3/4"	"	6.36	7.13	13.49
1"	"	9.81	7.60	17.41
Connector, steel compression				
1/2"	EA.	1.86	6.33	8.19
3/4"	"	3.55	6.33	9.88
1"	"	5.36	6.33	11.69
Coupling, steel, compression				
1/2"	EA.	3.16	4.22	7.38
3/4"	"	4.31	4.22	8.53
1"	"	6.51	4.22	10.73
1 hole strap, steel				
1/2"	EA.	0.20	2.85	3.05
3/4"	"	0.26	2.85	3.11

Basic Materials

16110.22 Emt Conduit *(Cont.)*

Basic Materials	UNIT	MAT.	INST.	TOTAL
1"	EA.	0.40	2.85	3.25
Connector, steel set screw				
1/2"	EA.	1.42	4.96	6.38
3/4"	"	2.28	4.96	7.24
1"	"	3.92	4.96	8.88
Insulated throat				
1/2"	EA.	1.88	4.96	6.84
3/4"	"	3.06	4.96	8.02
1"	"	5.05	4.96	10.01
Connector, die cast set screw				
1/2"	EA.	0.87	4.22	5.09
3/4"	"	1.49	4.22	5.71
1"	"	2.81	4.22	7.03
Insulated throat				
1/2"	EA.	1.87	4.22	6.09
3/4"	"	3.01	4.22	7.23
1"	"	5.24	4.22	9.46
Coupling, steel set screw				
1/2"	EA.	2.30	2.85	5.15
3/4"	"	3.47	2.85	6.32
1"	"	5.63	2.85	8.48
Diecast set screw				
1/2"	EA.	0.81	2.85	3.66
3/4"	"	1.30	2.85	4.15
1"	"	2.15	2.85	5.00
1 hole malleable straps				
1/2"	EA.	0.37	2.85	3.22
3/4"	"	0.51	2.85	3.36
1"	"	0.84	2.85	3.69
EMT to rigid compression coupling				
1/2"	EA.	4.34	7.13	11.47
3/4"	"	6.20	7.13	13.33
1"	"	9.45	10.75	20.20
Set screw couplings				
1/2"	EA.	1.13	7.13	8.26
3/4"	"	1.71	7.13	8.84
1"	"	2.86	10.25	13.11
Set screw offset connectors				
1/2"	EA.	2.55	7.13	9.68
3/4"	"	3.42	7.13	10.55
1"	"	6.21	10.25	16.46
Compression offset connectors				
1/2"	EA.	4.21	7.13	11.34
3/4"	"	5.32	7.13	12.45
1"	"	7.70	10.25	17.95
Type "LB" set screw condulets				
1/2"	EA.	12.00	16.25	28.25
3/4"	"	14.50	21.00	35.50
1"	"	22.25	27.25	49.50
Type "T" set screw condulets				
1/2"	EA.	15.00	21.00	36.00
3/4"	"	18.75	28.50	47.25
1"	"	27.00	31.75	58.75

Basic Materials

16110.22 Emt Conduit *(Cont.)*

Basic Materials	UNIT	MAT.	INST.	TOTAL
Type "C" set screw condulets				
1/2"	EA.	12.50	17.75	30.25
3/4"	"	15.75	21.00	36.75
1"	"	23.50	27.25	50.75
Type "LL" set screw condulets				
1/2"	EA.	12.50	17.75	30.25
3/4"	"	15.25	21.00	36.25
1"	"	23.25	27.25	50.50
Type "LR" set screw condulets				
1/2"	EA.	12.50	17.75	30.25
3/4"	"	15.25	21.00	36.25
1"	"	23.25	27.25	50.50
Type "LB" compression condulets				
1/2"	EA.	28.25	21.00	49.25
3/4"	"	41.75	35.75	77.50
1"	"	54.00	35.75	89.75
Type "T" compression condulets				
1/2"	EA.	37.75	28.50	66.25
3/4"	"	49.75	31.75	81.50
1"	"	77.00	43.75	121
Condulet covers				
1/2"	EA.	1.74	8.77	10.51
3/4"	"	2.11	8.77	10.88
1"	"	2.88	8.77	11.65
Clamp type entrance caps				
1/2"	EA.	9.33	17.75	27.08
3/4"	"	11.00	21.00	32.00
1"	"	13.00	28.50	41.50
Slip fitter type entrance caps				
1/2"	EA.	6.81	17.75	24.56
3/4"	"	8.14	21.00	29.14
1"	"	9.81	28.50	38.31

16110.23 Flexible Conduit

Basic Materials	UNIT	MAT.	INST.	TOTAL
Flexible conduit, steel				
3/8"	L.F.	0.75	2.15	2.90
1/2	"	0.86	2.15	3.01
3/4"	"	1.17	2.85	4.02
1"	"	2.23	2.85	5.08
Flexible conduit, liquid tight				
3/8"	L.F.	2.09	2.15	4.24
1/2"	"	2.37	2.15	4.52
3/4"	"	3.22	2.85	6.07
1"	"	4.86	2.85	7.71
Connector, straight				
3/8"	EA.	3.54	5.70	9.24
1/2"	"	3.79	5.70	9.49
3/4"	"	4.82	6.33	11.15
1"	"	8.60	7.13	15.73
Straight insulated throat connectors				
3/8"	EA.	4.29	8.77	13.06
1/2"	"	4.29	8.77	13.06
3/4"	"	6.29	10.25	16.54

Basic Materials

16110.23 Flexible Conduit *(Cont.)*

Basic Materials	UNIT	MAT.	INST.	TOTAL
1"	EA.	9.72	10.25	19.97
90 deg connectors				
3/8"	EA.	6.00	10.50	16.50
1/2"	"	6.00	10.50	16.50
3/4"	"	9.64	12.25	21.89
1"	"	18.50	13.00	31.50
90 degree insulated throat connectors				
3/8"	EA.	7.42	10.25	17.67
1/2"	"	7.42	10.25	17.67
3/4"	"	11.25	12.25	23.50
1"	"	21.25	12.75	34.00
Flexible aluminum conduit				
3/8"	L.F.	0.44	2.15	2.59
1/2"	"	0.52	2.15	2.67
3/4"	"	0.72	2.85	3.57
1"	"	1.36	2.85	4.21
Connector, straight				
3/8"	EA.	1.35	7.13	8.48
1/2"	"	1.87	7.13	9.00
3/4"	"	2.05	7.60	9.65
1"	"	7.52	8.77	16.29
Straight insulated throat connectors				
3/8"	EA.	1.31	6.33	7.64
1/2"	"	2.64	6.33	8.97
3/4"	"	2.81	6.33	9.14
1"	"	6.83	7.13	13.96
90 deg connectors				
3/8"	EA.	2.16	10.25	12.41
1/2"	"	3.65	10.25	13.90
3/4"	"	5.73	10.25	15.98
1"	"	9.94	12.25	22.19
90 deg insulated throat connectors				
3/8"	EA.	2.68	10.25	12.93
1/2"	"	4.05	10.25	14.30
3/4"	"	6.76	10.25	17.01
1"	"	11.00	12.25	23.25

16110.24 Galvanized Conduit

Basic Materials	UNIT	MAT.	INST.	TOTAL
Galvanized rigid steel conduit				
1/2"	L.F.	3.13	2.85	5.98
3/4"	"	3.47	3.56	7.03
1"	"	5.02	4.22	9.24
1-1/4"	"	6.93	5.70	12.63
1-1/2"	"	8.15	6.33	14.48
2"	"	10.25	7.13	17.38
90 degree ell				
1/2"	EA.	8.78	17.75	26.53
3/4"	"	9.19	22.00	31.19
1"	"	14.00	27.25	41.25
1-1/4"	"	19.25	31.75	51.00
1-1/2"	"	23.75	35.75	59.50
2"	"	34.50	38.00	72.50
Couplings, with set screws				

Basic Materials

16110.24 Galvanized Conduit (Cont.)

Basic Materials	UNIT	MAT.	INST.	TOTAL
1/2"	EA.	4.39	3.56	7.95
3/4"	"	5.79	4.22	10.01
1"	"	9.25	5.70	14.95
1-1/4"	"	15.75	7.13	22.88
1-1/2"	"	20.50	8.77	29.27
2"	"	45.75	10.25	56.00
Split couplings				
1/2"	EA.	3.74	13.50	17.24
3/4"	"	4.86	17.75	22.61
1"	"	6.82	19.75	26.57
1-1/4"	"	13.50	22.00	35.50
1-1/2"	"	17.50	27.25	44.75
2"	"	40.25	40.75	81.00
Erickson couplings				
1/2"	EA.	4.44	31.75	36.19
3/4"	"	5.42	35.75	41.17
1"	"	11.00	43.75	54.75
1-1/4"	"	19.75	63.00	82.75
1-1/2"	"	25.50	71.00	96.50
2"	"	49.25	95.00	144
Seal fittings				
1/2"	EA.	14.00	47.50	61.50
3/4"	"	15.50	57.00	72.50
1"	"	19.75	71.00	90.75
1-1/4"	"	23.25	81.00	104
1-1/2"	"	35.50	95.00	131
2"	"	45.00	110	155
Entrance fitting, (weather head), threaded				
1/2"	EA.	7.48	31.75	39.23
3/4"	"	9.16	35.75	44.91
1"	"	11.75	40.75	52.50
1-1/4"	"	15.25	52.00	67.25
1-1/2"	"	27.00	57.00	84.00
2"	"	41.25	63.00	104
Locknuts				
1/2"	EA.	0.19	3.56	3.75
3/4"	"	0.24	3.56	3.80
1"	"	0.38	3.56	3.94
1-1/4"	"	0.52	3.56	4.08
1-1/2"	"	0.86	4.22	5.08
2"	"	1.26	4.22	5.48
Plastic conduit bushings				
1/2"	EA.	0.43	8.77	9.20
3/4"	"	0.66	10.25	10.91
1"	"	0.94	13.50	14.44
1-1/4"	"	1.23	15.75	16.98
1-1/2"	"	1.67	17.75	19.42
2"	"	3.79	22.00	25.79
Conduit bushings, steel				
1/2"	EA.	0.58	8.77	9.35
3/4"	"	0.73	10.25	10.98
1"	"	1.11	13.50	14.61
1-1/4"	"	1.59	15.75	17.34

Basic Materials

16110.24 Galvanized Conduit *(Cont.)*

Basic Materials	UNIT	MAT.	INST.	TOTAL
1-1/2"	EA.	2.27	17.75	20.02
2"	"	4.62	22.00	26.62
Pipe cap				
1/2"	EA.	0.57	3.56	4.13
3/4"	"	0.61	3.56	4.17
1"	"	0.99	3.56	4.55
1-1/4"	"	1.69	5.70	7.39
1-1/2"	"	2.64	5.70	8.34
2"	"	2.97	5.70	8.67
GRS elbows, 36" radius				
2"	EA.	150	47.50	198
42" radius				
2"	EA.	160	58.00	218
48" radius				
2"	EA.	180	66.00	246
Threaded couplings				
1/2"	EA.	1.91	3.56	5.47
3/4"	"	2.34	4.22	6.56
1"	"	3.47	5.70	9.17
1-1/4"	"	4.34	6.33	10.67
1-1/2"	"	5.32	7.13	12.45
2"	"	7.24	7.60	14.84
Threadless couplings				
1/2"	EA.	6.70	7.13	13.83
3/4"	"	6.98	8.77	15.75
1"	"	9.05	10.25	19.30
1-1/4"	"	10.50	13.50	24.00
1-1/2"	"	12.50	17.75	30.25
2"	"	18.25	22.00	40.25
Threadless connectors				
1/2"	EA.	3.19	7.13	10.32
3/4"	"	5.09	8.77	13.86
1"	"	8.05	10.25	18.30
1-1/4"	"	13.75	13.50	27.25
1-1/2"	"	21.00	17.75	38.75
2"	"	40.00	22.00	62.00
Setscrew connectors				
1/2"	EA.	2.56	5.70	8.26
3/4"	"	3.55	6.33	9.88
1"	"	5.53	7.13	12.66
1-1/4"	"	9.82	8.77	18.59
1-1/2"	"	14.25	10.25	24.50
2"	"	28.25	13.50	41.75
Clamp type entrance caps				
1/2"	EA.	8.58	22.00	30.58
3/4"	"	10.00	27.25	37.25
1"	"	14.00	31.75	45.75
1-1/4"	"	16.50	35.75	52.25
1-1/2"	"	29.75	43.75	73.50
2"	"	35.75	52.00	87.75
"LB" condulets				
1/2"	EA.	9.67	22.00	31.67
3/4"	"	11.75	27.25	39.00

Basic Materials

16110.24 Galvanized Conduit (Cont.)

Basic Materials	UNIT	MAT.	INST.	TOTAL
1"	EA.	17.50	31.75	49.25
1-1/4"	"	30.25	35.75	66.00
1-1/2"	"	39.50	43.75	83.25
2"	"	65.00	52.00	117
"T" condulets				
1/2"	EA.	12.25	27.25	39.50
3/4"	"	14.50	31.75	46.25
1"	"	22.00	35.75	57.75
1-1/4"	"	32.00	40.75	72.75
1-1/2"	"	43.00	43.75	86.75
2"	"	66.00	52.00	118
"X" condulets				
1/2"	EA.	18.00	31.75	49.75
3/4"	"	19.25	35.75	55.00
1"	"	31.75	40.75	72.50
1-1/4"	"	41.75	43.75	85.50
1-1/2"	"	54.00	47.50	102
2"	"	110	63.00	173
Blank steel condulet covers				
1/2"	EA.	2.75	7.13	9.88
3/4"	"	3.41	7.13	10.54
1"	"	4.65	7.13	11.78
1-1/4"	"	5.69	8.77	14.46
1-1/2"	"	5.98	8.77	14.75
2"	"	10.00	8.77	18.77
Solid condulet gaskets				
1/2"	EA.	2.27	3.56	5.83
3/4"	"	2.46	3.56	6.02
1"	"	2.84	3.56	6.40
1-1/4"	"	3.54	5.70	9.24
1-1/2"	"	3.73	5.70	9.43
2"	"	4.17	5.70	9.87
One-hole malleable straps				
1/2"	EA.	0.37	2.85	3.22
3/4"	"	0.53	2.85	3.38
1"	"	0.75	2.85	3.60
1-1/4"	"	1.51	3.56	5.07
1-1/2"	"	1.74	3.56	5.30
2"	"	3.41	3.56	6.97
One-hole steel straps				
1/2"	EA.	0.10	2.85	2.95
3/4"	"	0.14	2.85	2.99
1"	"	0.24	2.85	3.09
1-1/4"	"	0.34	3.56	3.90
1-1/2"	"	0.45	3.56	4.01
2"	"	0.60	3.56	4.16
Grounding locknuts				
1/2"	EA.	2.09	5.70	7.79
3/4"	"	2.63	5.70	8.33
1"	"	3.81	5.70	9.51
1-1/4"	"	4.08	6.33	10.41
1-1/2"	"	4.27	6.33	10.60
2"	"	6.33	6.33	12.66

Basic Materials

16110.24 Galvanized Conduit *(Cont.)*

Basic Materials	UNIT	MAT.	INST.	TOTAL
Insulated grounding metal bushings				
1/2"	EA.	1.48	13.50	14.98
3/4"	"	2.19	15.75	17.94
1"	"	3.12	17.75	20.87
1-1/4"	"	4.98	22.00	26.98
1-1/2"	"	6.21	27.25	33.46
2"	"	9.00	31.75	40.75

16110.25 Plastic Conduit

Basic Materials	UNIT	MAT.	INST.	TOTAL
PVC conduit, schedule 40				
1/2"	L.F.	0.70	2.15	2.85
3/4"	"	0.87	2.15	3.02
1"	"	1.25	2.85	4.10
1-1/4"	"	1.74	2.85	4.59
1-1/2"	"	2.06	3.56	5.62
2"	"	2.64	3.56	6.20
Couplings				
1/2"	EA.	0.42	3.56	3.98
3/4"	"	0.51	3.56	4.07
1"	"	0.80	3.56	4.36
1-1/4"	"	1.05	4.22	5.27
1-1/2"	"	1.46	4.22	5.68
2"	"	1.92	4.22	6.14
90 degree elbows				
1/2"	EA.	1.66	7.13	8.79
3/4"	"	1.81	8.77	10.58
1"	"	2.87	8.77	11.64
1-1/4"	"	4.00	10.25	14.25
1-1/2"	"	5.42	13.50	18.92
2"	"	7.56	15.75	23.31
Terminal adapters				
1/2"	EA.	0.62	7.13	7.75
3/4"	"	1.01	7.13	8.14
1"	"	1.26	7.13	8.39
1-1/4"	"	1.59	11.50	13.09
1-1/2"	"	2.03	11.50	13.53
2"	"	2.80	11.50	14.30
End bells				
1"	EA.	4.21	7.13	11.34
1-1/4"	"	4.98	11.50	16.48
1-1/2"	"	5.18	11.50	16.68
2"	"	7.71	11.50	19.21
LB conduit body				
1/2"	EA.	5.43	13.50	18.93
3/4"	"	7.00	13.50	20.50
1	"	7.71	13.50	21.21
1-1/4"	"	11.75	22.00	33.75
1-1/2"	"	14.00	22.00	36.00
2"	"	25.00	22.00	47.00
"EB" and "DB" duct, 90 degree elbows				
1-1/2"	EA.	10.75	10.25	21.00
2"	"	11.75	16.25	28.00
45 degree elbows				

Basic Materials

16110.25 Plastic Conduit *(Cont.)*

Basic Materials	UNIT	MAT.	INST.	TOTAL
1-1/2"	EA.	12.75	16.25	29.00
2"	"	13.00	16.25	29.25
Couplings				
1-1/2"	EA.	1.04	4.22	5.26
2"	"	1.17	4.22	5.39
Bell ends				
1-1/2"	EA.	6.49	11.50	17.99
2"	"	8.25	11.50	19.75
Female adapters, 1-1/2"	"	1.98	14.25	16.23
5 degree couplings				
1-1/2"	EA.	8.32	4.96	13.28
2"	"	9.38	4.96	14.34
45 degree elbows				
1/2"	EA.	1.26	8.77	10.03
3/4"	"	1.59	10.25	11.84
1"	"	2.25	10.25	12.50
1-1/4"	"	3.24	13.00	16.24
1-1/2"	"	4.45	16.25	20.70
2"	"	6.60	19.00	25.60
Female adapters				
1/2"	EA.	0.66	8.77	9.43
3/4"	"	1.05	8.77	9.82
1"	"	1.32	8.77	10.09
1-1/4"	"	1.70	14.25	15.95
1-1/2"	"	1.87	14.25	16.12
2"	"	2.69	14.25	16.94
Expansion couplings				
1/2"	EA.	28.50	8.77	37.27
3/4"	"	28.00	8.77	36.77
1"	"	29.25	10.25	39.50
1-1/4"	"	29.50	14.25	43.75
1-1/2"	"	29.50	14.25	43.75
2"	"	32.50	14.25	46.75
PVC cement				
1 pint	EA.			15.00
1 quart	"			22.00
1 gallon	"			72.00
Type "T" condulets				
1/2"	EA.	7.42	21.00	28.42
3/4"	"	8.41	21.00	29.41
1"	"	9.57	21.00	30.57
1-1/4"	"	14.25	35.75	50.00
1-1/2"	"	18.75	35.75	54.50
2"	"	26.50	35.75	62.25

16110.27 Plastic Coated Conduit

Basic Materials	UNIT	MAT.	INST.	TOTAL
Rigid steel conduit, plastic coated				
1/2"	L.F.	5.82	3.56	9.38
3/4"	"	6.76	4.22	10.98
1"	"	8.75	5.70	14.45
1-1/4"	"	11.00	7.13	18.13
1-1/2"	"	13.50	8.77	22.27
2"	"	17.50	10.25	27.75

Basic Materials

16110.27 Plastic Coated Conduit *(Cont.)*

	UNIT	MAT.	INST.	TOTAL
90 degree elbows				
1/2"	EA.	23.00	22.00	45.00
3/4"	"	23.75	27.25	51.00
1"	"	27.25	31.75	59.00
1-1/4"	"	33.50	35.75	69.25
1-1/2"	"	41.25	43.75	85.00
2"	"	58.00	57.00	115
Couplings				
1/2"	EA.	6.54	4.22	10.76
3/4"	"	6.85	5.70	12.55
1"	"	9.07	6.33	15.40
1-1/4"	"	10.50	7.60	18.10
1-1/2"	"	14.75	8.77	23.52
2"	"	18.50	10.25	28.75
1 hole conduit straps				
3/4"	EA.	11.25	3.56	14.81
1"	"	11.50	3.56	15.06
1-1/4"	"	16.75	4.22	20.97
1-1/2"	"	17.75	4.22	21.97
2"	"	25.75	4.22	29.97
"L.B." condulets with covers				
1/2"	EA.	54.00	35.75	89.75
3/4"	"	60.00	35.75	95.75
1"	"	81.00	43.75	125
1-1/4"	"	120	52.00	172
1-1/2"	"	140	63.00	203
2"	"	210	71.00	281
"T" condulets with covers				
1/2"	EA.	63.00	40.75	104
3/4"	"	79.00	43.75	123
1"	"	94.00	47.50	142
1-1/4"	"	130	58.00	188
1-1/2"	"	170	67.00	237
2"	"	240	75.00	315

16110.28 Steel Conduit

	UNIT	MAT.	INST.	TOTAL
Intermediate metal conduit (IMC)				
1/2"	L.F.	2.03	2.15	4.18
3/4"	"	2.49	2.85	5.34
1"	"	3.77	3.56	7.33
1-1/4"	"	4.83	4.22	9.05
1-1/2"	"	6.04	5.70	11.74
2"	"	7.88	6.33	14.21
90 degree ell				
1/2"	EA.	15.50	17.75	33.25
3/4"	"	16.25	22.00	38.25
1"	"	25.00	27.25	52.25
1-1/4"	"	34.50	31.75	66.25
1-1/2"	"	42.50	35.75	78.25
2"	"	61.00	40.75	102
Couplings				

Basic Materials

	UNIT	MAT.	INST.	TOTAL
16110.28 Steel Conduit (Cont.)				
1/2"	EA.	3.80	3.56	7.36
3/4"	"	4.67	4.22	8.89
1"	"	6.92	5.70	12.62
1-1/4"	"	8.66	6.33	14.99
1-1/2"	"	11.00	7.13	18.13
2"	"	14.50	7.60	22.10
16110.35 Surface Mounted Raceway				
Single Raceway				
3/4" x 17/32" Conduit	L.F.	1.67	2.85	4.52
Mounting Strap	EA.	0.45	3.80	4.25
Connector	"	0.60	3.80	4.40
Elbow				
45 degree	EA.	7.62	3.56	11.18
90 degree	"	2.43	3.56	5.99
internal	"	3.05	3.56	6.61
external	"	2.82	3.56	6.38
Switch	"	19.75	28.50	48.25
Utility Box	"	13.25	28.50	41.75
Receptacle	"	23.50	28.50	52.00
3/4" x 21/32" Conduit	L.F.	1.90	2.85	4.75
Mounting Strap	EA.	0.70	3.80	4.50
Connector	"	0.72	3.80	4.52
Elbow				
45 degree	EA.	9.41	3.56	12.97
90 degree	"	2.59	3.56	6.15
internal	"	3.52	3.56	7.08
external	"	3.52	3.56	7.08
Switch	"	19.75	28.50	48.25
Utility Box	"	13.25	28.50	41.75
Receptacle	"	23.50	28.50	52.00
16120.41 Aluminum Conductors				
Type XHHW, stranded aluminum, 600v				
#8	L.F.	0.31	0.35	0.66
#6	"	0.33	0.43	0.76
#4	"	0.41	0.57	0.98
#2	"	0.56	0.64	1.20
1/0	"	0.90	0.78	1.68
2/0	"	1.17	0.85	2.02
3/0	"	1.45	1.00	2.45
4/0	"	1.62	1.06	2.68
THW, stranded				
#8	L.F.	0.31	0.35	0.66
#6	"	0.33	0.43	0.76
#4	"	0.41	0.57	0.98
#3	"	0.53	0.63	1.16
#1	"	0.90	0.71	1.61
1/0	"	0.99	0.78	1.77
2/0	"	1.17	0.84	2.01
3/0	"	1.45	0.84	2.29
4/0	"	1.62	1.06	2.68
XLP, stranded				

Basic Materials	UNIT	MAT.	INST.	TOTAL
16120.41 Aluminum Conductors *(Cont.)*				
#6	L.F.	0.39	0.35	0.74
#4	"	0.45	0.57	1.02
#2	"	0.62	0.64	1.26
#1	"	0.87	0.71	1.58
1/0	"	1.06	0.78	1.84
2/0	"	1.25	0.85	2.10
3/0	"	1.50	1.00	2.50
4/0	"	1.65	1.06	2.71
Bare stranded aluminum wire				
#4	L.F.	0.29	0.57	0.86
#2	"	0.41	0.64	1.05
1/0	"	0.55	0.78	1.33
2/0	"	0.68	0.85	1.53
3/0	"	0.85	1.00	1.85
4/0	"	1.07	1.06	2.13
Triplex XLP cable				
#4	L.F.	0.94	1.06	2.00
#2	"	1.16	1.42	2.58
1/0	"	1.85	2.15	4.00
4/0	"	3.40	3.45	6.85
Aluminum quadruplex XLP cable				
#4	L.F.	1.25	1.28	2.53
#2	"	1.62	1.62	3.24
1/0	"	2.56	2.28	4.84
2/0	"	3.07	3.00	6.07
4/0	"	4.42	4.56	8.98
Triplexed URD-XLP cable				
#6	L.F.	0.76	0.78	1.54
#4	"	1.08	1.00	2.08
#2	"	1.40	1.28	2.68
1/0	"	2.23	2.00	4.23
2/0	"	2.56	2.37	4.93
3/0	"	3.07	2.85	5.92
4/0	"	3.59	3.35	6.94
Type S.E.U. cable				
#8/3	L.F.	1.47	1.78	3.25
#6/3	"	1.47	2.00	3.47
#4/3	"	1.89	2.48	4.37
#2/3	"	2.52	2.71	5.23
#1/3	"	3.43	2.85	6.28
1/0-3	"	3.85	3.00	6.85
2/0-3	"	4.42	3.16	7.58
3/0-3	"	6.17	3.68	9.85
4/0-3	"	6.20	4.07	10.27
Type S.E.R. cable with ground				
#8/3	L.F.	1.79	2.00	3.79
#6/3	"	2.02	2.48	4.50
#4/3	"	2.26	2.71	4.97
#2/3	"	3.32	2.85	6.17
#1/3	"	4.33	3.16	7.49
1/0-3	"	5.04	3.56	8.60
2/0-3	"	5.94	3.93	9.87
3/0-3	"	7.32	4.22	11.54

Basic Materials

16120.41 Aluminum Conductors *(Cont.)*

	UNIT	MAT.	INST.	TOTAL
4/0-3	L.F.	8.47	4.75	13.22
#6/4	"	3.43	2.71	6.14
#4/4	"	3.87	3.16	7.03
#2/4	"	5.65	3.16	8.81
#1/4	"	7.32	3.56	10.88
1/0-4	"	8.54	3.68	12.22
2/0-4	"	10.00	4.07	14.07
3/0-4	"	12.50	4.56	17.06
4/0-4	"	14.50	5.43	19.93

16120.43 Copper Conductors

	UNIT	MAT.	INST.	TOTAL
Copper conductors, type THW, solid				
#14	L.F.	0.13	0.28	0.41
#12	"	0.20	0.35	0.55
#10	"	0.31	0.43	0.74
Stranded				
#14	L.F.	0.14	0.28	0.42
#12	"	0.18	0.35	0.53
#10	"	0.27	0.43	0.70
#8	"	0.45	0.57	1.02
#6	"	0.73	0.64	1.37
#4	"	1.16	0.71	1.87
#3	"	1.46	0.71	2.17
#2	"	1.83	0.85	2.68
#1	"	2.32	1.00	3.32
1/0	"	2.78	1.14	3.92
2/0	"	3.48	1.42	4.90
3/0	"	4.39	1.78	6.17
4/0	"	5.48	2.00	7.48
THHN-THWN, solid				
#14	L.F.	0.13	0.28	0.41
#12	"	0.20	0.35	0.55
#10	"	0.31	0.43	0.74
Stranded				
#14	L.F.	0.13	0.28	0.41
#12	"	0.20	0.35	0.55
#10	"	0.31	0.43	0.74
#8	"	0.54	0.57	1.11
#6	"	0.85	0.64	1.49
#4	"	1.35	0.71	2.06
#2	"	1.88	0.85	2.73
#1	"	2.38	1.00	3.38
1/0	"	2.94	1.14	4.08
2/0	"	3.63	1.42	5.05
3/0	"	4.56	1.78	6.34
4/0	"	5.69	2.00	7.69
XHHW				
#14	L.F.	0.22	0.28	0.50
#10	"	0.50	0.43	0.93
#8	"	0.73	0.57	1.30
#6	"	1.16	0.64	1.80
#4	"	1.80	0.64	2.44
#2	"	2.80	0.78	3.58

Basic Materials

16120.43 Copper Conductors *(Cont.)*

Basic Materials	UNIT	MAT.	INST.	TOTAL
#1	L.F.	3.58	1.00	4.58
1/0	"	4.22	1.14	5.36
2/0	"	5.28	1.35	6.63
3/0	"	6.59	1.78	8.37
XLP, 600v				
#12	L.F.	0.35	0.35	0.70
#10	"	0.50	0.43	0.93
#8	"	0.66	0.57	1.23
#6	"	1.02	0.64	1.66
#4	"	1.58	0.71	2.29
#3	"	1.97	0.78	2.75
#2	"	2.44	0.85	3.29
#1	"	3.14	1.00	4.14
1/0	"	3.54	1.14	4.68
2/0	"	4.42	1.42	5.84
3/0	"	5.55	1.84	7.39
4/0	"	6.94	2.00	8.94
Bare solid wire				
#14	L.F.	0.13	0.28	0.41
#12	"	0.22	0.35	0.57
#10	"	0.33	0.42	0.75
#8	"	0.45	0.57	1.02
#6	"	0.82	0.64	1.46
#4	"	1.35	0.71	2.06
#2	"	2.15	0.85	3.00
Bare stranded wire				
#8	L.F.	0.47	0.57	1.04
#6	"	0.78	0.71	1.49
#4	"	1.23	0.71	1.94
#2	"	1.96	0.78	2.74
#1	"	2.45	1.00	3.45
1/0	"	2.90	1.28	4.18
2/0	"	3.65	1.42	5.07
3/0	"	4.61	1.78	6.39
4/0	"	5.80	2.00	7.80
Type "BX" solid armored cable				
#14/2	L.F.	0.90	1.78	2.68
#14/3	"	1.42	2.00	3.42
#14/4	"	1.99	2.19	4.18
#12/2	"	0.93	2.00	2.93
#12/3	"	1.48	2.19	3.67
#12/4	"	2.05	2.48	4.53
#10/2	"	1.71	2.19	3.90
#10/3	"	2.45	2.48	4.93
#10/4	"	3.82	2.85	6.67
#8/2	"	3.42	2.48	5.90
#8/3	"	4.81	2.85	7.66
Steel type, metal clad cable, solid, with ground				
#14/2	L.F.	0.74	1.28	2.02
#14/3	"	1.14	1.42	2.56
#14/4	"	1.54	1.62	3.16
#12/2	"	0.77	1.42	2.19
#12/3	"	1.26	1.78	3.04

Basic Materials

Basic Materials	UNIT	MAT.	INST.	TOTAL
16120.43 Copper Conductors *(Cont.)*				
#12/4	L.F.	1.70	2.15	3.85
#10/2	"	1.58	1.62	3.20
#10/3	"	2.20	2.00	4.20
#10/4	"	3.42	2.37	5.79
Metal clad cable, stranded, with ground				
#8/2	L.F.	2.78	2.00	4.78
#8/3	"	3.98	2.48	6.46
#8/4	"	5.20	3.00	8.20
#6/2	"	3.79	2.15	5.94
#6/3	"	4.56	2.71	7.27
#6/4	"	5.44	3.16	8.60
#4/2	"	4.95	2.85	7.80
#4/3	"	5.61	3.16	8.77
#4/4	"	6.35	3.93	10.28
#3/3	"	6.50	3.56	10.06
#3/4	"	7.21	4.22	11.43
#2/3	"	5.22	4.07	9.29
#2/4	"	8.83	4.75	13.58
#1/3	"	9.18	5.43	14.61
#1/4	"	10.75	6.00	16.75
16120.45 Flat Conductor Cable				
Flat conductor cable, with shield, 3 conductor				
#12 awg	L.F.	6.82	4.22	11.04
#10 awg	"	7.99	4.22	12.21
4 conductor				
#12 awg	L.F.	9.25	5.70	14.95
#10 awg	"	10.50	5.70	16.20
Transition boxes				
#12 awg	L.F.	11.50	6.33	17.83
#10 awg	"	13.00	6.33	19.33
Flat conductor cable communication, with shield				
10 conductor	L.F.	4.48	4.22	8.70
16 conductor	"	5.18	4.96	10.14
24 conductor	"	5.79	7.13	12.92
Power and communication heads, duplex receptacle	EA.	61.00	57.00	118
Double duplex receptacle	"	68.00	68.00	136
Telephone	"	39.00	57.00	96.00
Receptacle and telephone	"	94.00	68.00	162
Blank cover	"	9.77	10.25	20.02
Transition boxes				
Surface	EA.	180	52.00	232
Flush	"	100	71.00	171
Flat conductor cable fittings				
End caps	EA.	1.87	10.25	12.12
Insulators	"	20.50	21.00	41.50
Splice connectors	"	1.40	31.75	33.15
Tap connectors	"	1.54	31.75	33.29
Cable connectors	"	1.72	31.75	33.47
Terminal blocks	"	11.75	43.75	55.50
Tape	"			18.25

Basic Materials	UNIT	MAT.	INST.	TOTAL
16120.47 Sheathed Cable				
Non-metallic sheathed cable				
Type NM cable with ground				
#14/2	L.F.	0.29	1.06	1.35
#12/2	"	0.45	1.14	1.59
#10/2	"	0.72	1.26	1.98
#8/2	"	1.18	1.42	2.60
#6/2	"	1.87	1.78	3.65
#14/3	"	0.42	1.84	2.26
#12/3	"	0.65	1.90	2.55
#10/3	"	1.03	1.93	2.96
#8/3	"	1.74	1.96	3.70
#6/3	"	2.82	2.00	4.82
#4/3	"	5.84	2.28	8.12
#2/3	"	8.78	2.48	11.26
Type U.F. cable with ground				
#14/2	L.F.	0.34	1.14	1.48
#12/2	"	0.52	1.35	1.87
#10/2	"	0.83	1.42	2.25
#8/2	"	1.43	1.62	3.05
#6/2	"	2.23	1.93	4.16
#14/3	"	0.48	1.42	1.90
#12/3	"	0.73	1.56	2.29
#10/3	"	1.15	1.78	2.93
#8/3	"	2.16	2.00	4.16
#6/3	"	3.50	2.28	5.78
Type S.F.U. cable, 3 conductor				
#8	L.F.	1.49	2.00	3.49
#6	"	2.60	2.19	4.79
#3	"	5.07	2.85	7.92
#2	"	6.30	3.16	9.46
#1	"	8.14	3.56	11.70
#1/0	"	10.00	3.93	13.93
#2/0	"	12.50	4.56	17.06
#3/0	"	15.75	4.96	20.71
#4/0	"	17.50	5.43	22.93
Type SER cable, 4 conductor				
#6	L.F.	3.73	2.59	6.32
#4	"	5.22	2.78	8.00
#3	"	7.04	3.16	10.20
#2	"	8.17	3.45	11.62
#1	"	10.25	3.93	14.18
#1/0	"	13.00	4.56	17.56
#2/0	"	16.25	4.75	21.00
#3/0	"	20.25	5.43	25.68
#4/0	"	25.25	6.00	31.25
Flexible cord, type STO cord				
#18/2	L.F.	0.63	0.28	0.91
#18/3	"	0.73	0.35	1.08
#18/4	"	1.02	0.42	1.44
#16/2	"	0.72	0.28	1.00
#16/3	"	0.61	0.31	0.92
#16/4	"	0.86	0.35	1.21
#14/2	"	1.14	0.35	1.49

Basic Materials

Basic Materials	UNIT	MAT.	INST.	TOTAL
16120.47 Sheathed Cable *(Cont.)*				
#14/3	L.F.	1.03	0.43	1.46
#14/4	"	1.29	0.49	1.78
#12/2	"	1.44	0.42	1.86
#12/3	"	1.09	0.47	1.56
#12/4	"	1.58	0.57	2.15
#10/2	"	1.79	0.49	2.28
#10/3	"	1.72	0.57	2.29
#10/4	"	2.66	0.64	3.30
#8/2	"	2.99	0.57	3.56
#8/3	"	3.31	0.63	3.94
#8/4	"	4.65	0.71	5.36
16130.10 Floor Boxes				
Aluminum round	EA.	53.00	52.00	105
1 gang	"	40.75	57.00	97.75
2 gang	"	49.50	68.00	118
3 gang	"	59.00	71.00	130
Steel plate single recept	"	13.50	10.25	23.75
Duplex receptacle	"	13.00	13.00	26.00
Twist lock receptacle	"	13.75	13.00	26.75
Plug, 3/4"	"	17.25	10.25	27.50
1" plug	"	16.00	10.25	26.25
Carpet flange	"	21.00	10.25	31.25
Adjustable bronze plates for round cast boxes				
1/2" plug	EA.	7.04	10.25	17.29
3/4" plug	"	7.04	10.25	17.29
1" plug	"	8.85	10.25	19.10
Combination plug	"	15.50	14.25	29.75
Duplex receptacle plug	"	26.25	14.25	40.50
Adjustable aluminum plates for round cast boxes				
1/2" plug	EA.	21.00	10.25	31.25
3/4" plug	"	21.25	10.25	31.50
1" plug	"	21.75	10.25	32.00
Combination plug	"	21.00	14.25	35.25
Duplex receptacle plug	"	35.75	14.25	50.00
Adjustable bronze plates for gang type boxes				
1/2" plug	EA.	22.00	10.25	32.25
3/4" plug	"	22.25	10.25	32.50
1" plug	"	22.50	10.25	32.75
Carpet plate				
1 gang	EA.	19.50	10.25	29.75
2 gang	"	29.25	10.25	39.50
3 gang	"	39.25	14.25	53.50
Adjustable aluminum plates for gang type boxes				
1/2" plug	EA.	19.75	10.25	30.00
3/4" plug	"	20.25	10.25	30.50
1" plug	"	20.50	10.25	30.75
Duplex recept	"	19.75	14.25	34.00
Carpet plate				
1 gang	EA.	43.25	10.25	53.50
2 gang	"	60.00	10.25	70.25
3 gang	"	95.00	14.25	109
4 gang carpet plate	"	32.00	40.75	72.75

Basic Materials

16130.10 Floor Boxes (Cont.)

Basic Materials	UNIT	MAT.	INST.	TOTAL
Telephone	EA.	29.25	35.75	65.00
Floor box nozzles, horizontal				
Duplex recept	EA.	49.50	38.00	87.50
Single recept	"	66.00	38.00	104
Double duplex recept	"	54.00	52.00	106
Vertical with duplex recept	"	44.00	43.75	87.75
Double duplex recept	"	46.75	52.00	98.75

16130.40 Boxes

Basic Materials	UNIT	MAT.	INST.	TOTAL
Round cast box, type SEH				
1/2"	EA.	20.00	24.75	44.75
3/4"	"	20.00	30.00	50.00
SEHC				
1/2"	EA.	24.00	24.75	48.75
3/4"	"	24.00	30.00	54.00
SEHL				
1/2"	EA.	24.50	24.75	49.25
3/4"	"	24.00	31.75	55.75
SEHT				
1/2"	EA.	26.25	30.00	56.25
3/4"	"	26.25	35.75	62.00
SEHX				
1/2"	EA.	28.50	35.75	64.25
3/4"	"	28.50	43.75	72.25
Blank cover	"	4.84	10.25	15.09
1/2", hub cover	"	4.62	10.25	14.87
Cover with gasket	"	5.06	12.75	17.81
Rectangle, type FS boxes				
1/2"	EA.	10.25	24.75	35.00
3/4"	"	11.00	28.50	39.50
1"	"	11.75	35.75	47.50
FSA				
1/2"	EA.	18.50	24.75	43.25
3/4"	"	17.25	28.50	45.75
FSC				
1/2"	EA.	11.50	24.75	36.25
3/4"	"	12.50	30.00	42.50
1"	"	15.75	35.75	51.50
FSL				
1/2"	EA.	18.25	24.75	43.00
3/4"	"	18.25	28.50	46.75
FSR				
1/2"	EA.	19.00	24.75	43.75
3/4"	"	19.50	28.50	48.00
FSS				
1/2"	EA.	11.50	24.75	36.25
3/4"	"	12.50	28.50	41.00
FSLA				
1/2"	EA.	7.83	24.75	32.58
3/4"	"	8.86	28.50	37.36
FSCA				
1/2"	EA.	23.00	24.75	47.75
3/4"	"	22.25	28.50	50.75

Basic Materials

16130.40 Boxes (Cont.)

Basic Materials	UNIT	MAT.	INST.	TOTAL
FSCC				
1/2"	EA.	14.00	28.50	42.50
3/4"	"	21.00	35.75	56.75
FSCT				
1/2"	EA.	14.00	28.50	42.50
3/4"	"	17.50	35.75	53.25
1"	"	14.25	40.75	55.00
FST				
1/2"	EA.	20.50	35.75	56.25
3/4"	"	20.50	40.75	61.25
FSX				
1/2"	EA.	23.50	43.75	67.25
3/4"	"	21.75	52.00	73.75
FSCD boxes				
1/2"	EA.	19.50	43.75	63.25
3/4"	"	20.50	52.00	72.50
Rectangle, type FS, 2 gang boxes				
1/2"	EA.	22.00	24.75	46.75
3/4"	"	22.50	28.50	51.00
1"	"	23.75	35.75	59.50
FSC, 2 gang boxes				
1/2"	EA.	23.25	24.75	48.00
3/4"	"	25.75	28.50	54.25
1"	"	31.25	35.75	67.00
FSS, 2 gang boxes				
3/4"	EA.	24.25	28.50	52.75
FS, tandem boxes				
1/2"	EA.	24.25	28.50	52.75
3/4"	"	25.00	31.75	56.75
FSC, tandem boxes				
1/2"	EA.	32.75	28.50	61.25
3/4"	"	35.00	31.75	66.75
FS, three gang boxes				
3/4"	EA.	35.75	31.75	67.50
1"	"	39.25	35.75	75.00
FSS, three gang boxes, 3/4"	"	46.00	35.75	81.75
Weatherproof cast aluminum boxes, 1 gang, 3 outlets				
1/2"	EA.	6.54	28.50	35.04
3/4"	"	7.09	35.75	42.84
2 gang, 3 outlets				
1/2"	EA.	12.50	35.75	48.25
3/4"	"	13.25	38.00	51.25
1 gang, 4 outlets				
1/2"	EA.	11.50	43.75	55.25
3/4"	"	12.50	52.00	64.50
2 gang, 4 outlets				
1/2"	EA.	12.00	43.75	55.75
3/4"	"	13.25	52.00	65.25
1 gang, 5 outlets				
1/2"	EA.	9.44	52.00	61.44
3/4"	"	11.25	57.00	68.25
2 gang, 5 outlets				
1/2"	EA.	17.00	52.00	69.00

Basic Materials

16130.40 Boxes (Cont.)

Basic Materials	UNIT	MAT.	INST.	TOTAL
3/4"	EA.	20.75	57.00	77.75
2 gang, 6 outlets				
1/2"	EA.	19.25	61.00	80.25
3/4"	"	20.75	64.00	84.75
2 gang, 7 outlets				
1/2"	EA.	20.50	71.00	91.50
3/4"	"	25.50	78.00	104
Weatherproof and type FS box covers, blank, 1 gang	"	2.98	10.25	13.23
Tumbler switch, 1 gang	"	6.11	10.25	16.36
1 gang, single recept	"	3.85	10.25	14.10
Duplex recept	"	4.91	10.25	15.16
Despard	"	4.93	10.25	15.18
Red pilot light	"	23.25	10.25	33.50
SW and				
Single recept	EA.	10.25	14.25	24.50
Duplex recept	"	8.47	14.25	22.72
2 gang				
Blank	EA.	3.10	13.00	16.10
Tumbler switch	"	4.07	13.00	17.07
Single recept	"	4.07	13.00	17.07
Duplex recept	"	4.07	13.00	17.07
3 gang				
Blank	EA.	7.09	14.25	21.34
Tumbler switch	"	8.80	14.25	23.05
4 gang				
Tumbler switch	EA.	11.25	17.75	29.00
Box covers				
Surface	EA.	15.50	14.25	29.75
Sealing	"	17.00	14.25	31.25
Dome	"	23.50	14.25	37.75
1/2" nipple	"	30.00	14.25	44.25
3/4" nipple	"	31.00	14.25	45.25

16130.60 Pull And Junction Boxes

Basic Materials	UNIT	MAT.	INST.	TOTAL
4"				
Octagon box	EA.	3.66	8.14	11.80
Box extension	"	6.17	4.22	10.39
Plaster ring	"	3.38	4.22	7.60
Cover blank	"	1.50	4.22	5.72
Square box	"	5.27	8.14	13.41
Box extension	"	5.16	4.22	9.38
Plaster ring	"	2.83	4.22	7.05
Cover blank	"	1.45	4.22	5.67
4-11/16"				
Square box	EA.	10.75	8.14	18.89
Box extension	"	11.50	4.22	15.72
Plaster ring	"	7.04	4.22	11.26
Cover blank	"	2.61	4.22	6.83
Switch and device boxes				
2 gang	EA.	16.00	8.14	24.14
3 gang	"	28.00	8.14	36.14
4 gang	"	37.50	11.50	49.00
Device covers				

Basic Materials

16130.60 Pull And Junction Boxes (Cont.)

Basic Materials	UNIT	MAT.	INST.	TOTAL
2 gang	EA.	12.75	4.22	16.97
3 gang	"	13.25	4.22	17.47
4 gang	"	17.75	4.22	21.97
Handy box	"	3.93	8.14	12.07
Extension	"	3.70	4.22	7.92
Switch cover	"	1.96	4.22	6.18
Switch box with knockout	"	5.90	10.25	16.15
Weatherproof cover, spring type	"	11.00	5.70	16.70
Cover plate, dryer receptacle 1 gang plastic	"	1.68	7.13	8.81
For 4" receptacle, 2 gang	"	2.98	7.13	10.11
Duplex receptacle cover plate, plastic	"	0.73	4.22	4.95
4", vertical bracket box, 1-1/2" with				
RMX clamps	EA.	7.59	10.25	17.84
BX clamps	"	8.15	10.25	18.40
4", octagon device cover				
1 switch	EA.	4.46	4.22	8.68
1 duplex recept	"	4.46	4.22	8.68
4", octagon swivel hanger box, 1/2" hub	"	12.00	4.22	16.22
3/4" hub	"	13.50	4.22	17.72
4" octagon adjustable bar hangers				
18-1/2"	EA.	5.54	3.56	9.10
26-1/2"	"	6.05	3.56	9.61
With clip				
18-1/2"	EA.	4.10	3.56	7.66
26-1/2"	"	4.59	3.56	8.15
4", square face bracket boxes, 1-1/2"				
RMX	EA.	9.07	10.25	19.32
BX	"	9.86	10.25	20.11
4" square to round plaster rings	"	3.02	4.22	7.24
2 gang device plaster rings	"	3.12	4.22	7.34
Surface covers				
1 gang switch	EA.	2.72	4.22	6.94
2 gang switch	"	2.78	4.22	7.00
1 single recept	"	4.10	4.22	8.32
1 20a twist lock recept	"	5.13	4.22	9.35
1 30a twist lock recept	"	6.57	4.22	10.79
1 duplex recept	"	2.54	4.22	6.76
2 duplex recept	"	2.54	4.22	6.76
Switch and duplex recept	"	4.23	4.22	8.45
4-11/16" square to round plaster rings	"	7.04	4.22	11.26
2 gang device plaster rings	"	5.80	4.22	10.02
Surface covers				
1 gang switch	EA.	7.81	4.22	12.03
2 gang switch	"	12.00	4.22	16.22
1 single recept	"	10.75	4.22	14.97
1 20a twist lock recept	"	10.75	4.22	14.97
1 30a twist lock recept	"	13.50	4.22	17.72
1 duplex recept	"	11.75	4.22	15.97
2 duplex recept	"	10.25	4.22	14.47
Switch and duplex recept	"	17.50	4.22	21.72
4" plastic round boxes, ground straps				
Box only	EA.	1.93	10.25	12.18
Box w/clamps	"	2.24	14.25	16.49

Basic Materials

16130.60 Pull And Junction Boxes *(Cont.)*

Basic Materials	UNIT	MAT.	INST.	TOTAL
Box w/16" bar	EA.	4.77	16.25	21.02
Box w/24" bar	"	4.76	17.75	22.51
4" plastic round box covers				
Blank cover	EA.	1.26	4.22	5.48
Plaster ring	"	2.06	4.22	6.28
4" plastic square boxes				
Box only	EA.	1.49	10.25	11.74
Box w/clamps	"	1.85	14.25	16.10
Box w/hanger	"	2.27	17.75	20.02
Box w/nails and clamp	"	3.26	17.75	21.01
4" plastic square box covers				
Blank cover	EA.	1.22	4.22	5.44
1 gang ring	"	1.49	4.22	5.71
2 gang ring	"	2.08	4.22	6.30
Round ring	"	1.66	4.22	5.88

16130.65 Pull Boxes And Cabinets

Basic Materials	UNIT	MAT.	INST.	TOTAL
Galvanized pull boxes, screw cover				
4x4x4	EA.	7.10	13.50	20.60
4x6x4	"	8.46	13.50	21.96
6x6x4	"	10.75	13.50	24.25
6x8x4	"	12.75	13.50	26.25
8x8x4	"	16.00	17.75	33.75

16130.80 Receptacles

Basic Materials	UNIT	MAT.	INST.	TOTAL
Contractor grade duplex receptacles, 15a 120v				
Duplex	EA.	1.60	14.25	15.85
125 volt, 20a, duplex, standard grade	"	12.00	14.25	26.25
Ground fault interrupter type	"	38.75	21.00	59.75
250 volt, 20a, 2 pole, single, ground type	"	20.00	14.25	34.25
120/208v, 4 pole, single receptacle, twist lock				
20a	EA.	23.75	24.75	48.50
50a	"	45.25	24.75	70.00
125/250v, 3 pole, flush receptacle				
30a	EA.	24.00	21.00	45.00
50a	"	29.75	21.00	50.75
60a	"	77.00	24.75	102
277v, 20a, 2 pole, grounding type, twist lock	"	13.00	14.25	27.25
Dryer receptacle, 250v, 30a/50a, 3 wire	"	18.00	21.00	39.00
Clock receptacle, 2 pole, grounding type	"	12.00	14.25	26.25
125v, 20a single recept. grounding type				
Standard grade	EA.	13.00	14.25	27.25
125/250v, 3 pole, 3 wire surface recepts				
30a	EA.	20.25	21.00	41.25
50a	"	22.50	21.00	43.50
60a	"	49.50	24.75	74.25
Cord set, 3 wire, 6' cord				
30a	EA.	18.25	21.00	39.25
50a	"	25.50	21.00	46.50
125/250v, 3 pole, 3 wire cap				
30a	EA.	18.00	28.50	46.50
50a	"	32.75	28.50	61.25
60a	"	42.00	31.75	73.75

Basic Materials

	UNIT	MAT.	INST.	TOTAL
16199.10 Utility Poles & Fittings				
Wood pole, creosoted				
25'	EA.	450	170	620
30'	"	540	210	750
Treated, wood preservative, 6"x6"				
8'	EA.	95.00	35.75	131
10'	"	140	57.00	197
12'	"	150	63.00	213
14'	"	180	95.00	275
16'	"	220	110	330
18'	"	250	140	390
20'	"	320	140	460
Aluminum, brushed, no base				
8'	EA.	590	140	730
10'	"	680	190	870
15'	"	760	200	960
20'	"	920	230	1,150
Steel, no base				
10'	EA.	690	180	870
15'	"	760	210	970
20'	"	1,010	270	1,280
Concrete, no base				
13'	EA.	890	390	1,280
16'	"	1,240	520	1,760
18'	"	1,490	630	2,120
Lightning arrester				
3kv	EA.	480	71.00	551
10kv	"	750	110	860
16350.10 Circuit Breakers				
Molded case, 240v, 15-60a, bolt-on				
1 pole	EA.	16.50	17.75	34.25
2 pole	"	35.25	24.75	60.00
70-100a, 2 pole	"	100	38.00	138
15-60a, 3 pole	"	120	28.50	149
70-100a, 3 pole	"	200	43.75	244
Load center circuit breakers, 240v				
1 pole, 10-60a	EA.	16.50	17.75	34.25
2 pole				
10-60a	EA.	38.50	28.50	67.00
70-100a	"	120	47.50	168
110-150a	"	250	52.00	302
Load center, G.F.I. breakers, 240v				
1 pole, 15-30a	EA.	140	21.00	161
Tandem breakers, 240v				
1 pole, 15-30a	EA.	31.25	28.50	59.75
2 pole, 15-30a	"	57.00	38.00	95.00
16365.10 Fuses				
Fuse, one-time, 250v				
30a	EA.	2.51	3.56	6.07
60a	"	4.25	3.56	7.81
100a	"	17.75	3.56	21.31
200a	"	43.00	3.56	46.56

Basic Materials

16395.10 Grounding

Basic Materials	UNIT	MAT.	INST.	TOTAL
Ground rods, copper clad, 1/2" x				
6'	EA.	14.75	47.50	62.25
8'	"	20.25	52.00	72.25
10'	"	25.25	71.00	96.25
5/8" x				
5'	EA.	18.25	43.75	62.00
6'	"	19.50	52.00	71.50
8'	"	25.25	71.00	96.25
10'	"	31.25	89.00	120
3/4" x				
8'	EA.	45.00	52.00	97.00
10'	"	49.25	57.00	106
Ground rod clamp				
5/8"	EA.	6.01	8.77	14.78
3/4"	"	8.50	8.77	17.27
Coupling, on threaded rods, 3/4"	"	17.75	3.56	21.31
Ground receptacles	"	22.25	17.75	40.00
Bus bar, copper, 2" x 1/4"	L.F.	6.48	10.25	16.73
Copper braid, 1" x 1/8", for door ground	EA.	5.03	7.13	12.16
Brazed connection for				
#6 wire	EA.	21.00	35.75	56.75
#2 wire	"	26.50	57.00	83.50
#2/0 wire	"	35.25	71.00	106
#4/0 wire	"	48.50	81.00	130
Ground rod couplings				
1/2"	EA.	11.00	7.13	18.13
5/8"	"	15.50	7.13	22.63
Ground rod, driving stud				
1/2"	EA.	8.83	7.13	15.96
5/8"	"	10.50	7.13	17.63
3/4"	"	11.75	7.13	18.88
Ground rod clamps, #8-2 to				
1" pipe	EA.	9.55	14.25	23.80
2" pipe	"	12.00	17.75	29.75
#4-4/0 to				
1" pipe	EA.	22.75	14.25	37.00
2" pipe	"	36.00	17.75	53.75

Service And Distribution

16430.20 Metering

Service And Distribution	UNIT	MAT.	INST.	TOTAL
Outdoor wp meter sockets, 1 gang, 240v, 1 phase				
Includes sealing ring, 100a	EA.	42.75	110	153
150a	"	57.00	130	187
200a	"	72.00	140	212
Die cast hubs, 1-1/4"	"	6.58	22.75	29.33

Service And Distribution

Service And Distribution	UNIT	MAT.	INST.	TOTAL
16430.20 Metering (Cont.)				
1-1/2"	EA.	7.56	22.75	30.31
2"	"	9.14	22.75	31.89
16470.10 Panelboards				
Indoor load center, 1 phase 240v main lug only				
30a - 2 spaces	EA.	29.25	140	169
100a - 8 spaces	"	93.00	170	263
150a - 16 spaces	"	240	210	450
200a - 24 spaces	"	500	250	750
200a - 42 spaces	"	520	290	810
Main circuit breaker				
100a - 8 spaces	EA.	300	170	470
100a - 16 spaces	"	320	200	520
150a - 16 spaces	"	520	210	730
150a - 24 spaces	"	620	230	850
200a - 24 spaces	"	580	250	830
200a - 42 spaces	"	830	260	1,090
120/208v, flush, 3 ph., 4 wire, main only				
100a				
12 circuits	EA.	930	360	1,290
20 circuits	"	1,280	450	1,730
30 circuits	"	1,900	500	2,400
225a				
30 circuits	EA.	1,930	550	2,480
42 circuits	"	2,440	680	3,120
16490.10 Switches				
Photo electric switches				
1000 watt				
105-135v	EA.	33.50	52.00	85.50
Dimmer switch and switch plate				
600w	EA.	30.75	22.00	52.75
1000w	"	51.00	24.75	75.75
Dimmer switch incandescent				
1500w	EA.	98.00	50.00	148
2000w	"	130	53.00	183
Fluorescent				
12 lamps	EA.	66.00	35.75	102
Time clocks with skip, 40a, 120v				
SPST	EA.	94.00	53.00	147
Contractor grade wall switch 15a, 120v				
Single pole	EA.	1.62	11.50	13.12
Three way	"	2.97	14.25	17.22
Four way	"	10.00	19.00	29.00
Specification grade toggle switches, 20a, 120-277v				
Single pole	EA.	3.57	14.25	17.82
Double pole	"	8.58	21.00	29.58
3 way	"	9.29	17.75	27.04
4 way	"	28.25	21.00	49.25
30a, 120-277v				
Single pole	EA.	23.25	14.25	37.50
Double pole	"	32.25	21.00	53.25
3 way	"	32.25	17.75	50.00

Service And Distribution

16490.10 Switches *(Cont.)*

Service And Distribution	UNIT	MAT.	INST.	TOTAL
Combination switch and pilot light, single pole	EA.	12.25	21.00	33.25
3 way	"	15.25	24.75	40.00
Combination switch and receptacle, single pole	"	17.75	21.00	38.75
3 way	"	21.75	21.00	42.75
Switch plates, plastic ivory				
1 gang	EA.	0.37	5.70	6.07
2 gang	"	0.88	7.13	8.01
3 gang	"	1.37	8.51	9.88
4 gang	"	3.52	10.25	13.77
5 gang	"	3.68	11.50	15.18
6 gang	"	4.34	13.00	17.34
Stainless steel				
1 gang	EA.	3.16	5.70	8.86
2 gang	"	4.40	7.13	11.53
3 gang	"	6.75	8.77	15.52
4 gang	"	11.50	10.25	21.75
5 gang	"	13.50	11.50	25.00
6 gang	"	17.00	13.00	30.00
Brass				
1 gang	EA.	5.90	5.70	11.60
2 gang	"	12.75	7.13	19.88
3 gang	"	19.50	8.77	28.27
4 gang	"	22.50	10.25	32.75
5 gang	"	28.00	11.50	39.50
6 gang	"	33.75	13.00	46.75

Lighting

16510.05 Interior Lighting

Lighting	UNIT	MAT.	INST.	TOTAL
Recessed fluorescent fixtures, 2'x2'				
2 lamp	EA.	69.00	52.00	121
4 lamp	"	93.00	52.00	145
1'x4'				
2 lamp	EA.	71.00	47.50	119
3 lamp	"	97.00	47.50	145
2 lamp w/flange	"	87.00	52.00	139
3 lamp w/flange	"	120	52.00	172
2'x4'				
2 lamp	EA.	87.00	52.00	139
3 lamp	"	110	52.00	162
4 lamp	"	97.00	52.00	149
2 lamp w/flange	"	110	71.00	181
3 lamp w/flange	"	120	71.00	191
4 lamp w/flange	"	120	71.00	191
Surface mounted incandescent fixtures				
40w	EA.	100	47.50	148

Lighting

16510.05 Interior Lighting (Cont.)

Lighting	UNIT	MAT.	INST.	TOTAL
75w	EA.	110	47.50	158
100w	"	120	47.50	168
150w	"	160	47.50	208
Pendant				
40w	EA.	86.00	57.00	143
75w	"	95.00	57.00	152
100w	"	110	57.00	167
150w	"	120	57.00	177
Recessed incandescent fixtures				
40w	EA.	150	110	260
75w	"	160	110	270
100w	"	170	110	280
150w	"	180	110	290
Light track single circuit				
2'	EA.	42.50	35.75	78.25
4'	"	50.00	35.75	85.75
8'	"	68.00	71.00	139
12'	"	97.00	110	207
Fittings and accessories				
Dead end	EA.	16.75	10.25	27.00
Starter kit	"	22.50	17.75	40.25
Conduit feed	"	21.75	10.25	32.00
Straight connector	"	19.25	10.25	29.50
Center feed	"	30.75	10.25	41.00
L-connector	"	21.75	10.25	32.00
T-connector	"	29.25	10.25	39.50
X-connector	"	35.25	14.25	49.50
Cord and plug	"	35.25	7.13	42.38
Rigid corner	"	46.50	10.25	56.75
Flex connector	"	36.25	10.25	46.50
2 way connector	"	100	14.25	114
Spacer clip	"	1.57	3.56	5.13
Grid box	"	8.77	10.25	19.02
T-bar clip	"	2.34	3.56	5.90
Utility hook	"	6.80	10.25	17.05
Fixtures, square				
R-20	EA.	43.00	10.25	53.25
R-30	"	67.00	10.25	77.25
40w flood	"	110	10.25	120
40w spot	"	110	10.25	120
100w flood	"	120	10.25	130
100w spot	"	97.00	10.25	107
Mini spot	"	41.00	10.25	51.25
Mini flood	"	94.00	10.25	104
Quartz, 500w	"	240	10.25	250
R-20 sphere	"	72.00	10.25	82.25
R-30 sphere	"	38.00	10.25	48.25
R-20 cylinder	"	51.00	10.25	61.25
R-30 cylinder	"	59.00	10.25	69.25
R-40 cylinder	"	60.00	10.25	70.25
R-30 wall wash	"	94.00	10.25	104
R-40 wall wash	"	120	10.25	130

16 ELECTRICAL

Lighting

Lighting	UNIT	MAT.	INST.	TOTAL
16510.10 Lighting Industrial				
Surface mounted fluorescent, wrap around lens				
1 lamp	EA.	86.00	57.00	143
2 lamps	"	140	63.00	203
Wall mounted fluorescent				
2-20w lamps	EA.	88.00	35.75	124
2-30w lamps	"	100	35.75	136
2-40w lamps	"	100	47.50	148
Strip fluorescent				
4'				
1 lamp	EA.	43.25	47.50	90.75
2 lamps	"	53.00	47.50	101
8'				
1 lamp	EA.	63.00	52.00	115
2 lamps	"	95.00	63.00	158
Compact fluorescent				
2-7w	EA.	150	71.00	221
2-13w	"	180	95.00	275
16670.10 Lightning Protection				
Lightning protection				
Copper point, nickel plated, 12'				
1/2" dia.	EA.	44.00	71.00	115
5/8" dia.	"	49.50	71.00	121
16720.50 Security Systems				
Sensors				
Balanced magnetic door switch, surface mounted	EA.	150	35.75	186
With remote test	"	200	71.00	271
Flush mounted	"	140	130	270
Mounted bracket	"	11.00	24.75	35.75
Mounted bracket spacer	"	9.84	24.75	34.59
Photoelectric sensor, for fence				
6 beam	EA.	16,060	200	16,260
9 beam	"	19,630	300	19,930
Photoelectric sensor, 12 volt dc				
500' range	EA.	470	110	580
800' range	"	530	140	670
Vibration sensor, 30 max per zone	"	200	35.75	236
Audio sensor, 30 max per zone	"	210	35.75	246
Inertia sensor				
Outdoor	EA.	150	52.00	202
Indoor	"	98.00	35.75	134
Monitor panel, with access/secure tone, standard	"	590	120	710
High security	"	870	140	1,010
Emergency power indicator	"	350	35.75	386
Monitor rack with 115v power supply				
1 zone	EA.	500	71.00	571
10 zone	"	2,520	180	2,700
Monitor cabinet, wall mounted				
1 zone	EA.	760	71.00	831
5 zone	"	2,740	110	2,850
10 zone	"	1,250	120	1,370
Audible alarm	"	110	35.75	146

Lighting

	UNIT	MAT.	INST.	TOTAL
16720.50 Security Systems *(Cont.)*				
Audible alarm control	EA.	460	24.75	485
16750.20 Signaling Systems				
Contractor grade doorbell chime kit				
Chime	EA.	36.75	71.00	108
Doorbutton	"	5.11	22.75	27.86
Transformer	"	16.25	35.75	52.00
16770.30 Sound Systems				
Power amplifiers	EA.	950	250	1,200
Tuner	"	490	100	590
Equalizer	"	1,230	110	1,340
Mixer	"	500	160	660
Cassette Player	"	810	150	960
Record player	"	71.00	140	211
Equipment rack	"	92.00	92.00	184
Speaker				
Wall	EA.	500	290	790
Paging	"	180	57.00	237
Column	"	270	38.00	308
Single	"	61.00	43.75	105
Double	"	190	320	510
Volume control	"	61.00	38.00	99.00
Plug-in	"	190	57.00	247
Desk	"	150	28.50	179
Outlet	"	30.50	28.50	59.00
Stand	"	61.00	21.00	82.00
Console	"	14,330	570	14,900
Power supply	"	270	92.00	362
16780.50 Television Systems				
TV outlet, self terminating, w/cover plate	EA.	5.97	22.00	27.97
Thru splitter	"	13.00	110	123
End of line	"	10.75	95.00	106
In line splitter multitap				
4 way	EA.	21.75	130	152
2 way	"	16.25	120	136
Equipment cabinet	"	54.00	110	164
Antenna				
Broad band uhf	EA.	110	250	360
Lightning arrester	"	33.00	52.00	85.00
TV cable	L.F.	0.49	0.35	0.84
Coaxial cable rg	"	0.33	0.35	0.68
Cable drill, with replacement tip	EA.	5.44	35.75	41.19
Cable blocks for in-line taps	"	10.75	52.00	62.75
In-line taps ptu-series 36 tv system	"	13.00	81.00	94.00
Control receptacles	"	8.52	32.00	40.52
Coupler	"	16.25	170	186
Head end equipment	"	2,050	480	2,530
TV camera	"	1,110	120	1,230
TV power bracket	"	99.00	57.00	156
TV monitor	"	850	100	950
Video recorder	"	1,650	150	1,800

Lighting

16780.50 Television Systems (Cont.)	UNIT	MAT.	INST.	TOTAL
Console	EA.	3,420	610	4,030
Selector switch	"	530	98.00	628
TV controller	"	250	100	350

Resistance Heating

16850.10 Electric Heating	UNIT	MAT.	INST.	TOTAL
Baseboard heater				
2', 375w	EA.	41.75	71.00	113
3', 500w	"	49.50	71.00	121
4', 750w	"	55.00	81.00	136
5', 935w	"	78.00	95.00	173
6', 1125w	"	92.00	110	202
7', 1310w	"	100	130	230
8', 1500w	"	120	140	260
9', 1680w	"	130	160	290
10', 1875w	"	180	160	340
Unit heater, wall mounted				
750w	EA.	160	110	270
1500w	"	220	120	340
2000w	"	230	120	350
2500w	"	240	130	370
3000w	"	280	140	420
4000w	"	320	160	480
Thermostat				
Integral	EA.	37.50	35.75	73.25
Line voltage	"	38.50	35.75	74.25
Electric heater connection	"	1.65	17.75	19.40
Fittings				
Inside corner	EA.	24.25	28.50	52.75
Outside corner	"	26.50	28.50	55.00
Receptacle section	"	27.50	28.50	56.00
Blank section	"	34.00	28.50	62.50
Infrared heaters				
600w	EA.	150	71.00	221
Radiant ceiling heater panels				
500w	EA.	290	71.00	361
750w	"	320	71.00	391

Controls

16910.40 Control Cable

Controls	UNIT	MAT.	INST.	TOTAL
Control cable, 600v, #14 THWN, PVC jacket				
2 wire	L.F.	0.35	0.57	0.92
4 wire	"	0.59	0.71	1.30
Audio cables, shielded, #24 gauge				
3 conductor	L.F.	0.29	0.28	0.57
4 conductor	"	0.35	0.42	0.77
5 conductor	"	0.41	0.49	0.90
6 conductor	"	0.45	0.64	1.09
#22 gauge				
3 conductor	L.F.	0.42	0.28	0.70
4 conductor	"	0.54	0.42	0.96
#20 gauge				
3 conductor	L.F.	0.31	0.28	0.59
10 conductor	"	1.01	1.06	2.07
#18 gauge				
3 conductor	L.F.	0.41	0.28	0.69
4 conductor	"	0.60	0.42	1.02
Computer cables shielded, #24 gauge				
1 pair	L.F.	0.26	0.28	0.54
2 pair	"	0.36	0.28	0.64
3 pair	"	0.44	0.42	0.86
4 pair	"	0.52	0.49	1.01
Coaxial cables				
RG 6/u	L.F.	0.41	0.42	0.83
RG 6a/u	"	0.64	0.42	1.06
RG 8/u	"	0.73	0.42	1.15
RG 8a/u	"	0.89	0.42	1.31
MATV and CCTV camera cables				
1 conductor	L.F.	0.41	0.28	0.69
2 conductor	"	0.52	0.35	0.87
4 conductor	"	1.11	0.42	1.53
Fire alarm cables, #22 gauge				
6 conductor	L.F.	2.03	0.71	2.74
9 conductor	"	2.61	1.06	3.67
12 conductor	"	3.00	1.14	4.14
#18 gauge				
2 conductor	L.F.	2.03	0.35	2.38
4 conductor	"	2.61	0.49	3.10
#16 gauge				
2 conductor	L.F.	2.03	0.49	2.52
4 conductor	"	2.80	0.57	3.37
#14 gauge				
2 conductor	L.F.	3.00	0.57	3.57
#12 gauge				
2 conductor	L.F.	3.67	0.71	4.38
Plastic jacketed thermostat cable				
2 conductor	L.F.	0.13	0.28	0.41
3 conductor	"	0.18	0.35	0.53

BNi® Building News

Man-Hour Tables

The man-hour productivities used to develop the labor costs are listed in the following section of this book. These productivities represent typical installation labor for thousands of construction items. The data takes into account all activities involved in normal construction under commonly experienced working conditions. As with the Costbook pages, these items are listed according to the 16 divisions. In order to best use the information in this book, please review this sample page and read the "Features in this Book" section.

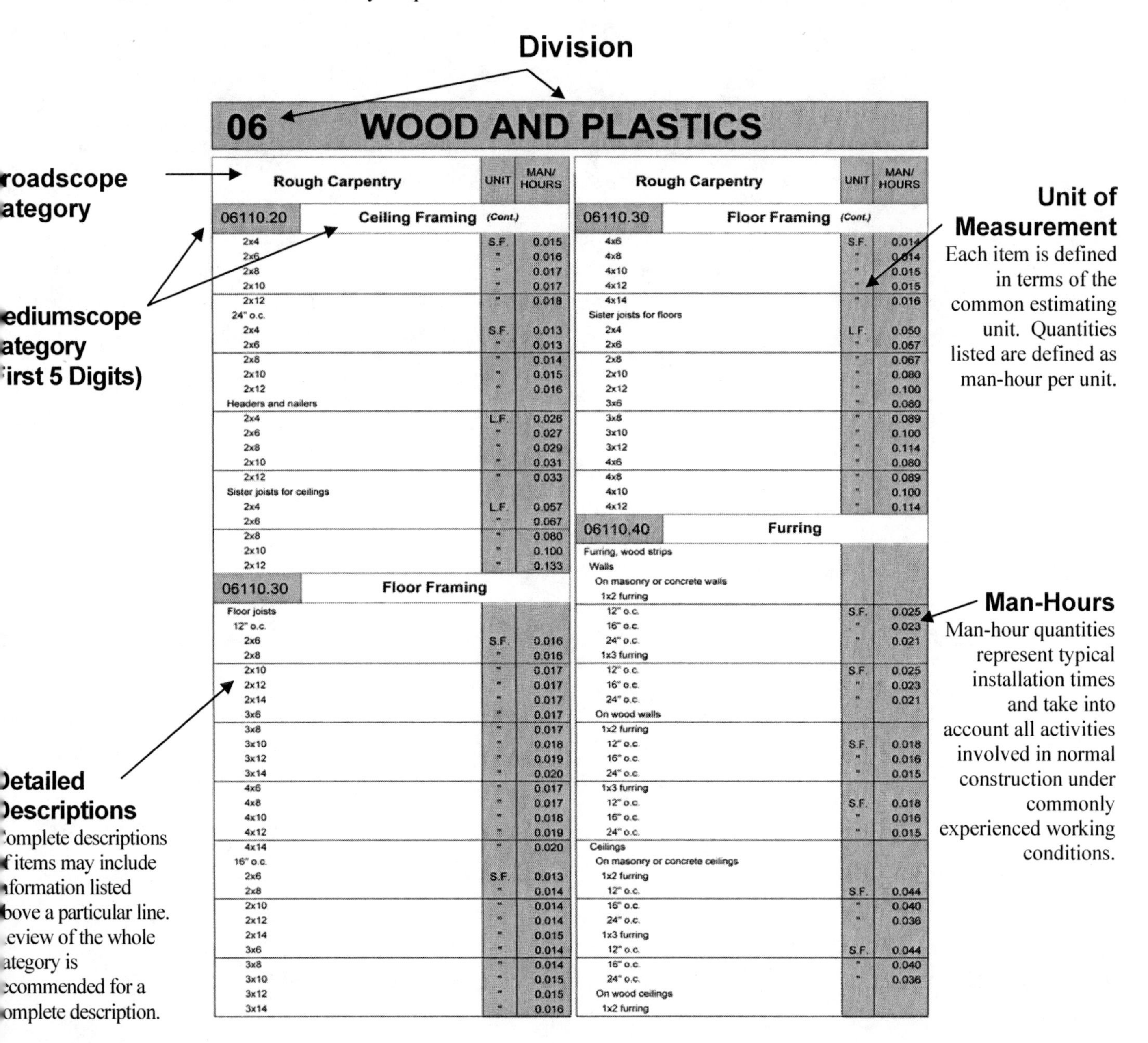

06 WOOD AND PLASTICS

Rough Carpentry	UNIT	MAN/ HOURS
06110.20 Ceiling Framing *(Cont.)*		
2x4	S.F.	0.015
2x6	"	0.016
2x8	"	0.017
2x10	"	0.017
2x12	"	0.018
24" o.c.		
2x4	S.F.	0.013
2x6	"	0.013
2x8	"	0.014
2x10	"	0.015
2x12	"	0.016
Headers and nailers		
2x4	L.F.	0.026
2x6	"	0.027
2x8	"	0.029
2x10	"	0.031
2x12	"	0.033
Sister joists for ceilings		
2x4	L.F.	0.057
2x6	"	0.067
2x8	"	0.080
2x10	"	0.100
2x12	"	0.133
06110.30 Floor Framing		
Floor joists		
12" o.c.		
2x6	S.F.	0.016
2x8	"	0.016
2x10	"	0.017
2x12	"	0.017
2x14	"	0.017
3x6	"	0.017
3x8	"	0.017
3x10	"	0.018
3x12	"	0.019
3x14	"	0.020
4x6	"	0.017
4x8	"	0.017
4x10	"	0.018
4x12	"	0.019
4x14	"	0.020
16" o.c.		
2x6	S.F.	0.013
2x8	"	0.014
2x10	"	0.014
2x12	"	0.014
2x14	"	0.015
3x6	"	0.014
3x8	"	0.014
3x10	"	0.015
3x12	"	0.015
3x14	"	0.016

Rough Carpentry	UNIT	MAN/ HOURS
06110.30 Floor Framing *(Cont.)*		
4x6	S.F.	0.014
4x8	"	0.014
4x10	"	0.015
4x12	"	0.015
4x14	"	0.016
Sister joists for floors		
2x4	L.F.	0.050
2x6	"	0.057
2x8	"	0.067
2x10	"	0.080
2x12	"	0.100
3x6	"	0.080
3x8	"	0.089
3x10	"	0.100
3x12	"	0.114
4x6	"	0.080
4x8	"	0.089
4x10	"	0.100
4x12	"	0.114
06110.40 Furring		
Furring, wood strips		
Walls		
On masonry or concrete walls		
1x2 furring		
12" o.c.	S.F.	0.025
16" o.c.	"	0.023
24" o.c.	"	0.021
1x3 furring		
12" o.c.	S.F.	0.025
16" o.c.	"	0.023
24" o.c.	"	0.021
On wood walls		
1x2 furring		
12" o.c.	S.F.	0.018
16" o.c.	"	0.016
24" o.c.	"	0.015
1x3 furring		
12" o.c.	S.F.	0.018
16" o.c.	"	0.016
24" o.c.	"	0.015
Ceilings		
On masonry or concrete ceilings		
1x2 furring		
12" o.c.	S.F.	0.044
16" o.c.	"	0.040
24" o.c.	"	0.036
1x3 furring		
12" o.c.	S.F.	0.044
16" o.c.	"	0.040
24" o.c.	"	0.036
On wood ceilings		
1x2 furring		

BNi® Building News

Site Remediation

02115.60 Underground Storage Tank	UNIT	MAN/ HOURS
Remove underground storage tank, and backfill		
50 to 250 gals	EA.	8.000
600 gals	"	8.000
1000 gals	"	12.000

02115.66 Septic Tank Removal	UNIT	MAN/ HOURS
Remove septic tank		
1000 gals	EA.	2.000
2000 gals	"	2.400

Site Preparation

02210.10 Soil Boring	UNIT	MAN/ HOURS
Borings, uncased, stable earth		
2-1/2" dia.	L.F.	0.300

Demolition

02220.10 Complete Building Demolition	UNIT	MAN/ HOURS
Wood frame	C.F.	0.003
Concrete	"	0.004
Steel frame	"	0.005

02220.15 Selective Building Demolition	UNIT	MAN/ HOURS
Partition removal		
Concrete block partitions		
4" thick	S.F.	0.040
8" thick	"	0.053
12" thick	"	0.073
Brick masonry partitions		
4" thick	S.F.	0.040
8" thick	"	0.050
12" thick	"	0.067
16" thick	"	0.100
Cast in place concrete partitions		
Unreinforced		
6" thick	S.F.	0.160
8" thick	"	0.171
10" thick	"	0.200

Demolition

02220.15 Selective Building Demolition (Cont.)	UNIT	MAN/ HOURS
12" thick	S.F.	0.240
Reinforced		
6" thick	S.F.	0.185
8" thick	"	0.240
10" thick	"	0.267
12" thick	"	0.320
Terra cotta		
To 6" thick	S.F.	0.040
Stud partitions		
Metal or wood, with drywall both sides	S.F.	0.040
Metal studs, both sides, lath and plaster	"	0.053
Door and frame removal		
Hollow metal in masonry wall		
Single		
2'6"x6'8"	EA.	1.000
3'x7'	"	1.333
Double		
3'x7'	EA.	1.600
4'x8'	"	1.600
Wood in framed wall		
Single		
2'6"x6'8"	EA.	0.571
3'x6'8"	"	0.667
Double		
2'6"x6'8"	EA.	0.800
3'x6'8"	"	0.889
Remove for re-use		
Hollow metal	EA.	2.000
Wood	"	1.333
Floor removal		
Brick flooring	S.F.	0.032
Ceramic or quarry tile	"	0.018
Terrazzo	"	0.036
Heavy wood	"	0.021
Residential wood	"	0.023
Resilient tile or linoleum	"	0.008
Ceiling removal		
Acoustical tile ceiling		
Adhesive fastened	S.F.	0.008
Furred and glued	"	0.007
Suspended grid	"	0.005
Drywall ceiling		
Furred and nailed	S.F.	0.009
Nailed to framing	"	0.008
Plastered ceiling		
Furred on framing	S.F.	0.020
Suspended system	"	0.027
Roofing removal		
Steel frame		
Corrugated metal roofing	S.F.	0.016
Built-up roof on metal deck	"	0.027
Wood frame		
Built up roof on wood deck	S.F.	0.025

Demolition	UNIT	MAN/ HOURS
02220.15 Selective Building Demolition *(Cont.)*		
Roof shingles	S.F.	0.013
Roof tiles	"	0.027
Concrete frame	C.F.	0.053
Concrete plank	S.F.	0.040
Built-up roof on concrete	"	0.023
Cut-outs		
Concrete, elevated slabs, mesh reinforcing		
Under 5 cf	C.F.	0.800
Over 5 cf	"	0.667
Bar reinforcing		
Under 5 cf	C.F.	1.333
Over 5 cf	"	1.000
Window removal		
Metal windows, trim included		
2'x3'	EA.	0.800
2'x4'	"	0.889
2'x6'	"	1.000
3'x4'	"	1.000
3'x6'	"	1.143
3'x8'	"	1.333
4'x4'	"	1.333
4'x6'	"	1.600
4'x8'	"	2.000
Wood windows, trim included		
2'x3'	EA.	0.444
2'x4'	"	0.471
2'x6'	"	0.500
3'x4'	"	0.533
3'x6'	"	0.571
3'x8'	"	0.615
6'x4'	"	0.667
6'x6'	"	0.727
6'x8'	"	0.800
Walls, concrete, bar reinforcing		
Small jobs	C.F.	0.533
Large jobs	"	0.444
Brick walls, not including toothing		
4" thick	S.F.	0.040
8" thick	"	0.050
12" thick	"	0.067
16" thick	"	0.100
Concrete block walls, not including toothing		
4" thick	S.F.	0.044
6" thick	"	0.047
8" thick	"	0.050
10" thick	"	0.057
12" thick	"	0.067
Rubbish handling		
Load in dumpster or truck		
Minimum	C.F.	0.018
Maximum	"	0.027
Rubbish hauling		
Hand loaded on trucks, 2 mile trip	C.Y.	0.320

Demolition	UNIT	MAN/ HOURS
02220.15 Selective Building Demolition *(Cont.)*		
Machine loaded on trucks, 2 mile trip	C.Y.	0.240

Selective Site Demolition	UNIT	MAN/ HOURS
02225.13 Core Drilling		
Concrete		
6" thick		
3" dia.	EA.	0.571
8" thick		
3" dia.	EA.	0.800
02225.20 Fence Demolition		
Remove fencing		
Chain link, 8' high		
For disposal	L.F.	0.040
For reuse	"	0.100
Wood		
4' high	S.F.	0.027
6' high	"	0.032
8' high	"	0.040
Masonry		
8" thick		
4' high	S.F.	0.080
6' high	"	0.100
8' high	"	0.114
02225.42 Drainage Piping Demolition		
Remove drainage pipe, not including excavation		
12" dia.	L.F.	0.100
18" dia.	"	0.126
02225.43 Gas Piping Demolition		
Remove welded steel pipe, not including excavation		
4" dia.	L.F.	0.150
5" dia.	"	0.240
02225.45 Sanitary Piping Demolition		
Remove sewer pipe, not including excavation		
4" dia.	L.F.	0.096
6" dia.	"	0.109

02 SITE CONSTRUCTION

Selective Site Demolition

Selective Site Demolition	UNIT	MAN/ HOURS
02225.48 Water Piping Demolition		
Remove water pipe, not including excavation		
4" dia.	L.F.	0.109
6" dia.	"	0.114
02225.50 Saw Cutting Pavement		
Pavement, bituminous		
2" thick	L.F.	0.016
3" thick	"	0.020
Concrete pavement, with wire mesh		
4" thick	L.F.	0.031
5" thick	"	0.033
Plain concrete, unreinforced		
4" thick	L.F.	0.027
5" thick	"	0.031
02225.80 Wall, Exterior, Demolition		
Concrete wall		
Light reinforcing		
6" thick	S.F.	0.120
8" thick	"	0.126
Medium reinforcing		
6" thick	S.F.	0.126
8" thick	"	0.133
Heavy reinforcing		
6" thick	S.F.	0.141
8" thick	"	0.150
Masonry		
No reinforcing		
8" thick	S.F.	0.053
12" thick	"	0.060
Horizontal reinforcing		
8" thick	S.F.	0.060
12" thick	"	0.065
Vertical reinforcing		
8" thick	S.F.	0.077
12" thick	"	0.089

Site Clearing

Site Clearing	UNIT	MAN/ HOURS
02230.50 Tree Cutting & Clearing		
Cut trees and clear out stumps		
9" to 12" dia.	EA.	4.800
Loading and trucking		
For machine load, per load, round trip		
1 mile	EA.	0.960
10 mile	"	1.600
02230.50 Tree Cutting & Clearing *(Cont.)*		
Hand loaded, round trip		
1 mile	EA.	2.000
10 mile	"	3.200

Earthwork, Excavation & Fill

Earthwork, Excavation & Fill	UNIT	MAN/ HOURS
02315.10 Base Course		
Base course, crushed stone		
3" thick	S.Y.	0.004
4" thick	"	0.004
6" thick	"	0.005
Base course, bank run gravel		
4" deep	S.Y.	0.004
6" deep	"	0.005
Prepare and roll sub base		
Minimum	S.Y.	0.004
Average	"	0.005
Maximum	"	0.007
02315.20 Borrow		
Borrow fill, F.O.B. at pit		
Sand, haul to site, round trip		
10 mile	C.Y.	0.080
02315.30 Bulk Excavation		
Excavation, by small dozer		
Small areas	C.Y.	0.027
Trim banks	"	0.040
Hydraulic excavator		
1 cy capacity		
Light material	C.Y.	0.040
Medium material	"	0.048
Wet material	"	0.060
Blasted rock	"	0.069
Wheel mounted front-end loader		
7/8 cy capacity		
Light material	C.Y.	0.020
Medium material	"	0.023
Wet material	"	0.027
02315.40 Building Excavation		
Structural excavation, unclassified earth		
3/8 cy backhoe	C.Y.	0.107
3/4 cy backhoe	"	0.080
1 cy backhoe	"	0.067
Foundation backfill and compaction by machine	"	0.160

02 SITE CONSTRUCTION

Earthwork, Excavation & Fill	UNIT	MAN/HOURS
02315.45 Hand Excavation		
Excavation		
To 2' deep		
Normal soil	C.Y.	0.889
Sand and gravel	"	0.800
Medium clay	"	1.000
Heavy clay	"	1.143
Loose rock	"	1.333
To 6' deep		
Normal soil	C.Y.	1.143
Sand and gravel	"	1.000
Medium clay	"	1.333
Heavy clay	"	1.600
Loose rock	"	2.000
Backfilling foundation without compaction, 6" lifts	"	0.500
Compaction of backfill around structures or in trench		
By hand with air tamper	C.Y.	0.571
By hand with vibrating plate tamper	"	0.533
1 ton roller	"	0.400
Miscellaneous hand labor		
Excavation around obstructions and services	C.Y.	2.667
02315.60 Trenching		
Trenching and continuous footing excavation		
By gradall		
1 cy capacity		
Light soil	C.Y.	0.023
Medium soil	"	0.025
Heavy/wet soil	"	0.027
Loose rock	"	0.029
Blasted rock	"	0.031
By hydraulic excavator		
1/2 cy capacity		
Light soil	C.Y.	0.027
Medium soil	"	0.029
Heavy/wet soil	"	0.032
Loose rock	"	0.036
Blasted rock	"	0.040
Hand excavation		
Bulk, wheeled 100'		
Normal soil	C.Y.	0.889
Sand or gravel	"	0.800
Medium clay	"	1.143
Heavy clay	"	1.600
Loose rock	"	2.000
Trenches, up to 2' deep		
Normal soil	C.Y.	1.000
Sand or gravel	"	0.889
Medium clay	"	1.333
Heavy clay	"	2.000
Loose rock	"	2.667
Backfill trenches		
With compaction		
By hand	C.Y.	0.667

Earthwork, Excavation & Fill	UNIT	MAN/HOURS
02315.60 Trenching *(Cont.)*		
By 60 hp tracked dozer	C.Y.	0.020
By small front-end loader	"	0.023
Spread dumped fill or gravel, no compaction		
6" layers	S.Y.	0.013
Compaction in 6" layers		
By hand with air tamper	S.Y.	0.016
02315.70 Utility Excavation		
Trencher, sandy clay, 8" wide trench		
18" deep	L.F.	0.018
24" deep	"	0.020
36" deep	"	0.023
Trench backfill, 95% compaction		
Tamp by hand	C.Y.	0.500
Vibratory compaction	"	0.400
Trench backfilling, with borrow sand, place & compact	"	0.400
02315.80 Hauling Material		
Haul material by 10 cy dump truck, round trip distance		
1 mile	C.Y.	0.044
5 mile	"	0.073
Spread topsoil by equipment on site	"	0.036

Soil Stabilization & Treatment	UNIT	MAN/HOURS
02340.05 Soil Stabilization		
Straw bale secured with rebar	L.F.	0.027
Filter barrier, 18" high filter fabric	"	0.080
Sediment fence, 36" fabric with 6" mesh	"	0.100
Soil stabilization with tar paper, burlap, straw and stakes	S.F.	0.001
02360.20 Soil Treatment		
Soil treatment, termite control pretreatment		
Under slabs	S.F.	0.004
By walls	"	0.005
02445.10 Pipe Jacking		
Pipe casing, horizontal jacking		
18" dia.	L.F.	0.711

Piles And Caissons

02455.65 Steel Pipe Piles	UNIT	MAN/ HOURS
Concrete filled, 3000# concrete, up to 40'		
8" dia.	L.F.	0.069
10" dia.	"	0.071
12" dia.	"	0.074
Pipe piles, non-filled		
8" dia.	L.F.	0.053
10" dia.	"	0.055
12" dia.	"	0.056

02455.80 Wood And Timber Piles	UNIT	MAN/ HOURS
Treated wood piles, 12" butt, 8" tip		
25' long	L.F.	0.096
30' long	"	0.080

02465.50 Prestressed Piling	UNIT	MAN/ HOURS
Prestressed concrete piling, less than 60' long		
10" sq.	L.F.	0.040
12" sq.	"	0.042
Straight cylinder, less than 60' long		
12" dia.	L.F.	0.044
14" dia.	"	0.045

02475.10 Caissons, Includes Casing	UNIT	MAN/ HOURS
Caisson, 3000# conc., 60 # reinf./CY, stable ground		
18" dia., 0.065 CY/ LF	L.F.	0.192
24" dia., 0.116 CY/ LF	"	0.200
Wet ground, casing required but pulled		
18" dia.	L.F.	0.240
24" dia.	"	0.267
Soft rock		
18" dia.	L.F.	0.686
24" dia.	"	1.200

Utility Services

02510.10 Wells	UNIT	MAN/ HOURS
Domestic water, drilled and cased		
4" dia.	L.F.	0.480
6" dia.	"	0.533

02510.15 Water Meters	UNIT	MAN/ HOURS
Water meter, displacement type		
1"	EA.	0.800

Utility Services

02510.20 Tapping Saddles & Sleeves	UNIT	MAN/ HOURS
Tapping saddle, tap size to 2"		
4" saddle	EA.	0.400
Tap hole in pipe		
4" hole	EA.	1.000

02510.25 Valve Boxes	UNIT	MAN/ HOURS
Valve box, adjustable, for valves up to 20"		
3' deep	EA.	0.267
4' deep	"	0.320
5' deep	"	0.400

02510.40 Ductile Iron Pipe	UNIT	MAN/ HOURS
Ductile iron pipe, cement lined, slip-on joints		
4"	L.F.	0.067
6"	"	0.071
Mechanical joint pipe		
4"	L.F.	0.092
6"	"	0.100

02510.60 Plastic Pipe	UNIT	MAN/ HOURS
PVC, class 150 pipe		
4" dia.	L.F.	0.060
6" dia.	"	0.065
Schedule 40 pipe		
1-1/2" dia.	L.F.	0.047
2" dia.	"	0.050
2-1/2" dia.	"	0.053
3" dia.	"	0.057
4" dia.	"	0.067
6" dia.	"	0.080
Drainage pipe		
PVC schedule 80		
1" dia.	L.F.	0.047
1-1/2" dia.	"	0.047
ABS, 2" dia.	"	0.050
2-1/2" dia.	"	0.053
3" dia.	"	0.057
4" dia.	"	0.067
6" dia.	"	0.080
90 degree elbows		
1"	EA.	0.133
1-1/2"	"	0.133
2"	"	0.145
2-1/2"	"	0.160
3"	"	0.178
4"	"	0.200
6"	"	0.267
45 degree elbows		
1"	EA.	0.133
1-1/2"	"	0.133
2"	"	0.145
2-1/2"	"	0.160
3"	"	0.178

Utility Services

02510.60 Plastic Pipe (Cont.)	UNIT	MAN/HOURS
4"	EA.	0.200
6"	"	0.267
Tees		
1"	EA.	0.160
1-1/2"	"	0.160
2"	"	0.178
2-1/2"	"	0.200
3"	"	0.229
4"	"	0.267
6"	"	0.320
Couplings		
1"	EA.	0.133
1-1/2"	"	0.133
2"	"	0.145
2-1/2"	"	0.160
3"	"	0.178
4"	"	0.200
6"	"	0.267

Sanitary Sewer

02530.20 Vitrified Clay Pipe	UNIT	MAN/HOURS
Vitrified clay pipe, extra strength		
6" dia.	L.F.	0.109

02530.40 Sanitary Sewers	UNIT	MAN/HOURS
Clay		
6" pipe	L.F.	0.080
PVC		
4" pipe	L.F.	0.060
6" pipe	"	0.063
Connect new sewer line		
To existing manhole	EA.	2.667
To new manhole	"	1.600

02540.10 Drainage Fields	UNIT	MAN/HOURS
Perforated PVC pipe, for drain field		
4" pipe	L.F.	0.053
6" pipe	"	0.057

02540.50 Septic Tanks	UNIT	MAN/HOURS
Septic tank, precast concrete		
1000 gals	EA.	4.000
2000 gals	"	6.000
Leaching pit, precast concrete, 72" diameter		
3' deep	EA.	3.000

Sanitary Sewer

02540.50 Septic Tanks (Cont.)	UNIT	MAN/HOURS
6' deep	EA.	3.429
8' deep	"	4.000

02630.70 Underdrain	UNIT	MAN/HOURS
Drain tile, clay		
6" pipe	L.F.	0.053
8" pipe	"	0.056
Porous concrete, standard strength		
6" pipe	L.F.	0.053
8" pipe	"	0.056
Corrugated metal pipe, perforated type		
6" pipe	L.F.	0.060
8" pipe	"	0.063
Perforated clay pipe		
6" pipe	L.F.	0.069
8" pipe	"	0.071
Drain tile, concrete		
6" pipe	L.F.	0.053
8" pipe	"	0.056
Perforated rigid PVC underdrain pipe		
4" pipe	L.F.	0.040
6" pipe	"	0.048
8" pipe	"	0.053
Underslab drainage, crushed stone		
3" thick	S.F.	0.008
4" thick	"	0.009
6" thick	"	0.010

Flexible Surfaces

02740.10 Asphalt Repair	UNIT	MAN/HOURS
Coal tar seal coat, rubber add., fuel resist.	S.Y.	0.011
Bituminous surface treatment, single	"	0.008
Double	"	0.001
Bituminous prime coat	"	0.001
Tack coat	"	0.001
Crack sealing, concrete paving	L.F.	0.005
Rubberized asphalt	S.Y.	0.073
Asphalt slurry seal	"	0.047

02740.20 Asphalt Surfaces	UNIT	MAN/HOURS
Asphalt wearing surface, flexible pavement		
1" thick	S.Y.	0.016
1-1/2" thick	"	0.019

Flexible Surfaces	UNIT	MAN/ HOURS
02740.20 Asphalt Surfaces *(Cont.)*		
Binder course		
1-1/2" thick	S.Y.	0.018
2" thick	"	0.022
Bituminous sidewalk, no base		
2" thick	S.Y.	0.028
3" thick	"	0.030

Rigid Pavement	UNIT	MAN/ HOURS
02750.10 Concrete Paving		
Concrete paving, reinforced, 5000 psi concrete		
6" thick	S.Y.	0.150
8" thick	"	0.171
02760.10 Pavement Markings		
Pavement line marking, paint		
4" wide	L.F.	0.002
Directional arrows, reflective preformed tape	EA.	0.800
Messages, reflective preformed tape (per letter)	"	0.400
Handicap symbol, preformed tape	"	0.800
Parking stall painting	"	0.160

Site Improvements	UNIT	MAN/ HOURS
02810.40 Lawn Irrigation		
02820.10 Chain Link Fence		
Chain link fence, 9 ga., galvanized, with posts 10' o.c.		
4' high	L.F.	0.057
5' high	"	0.073
6' high	"	0.100
Corner or gate post, 3" post		
4' high	EA.	0.267
5' high	"	0.296
6' high	"	0.348
Gate with gate posts, galvanized, 3' wide		
4' high	EA.	2.000
5' high	"	2.667
6' high	"	2.667

Site Improvements	UNIT	MAN/ HOURS
02820.10 Chain Link Fence *(Cont.)*		
Fabric, galvanized chain link, 2" mesh, 9 ga.		
4' high	L.F.	0.027
5' high	"	0.032
6' high	"	0.040
Line post, no rail fitting, galvanized, 2-1/2" dia.		
4' high	EA.	0.229
5' high	"	0.250
6' high	"	0.267
Vinyl coated, 9 ga., with posts 10' o.c.		
4' high	L.F.	0.057
5' high	"	0.073
6' high	"	0.100
Gate, with posts, 3' wide		
4' high	EA.	2.000
5' high	"	2.667
6' high	"	2.667
Fabric, vinyl, chain link, 2" mesh, 9 ga.		
4' high	L.F.	0.027
5' high	"	0.032
6' high	"	0.040
Swing gates, galvanized, 4' high		
Single gate		
3' wide	EA.	2.000
4' wide	"	2.000
6' high		
Single gate		
3' wide	EA.	2.667
4' wide	"	2.667
02880.70 Recreational Courts		
Walls, galvanized steel		
8' high	L.F.	0.160
10' high	"	0.178
12' high	"	0.211
Vinyl coated		
8' high	L.F.	0.160
10' high	"	0.178
12' high	"	0.211
Gates, galvanized steel		
Single, 3' transom		
3'x7'	EA.	4.000
4'x7'	"	4.571
5'x7'	"	5.333
6'x7'	"	6.400
Vinyl coated		
Single, 3' transom		
3'x7'	EA.	4.000
4'x7'	"	4.571
5'x7'	"	5.333
6'x7'	"	6.400

Planting	UNIT	MAN/ HOURS
02910.10 Topsoil		
Spread topsoil, with equipment		
Minimum	C.Y.	0.080
Maximum	"	0.100
By hand		
Minimum	C.Y.	0.800
Maximum	"	1.000
Area prep. seeding (grade, rake and clean)		
Square yard	S.Y.	0.006
Remove topsoil and stockpile on site		
4" deep	C.Y.	0.067
6" deep	"	0.062
Spreading topsoil from stock pile		
By loader	C.Y.	0.073
By hand	"	0.800
Top dress by hand	S.Y.	0.008
Place imported top soil		
By loader		
4" deep	S.Y.	0.008
6" deep	"	0.009
By hand		
4" deep	S.Y.	0.089
6" deep	"	0.100
Plant bed preparation, 18" deep		
With backhoe/loader	S.Y.	0.020
By hand	"	0.133
02920.10 Fertilizing		
Fertilizing (23#/1000 sf)		
By square yard	S.Y.	0.002
Liming (70#/1000 sf)		
By square yard	S.Y.	0.003
02920.30 Seeding		
Seeding by hand, 10 lb per 100 s.y.		
By square yard	S.Y.	0.003
Reseed disturbed areas	S.F.	0.004
02935.10 Shrub & Tree Maintenance		
Moving shrubs on site		
12" ball	EA.	1.000
24" ball	"	1.333
3' high	"	0.800
4' high	"	0.889
5' high	"	1.000
Moving trees on site		
24" ball	EA.	1.200
48" ball	"	1.600
Trees		
3' high	EA.	0.480
6' high	"	0.533
8' high	"	0.600
10' high	"	0.800
Palm trees		

Planting	UNIT	MAN/ HOURS
02935.10 Shrub & Tree Maintenance *(Cont.)*		
7' high	EA.	0.600
10' high	"	0.800
20' high	"	2.400
40' high	"	4.800
Guying trees		
4" dia.	EA.	0.400
8" dia.	"	0.500
02935.30 Weed Control		
Weed control, bromicil, 15 lb./acre, wettable powder	ACRE	4.000
Vegetation control, by application of plant killer	S.Y.	0.003
Weed killer, lawns and fields	"	0.002
02945.10 Prefabricated Planters		
Concrete precast, circular		
24" dia., 18" high	EA.	0.800
Fiberglass, circular		
36" dia., 27" high	EA.	0.400
02945.20 Landscape Accessories		
Steel edging, 3/16" x 4"	L.F.	0.010
Landscaping stepping stones, 15"x15", white	EA.	0.040
Wood chip mulch	C.Y.	0.533
2" thick	S.Y.	0.016
4" thick	"	0.023
6" thick	"	0.029
Gravel mulch, 3/4" stone	C.Y.	0.800
White marble chips, 1" deep	S.F.	0.008
Peat moss		
2" thick	S.Y.	0.018
4" thick	"	0.027
6" thick	"	0.033
Landscaping timbers, treated lumber		
4" x 4"	L.F.	0.027
6" x 6"	"	0.029
8" x 8"	"	0.033

Formwork	UNIT	MAN/ HOURS
03110.05 Beam Formwork		
Beam forms, job built		
Beam bottoms		
1 use	S.F.	0.133
5 uses	"	0.114
Beam sides		
1 use	S.F.	0.089
5 uses	"	0.073
03110.15 Column Formwork		
Column, square forms, job built		
8" x 8" columns		
1 use	S.F.	0.160
5 uses	"	0.138
12" x 12" columns		
1 use	S.F.	0.145
5 uses	"	0.127
Round fiber forms, 1 use		
10" dia.	L.F.	0.160
12" dia.	"	0.163
03110.18 Curb Formwork		
Curb forms		
Straight, 6" high		
1 use	L.F.	0.080
5 uses	"	0.067
Curved, 6" high		
1 use	L.F.	0.100
5 uses	"	0.082
03110.25 Equipment Pad Formwork		
Equipment pad, job built		
1 use	S.F.	0.100
2 uses	"	0.094
3 uses	"	0.089
4 uses	"	0.084
5 uses	"	0.080
03110.35 Footing Formwork		
Wall footings, job built, continuous		
1 use	S.F.	0.080
5 uses	"	0.067
Column footings, spread		
1 use	S.F.	0.100
5 uses	"	0.080
03110.50 Grade Beam Formwork		
Grade beams, job built		
1 use	S.F.	0.080
5 uses	"	0.067

Formwork	UNIT	MAN/ HOURS
03110.55 Slab / Mat Formwork		
Mat foundations, job built		
1 use	S.F.	0.100
5 uses	"	0.080
Edge forms		
6" high		
1 use	L.F.	0.073
5 uses	"	0.062
5 uses	"	0.067
Formwork for openings		
1 use	S.F.	0.160
5 uses	"	0.114
03110.60 Stair Formwork		
Stairway forms, job built		
1 use	S.F.	0.160
5 uses	"	0.114
Stairs, elevated		
1 use	S.F.	0.160
5 uses	"	0.100
03110.65 Wall Formwork		
Wall forms, exterior, job built		
Up to 8' high wall		
1 use	S.F.	0.080
5 uses	"	0.067
Over 8' high wall		
5 uses	S.F.	0.080
Radial wall forms		
1 use	S.F.	0.123
5 uses	"	0.094
Retaining wall forms		
1 use	S.F.	0.089
5 uses	"	0.073
Radial retaining wall forms		
1 use	S.F.	0.133
5 uses	"	0.100
Column pier and pilaster		
1 use	S.F.	0.160
5 uses	"	0.114
Interior wall forms		
Up to 8' high		
1 use	S.F.	0.073
5 uses	"	0.062
Radial wall forms		
1 use	S.F.	0.107
5 uses	"	0.084
Curved wall forms, 24" sections		
1 use	S.F.	0.160
5 uses	"	0.114
PVC form liner, per side, smooth finish		
1 use	S.F.	0.067
5 uses	"	0.053

Formwork	UNIT	MAN/ HOURS
03110.90 Miscellaneous Formwork		
Keyway forms (5 uses)		
2 x 4	L.F.	0.040
2 x 6	"	0.044
Bulkheads		
Walls, with keyways		
2 piece	L.F.	0.073
3 piece	"	0.080
Ground slab, with keyway		
2 piece	L.F.	0.057
3 piece	"	0.062
Chamfer strips		
Wood		
1/2" wide	L.F.	0.018
3/4" wide	"	0.018
1" wide	"	0.018
PVC		
1/2" wide	L.F.	0.018
3/4" wide	"	0.018
1" wide	"	0.018
Radius		
1"	L.F.	0.019
1-1/2"	"	0.019
Reglets		
Galvanized steel, 24 ga.	L.F.	0.032
Metal formwork		
Straight edge forms		
4" high	L.F.	0.050
6" high	"	0.053
8" high	"	0.057
Curb form, S-shape		
12" x		
1'-6"	L.F.	0.114
2'	"	0.107

Reinforcement	UNIT	MAN/ HOURS
03210.05 Beam Reinforcing		
Beam-girders		
#3 - #4	TON	20.000
#5 - #6	"	16.000
Galvanized		
#3 - #4	TON	20.000
#5 - #6	"	16.000
Bond Beams		
#3 - #4	TON	26.667
#5 - #6	"	20.000

Reinforcement	UNIT	MAN/ HOURS
03210.05 Beam Reinforcing *(Cont.)*		
Galvanized		
#3 - #4	TON	26.667
#5 - #6	"	20.000
03210.15 Column Reinforcing		
Columns		
#3 - #4	TON	22.857
#5 - #6	"	17.778
Galvanized		
#3 - #4	TON	22.857
#5 - #6	"	17.778
03210.25 Equip. Pad Reinforcing		
Equipment pad		
#3 - #4	TON	16.000
#5 - #6	"	14.545
03210.35 Footing Reinforcing		
Footings		
#3 - #4	TON	13.333
#5 - #6	"	11.429
#7 - #8	"	10.000
Straight dowels, 24" long		
3/4" dia. (#6)	EA.	0.080
5/8" dia. (#5)	"	0.067
1/2" dia. (#4)	"	0.057
03210.45 Foundation Reinforcing		
Foundations		
#3 - #4	TON	13.333
#5 - #6	"	11.429
Galvanized		
#3 - #4	TON	13.333
#5 - #6	"	11.429
03210.50 Grade Beam Reinforcing		
Grade beams		
#3 - #4	TON	12.308
#5 - #6	"	10.667
Galvanized		
#3 - #4	TON	12.308
#5 - #6	"	10.667
03210.55 Slab / Mat Reinforcing		
Bars, slabs		
#3 - #4	TON	13.333
#5 - #6	"	11.429
Galvanized		
#3 - #4	TON	13.333
#5 - #6	"	11.429
Wire mesh, slabs		
Galvanized		

Reinforcement

03210.55 Slab / Mat Reinforcing *(Cont.)*	UNIT	MAN/HOURS
4x4		
W1.4xW1.4	S.F.	0.005
W4.0xW4.0	"	0.007
6x6		
W1.4xW1.4	S.F.	0.004
W4.0xW4.0	"	0.005

03210.60 Stair Reinforcing	UNIT	MAN/HOURS
Stairs		
#3 - #4	TON	16.000
#5 - #6	"	13.333
Galvanized		
#3 - #4	TON	16.000
#5 - #6	"	13.333

03210.65 Wall Reinforcing	UNIT	MAN/HOURS
Walls		
#3 - #4	TON	11.429
#5 - #6	"	10.000
Galvanized		
#3 - #4	TON	11.429
#5 - #6	"	10.000
Masonry wall (horizontal)		
#3 - #4	TON	32.000
#5 - #6	"	26.667
Galvanized		
#3 - #4	TON	32.000
#5 - #6	"	26.667
Masonry wall (vertical)		
#3 - #4	TON	40.000
#5 - #6	"	32.000
Galvanized		
#3 - #4	TON	40.000
#5 - #6	"	32.000

Accessories

03250.40 Concrete Accessories	UNIT	MAN/HOURS
Expansion joint, poured		
Asphalt		
1/2" x 1"	L.F.	0.016
1" x 2"	"	0.017
Vapor barrier		
4 mil polyethylene	S.F.	0.003
6 mil polyethylene	"	0.003
Gravel porous fill, under floor slabs, 3/4" stone	C.Y.	1.333

Cast-in-place Concrete

03300.10 Concrete Admixtures	UNIT	MAN/HOURS
Floor finishes		
Broom	S.F.	0.011
Screed	"	0.010
Darby	"	0.010
Steel float	"	0.013
Break ties and patch holes	"	0.016
Carborundum		
Dry rub	S.F.	0.027
Wet rub	"	0.040

03360.10 Pneumatic Concrete	UNIT	MAN/HOURS
Pneumatic applied concrete (gunite)		
2" thick	S.F.	0.030
3" thick	"	0.040
4" thick	"	0.048
Finish surface		
Minimum	S.F.	0.040
Maximum	"	0.080

03370.10 Curing Concrete	UNIT	MAN/HOURS
Sprayed membrane		
Slabs	S.F.	0.002

Placing Concrete

03380.05 Beam Concrete	UNIT	MAN/HOURS
Beams and girders		
2500# or 3000# concrete		
By pump	C.Y.	0.873
By hand buggy	"	0.800
3500# or 4000# concrete		
By pump	C.Y.	0.873
By hand buggy	"	0.800

03380.15 Column Concrete	UNIT	MAN/HOURS
Columns		
2500# or 3000# concrete		
By pump	C.Y.	0.800
3500# or 4000# concrete		
By pump	C.Y.	0.800

Placing Concrete

Placing Concrete	UNIT	MAN/ HOURS
03380.25 Equipment Pad Concrete		
Equipment pad		
2500# or 3000# concrete		
By chute	C.Y.	0.267
By pump	"	0.686
3500# or 4000# concrete		
By chute	C.Y.	0.267
By pump	"	0.686
03380.35 Footing Concrete		
Continuous footing		
2500# or 3000# concrete		
By chute	C.Y.	0.267
By pump	"	0.600
Spread footing		
2500# or 3000# concrete		
By chute	C.Y.	0.267
By pump	"	0.640
03380.50 Grade Beam Concrete		
Grade beam		
2500# or 3000# concrete		
By chute	C.Y.	0.267
By pump	"	0.600
By hand buggy	"	0.800
03380.53 Pile Cap Concrete		
Pile cap		
2500# or 3000 concrete		
By chute	C.Y.	0.267
By pump	"	0.686
By hand buggy	"	0.800
3500# or 4000# concrete		
By chute	C.Y.	0.267
By pump	"	0.686
By hand buggy	"	0.800
03380.55 Slab / Mat Concrete		
Slab on grade		
2500# or 3000# concrete		
By chute	C.Y.	0.200
By pump	"	0.343
By hand buggy	"	0.533
03380.58 Sidewalks		
Walks, cast in place with wire mesh, base not incl.		
4" thick	S.F.	0.027
5" thick	"	0.032
6" thick	"	0.040

Placing Concrete	UNIT	MAN/ HOURS
03380.60 Stair Concrete		
Stairs		
2500# or 3000# concrete		
By chute	C.Y.	0.267
By pump	"	0.686
By hand buggy	"	0.800
3500# or 4000# concrete		
By chute	C.Y.	0.267
By pump	"	0.686
By hand buggy	"	0.800
03380.65 Wall Concrete		
Walls		
2500# or 3000# concrete		
To 4'		
By chute	C.Y.	0.229
By pump	"	0.738
To 8'		
By pump	C.Y.	0.800
Filled block (CMU)		
3000# concrete, by pump		
4" wide	S.F.	0.034
6" wide	"	0.040
8" wide	"	0.048
Pilasters, 3000# concrete	C.F.	0.960
Wall cavity, 2" thick, 3000# concrete	S.F.	0.032
03400.90 Precast Specialties		
Precast concrete, coping, 4' to 8' long		
12" wide	L.F.	0.060
10" wide	"	0.069
Splash block, 30"x12"x4"	EA.	0.400
Stair unit, per riser	"	0.400
Sun screen and trellis, 8' long, 12" high		
4" thick blades	EA.	0.300

Concrete Restoration

Concrete Restoration	UNIT	MAN/ HOURS
03730.10 Concrete Repair		
Epoxy grout floor patch, 1/4" thick	S.F.	0.080
Grout crack seal, 2 component	C.F.	0.800
Concrete, epoxy modified		
Sand mix	C.F.	0.320
Gravel mix	"	0.296
Concrete repair		
Edge repair		
2" spall	L.F.	0.200

Concrete Restoration	UNIT	MAN/ HOURS
03730.10 Concrete Repair *(Cont.)*		
3" spall	L.F.	0.211
8" spall	"	0.235
Crack repair, 1/8" crack	"	0.080
Reinforcing steel repair		
1 bar, 4 ft		
#4 bar	L.F.	0.100
#5 bar	"	0.100
#6 bar	"	0.107

Mortar And Grout	UNIT	MAN/ HOURS
04100.10 Masonry Grout		
Grout, non shrink, non-metallic, trowelable	C.F.	0.016
Grout door frame, hollow metal		
Single	EA.	0.600
Double	"	0.632
Grout-filled concrete block (CMU)		
4" wide	S.F.	0.020
6" wide	"	0.022
8" wide	"	0.024
12" wide	"	0.025
Grout-filled individual CMU cells		
4" wide	L.F.	0.012
6" wide	"	0.012
8" wide	"	0.012
10" wide	"	0.014
12" wide	"	0.014
Bond beams or lintels, 8" deep		
6" thick	L.F.	0.022
8" thick	"	0.024
10" thick	"	0.027
12" thick	"	0.030
Cavity walls		
2" thick	S.F.	0.032
3" thick	"	0.032
4" thick	"	0.034
6" thick	"	0.040
04150.10 Masonry Accessories		
Foundation vents	EA.	0.320
Bar reinforcing		
Horizontal		
#3 - #4	Lb.	0.032
#5 - #6	"	0.027
Vertical		
#3 - #4	Lb.	0.040
#5 - #6	"	0.032
Horizontal joint reinforcing		
Truss type		
4" wide, 6" wall	L.F.	0.003
6" wide, 8" wall	"	0.003
8" wide, 10" wall	"	0.003
10" wide, 12" wall	"	0.004
12" wide, 14" wall	"	0.004
Ladder type		
4" wide, 6" wall	L.F.	0.003
6" wide, 8" wall	"	0.003
8" wide, 10" wall	"	0.003
10" wide, 12" wall	"	0.003
Rectangular wall ties		
3/16" dia., galvanized		
2" x 6"	EA.	0.013
2" x 8"	"	0.013
2" x 10"	"	0.013
2" x 12"	"	0.013

Mortar And Grout	UNIT	MAN/ HOURS
04150.10 Masonry Accessories (Cont.)		
4" x 6"	EA.	0.016
4" x 8"	"	0.016
4" x 10"	"	0.016
4" x 12"	"	0.016
1/4" dia., galvanized		
2" x 6"	EA.	0.013
2" x 8"	"	0.013
2" x 10"	"	0.013
2" x 12"	"	0.013
4" x 6"	"	0.016
4" x 8"	"	0.016
4" x 10"	"	0.016
4" x 12"	"	0.016
"Z" type wall ties, galvanized		
6" long		
1/8" dia.	EA.	0.013
3/16" dia.	"	0.013
1/4" dia.	"	0.013
8" long		
1/8" dia.	EA.	0.013
3/16" dia.	"	0.013
1/4" dia.	"	0.013
10" long		
1/8" dia.	EA.	0.013
3/16" dia.	"	0.013
1/4" dia.	"	0.013
Dovetail anchor slots		
Galvanized steel, filled		
24 ga.	L.F.	0.020
20 ga.	"	0.020
16 oz. copper, foam filled	"	0.020
Dovetail anchors		
16 ga.		
3-1/2" long	EA.	0.013
5-1/2" long	"	0.013
12 ga.		
3-1/2" long	EA.	0.013
5-1/2" long	"	0.013
Dovetail, triangular galvanized ties, 12 ga.		
3" x 3"	EA.	0.013
5" x 5"	"	0.013
7" x 7"	"	0.013
7" x 9"	"	0.013
Brick anchors		
Corrugated, 3-1/2" long		
16 ga.	EA.	0.013
12 ga.	"	0.013
Non-corrugated, 3-1/2" long		
16 ga.	EA.	0.013
12 ga.	"	0.013
Cavity wall anchors, corrugated, galvanized		
5" long		
16 ga.	EA.	0.013

Mortar And Grout

04150.10 Masonry Accessories (Cont.)

Description	UNIT	MAN/ HOURS
12 ga.	EA.	0.013
7" long		
28 ga.	EA.	0.013
24 ga.	"	0.013
22 ga.	"	0.013
16 ga.	"	0.013
Mesh ties, 16 ga., 3" wide		
8" long	EA.	0.013
12" long	"	0.013
20" long	"	0.013
24" long	"	0.013

04150.20 Masonry Control Joints

Description	UNIT	MAN/ HOURS
Control joint, cross shaped PVC	L.F.	0.020
Closed cell joint filler		
1/2"	L.F.	0.020
3/4"	"	0.020
Rubber, for		
4" wall	L.F.	0.020
6" wall	"	0.021
8" wall	"	0.022
PVC, for		
4" wall	L.F.	0.020
6" wall	"	0.021
8" wall	"	0.022

04150.50 Masonry Flashing

Description	UNIT	MAN/ HOURS
Through-wall flashing		
5 oz. coated copper	S.F.	0.067
0.030" elastomeric	"	0.053

Unit Masonry

04210.10 Brick Masonry

Description	UNIT	MAN/ HOURS
Standard size brick, running bond		
Face brick, red (6.4/sf)		
Veneer	S.F.	0.133
Cavity wall	"	0.114
9" solid wall	"	0.229
Common brick (6.4/sf)		
Select common for veneers	S.F.	0.133
Back-up		
4" thick	S.F.	0.100
8" thick	"	0.160
Firewall		
12" thick	S.F.	0.267

Unit Masonry

04210.10 Brick Masonry (Cont.)

Description	UNIT	MAN/ HOURS
16" thick	S.F.	0.364
Glazed brick (7.4/sf)		
Veneer	S.F.	0.145
Buff or gray face brick (6.4/sf)		
Veneer	S.F.	0.133
Cavity wall	"	0.114
Jumbo or oversize brick (3/sf)		
4" veneer	S.F.	0.080
4" back-up	"	0.067
8" back-up	"	0.114
12" firewall	"	0.200
16" firewall	"	0.267
Norman brick, red face, (4.5/sf)		
4" veneer	S.F.	0.100
Cavity wall	"	0.089
Chimney, standard brick, including flue		
16" x 16"	L.F.	0.800
16" x 20"	"	0.800
16" x 24"	"	0.800
20" x 20"	"	1.000
20" x 24"	"	1.000
20" x 32"	"	1.143
Window sill, face brick on edge	"	0.200

04210.60 Pavers, Masonry

Description	UNIT	MAN/ HOURS
Brick walk laid on sand, sand joints		
Laid flat, (4.5 per sf)	S.F.	0.089
Laid on edge, (7.2 per sf)	"	0.133
Precast concrete patio blocks		
2" thick		
Natural	S.F.	0.027
Colors	"	0.027
Exposed aggregates, local aggregate		
Natural	S.F.	0.027
Colors	"	0.027
Granite or limestone aggregate	"	0.027
White tumblestone aggregate	"	0.027
Stone pavers, set in mortar		
Bluestone		
1" thick		
Irregular	S.F.	0.200
Snapped rectangular	"	0.160
1-1/2" thick, random rectangular	"	0.200
2" thick, random rectangular	"	0.229
Slate		
Natural cleft		
Irregular, 3/4" thick	S.F.	0.229
Random rectangular		
1-1/4" thick	S.F.	0.200
1-1/2" thick	"	0.222
Granite blocks		
3" thick, 3" to 6" wide		
4" to 12" long	S.F.	0.267

Unit Masonry	UNIT	MAN/ HOURS
04210.60 Pavers, Masonry *(Cont.)*		
6" to 15" long	S.F.	0.229
Crushed stone, white marble, 3" thick	"	0.016
04220.10 Concrete Masonry Units		
Hollow, load bearing		
4"	S.F.	0.059
6"	"	0.062
8"	"	0.067
Solid, load bearing		
4"	S.F.	0.059
6"	"	0.062
8"	"	0.067
Back-up block, 8" x 16"		
2"	S.F.	0.046
4"	"	0.047
6"	"	0.050
8"	"	0.053
Foundation wall, 8" x 16"		
6"	S.F.	0.057
8"	"	0.062
10"	"	0.067
12"	"	0.073
Solid		
6"	S.F.	0.062
8"	"	0.067
10"	"	0.073
12"	"	0.080
Filled cavities		
4"	S.F.	0.089
6"	"	0.094
8"	"	0.100
Gypsum unit masonry		
Partition blocks (12"x30")		
Solid		
2"	S.F.	0.032
Hollow		
3"	S.F.	0.032
4"	"	0.033
6"	"	0.036
Vertical reinforcing		
4' o.c., add 5% to labor		
2'8" o.c., add 15% to labor		
Interior partitions, add 10% to labor		
04220.90 Bond Beams & Lintels		
Bond beam, no grout or reinforcement		
8" x 16" x		
4" thick	L.F.	0.062
6" thick	"	0.064
8" thick	"	0.067
Beam lintel, no grout or reinforcement		
8" x 16" x		
Precast masonry lintel		

Unit Masonry	UNIT	MAN/ HOURS
04220.90 Bond Beams & Lintels *(Cont.)*		
6 lf, 8" high x		
4" thick	L.F.	0.133
6" thick	"	0.133
8" thick	"	0.145
Steel angles and plates		
Minimum	Lb.	0.011
Maximum	"	0.020
Various size angle lintels		
1/4" stock		
3" x 3"	L.F.	0.050
3" x 3-1/2"	"	0.050
3/8" stock		
3" x 4"	L.F.	0.050
3-1/2" x 4"	"	0.050
4" x 4"	"	0.050
5" x 3-1/2"	"	0.050
6" x 3-1/2"	"	0.050
1/2" stock		
6" x 4"	L.F.	0.050
04240.10 Clay Tile		
Hollow clay tile, for back-up, 12" x 12"		
Scored face		
Load bearing		
4" thick	S.F.	0.057
6" thick	"	0.059
8" thick	"	0.062
Non-load bearing		
3" thick	S.F.	0.055
4" thick	"	0.057
6" thick	"	0.059
8" thick	"	0.062
Clay tile floors		
4" thick	S.F.	0.044
6" thick	"	0.047
8" thick	"	0.050
Terra cotta		
Coping, 10" or 12" wide, 3" thick	L.F.	0.160
04270.10 Glass Block		
Glass block, 4" thick		
6" x 6"	S.F.	0.267
8" x 8"	"	0.200
12" x 12"	"	0.160
Replacement glass blocks, 4" x 8" x 8"		
Minimum	S.F.	0.800
Maximum	"	1.600

04 MASONRY

Unit Masonry

04295.10 Parging / Masonry Plaster	UNIT	MAN/ HOURS
Parging		
1/2" thick	S.F.	0.053
3/4" thick	"	0.067
1" thick	"	0.080

Stone

04400.10 Stone	UNIT	MAN/ HOURS
Rubble stone		
Walls set in mortar		
8" thick	S.F.	0.200
12" thick	"	0.320
18" thick	"	0.400
24" thick	"	0.533
Dry set wall		
8" thick	S.F.	0.133
12" thick	"	0.200
18" thick	"	0.267
24" thick	"	0.320
Cut stone		
Imported marble		
Facing panels		
3/4" thick	S.F.	0.320
1-1/2" thick	"	0.364
Flooring, travertine, minimum	"	0.123
Average	"	0.160
Maximum	"	0.178
Domestic marble		
Stairs		
12" treads	L.F.	0.400
6" risers	"	0.267
Thresholds, 7/8" thick, 3' long, 4" to 6" wide		
Plain	EA.	0.667
Beveled	"	0.667
Window sill		
6" wide, 2" thick	L.F.	0.320
Stools		
5" wide, 7/8" thick	L.F.	0.320
Limestone panels up to 12' x 5', smooth finish		
2" thick	S.F.	0.096
Miscellaneous limestone items		
Steps, 14" wide, 6" deep	L.F.	0.533
Coping, smooth finish	C.F.	0.267
Sills, lintels, jambs, smooth finish	"	0.320
Granite veneer facing panels, polished		
7/8" thick		

Stone

04400.10 Stone (Cont.)	UNIT	MAN/ HOURS
Black	S.F.	0.320
Gray	"	0.320
Slate, panels		
1" thick	S.F.	0.320
Sills or stools		
1" thick		
6" wide	L.F.	0.320

Masonry Restoration

04520.10 Restoration And Cleaning	UNIT	MAN/ HOURS
Masonry cleaning		
Washing brick		
Smooth surface	S.F.	0.013
Rough surface	"	0.018
Steam clean masonry		
Smooth face		
Minimum	S.F.	0.010
Maximum	"	0.015
Rough face		
Minimum	S.F.	0.013
Maximum	"	0.020
Sandblast masonry		
Minimum	S.F.	0.016
Maximum	"	0.027
Pointing masonry		
Brick	S.F.	0.032
Concrete block	"	0.023
Cut and repoint		
Brick		
Minimum	S.F.	0.040
Maximum	"	0.080
Stone work	L.F.	0.062
Cut and recaulk		
Oil base caulks	L.F.	0.053
Butyl caulks	"	0.053
Polysulfides and acrylics	"	0.053
Silicones	"	0.053
Cement and sand grout on walls, to 1/8" thick		
Minimum	S.F.	0.032
Maximum	"	0.040
Brick removal and replacement		
Minimum	EA.	0.100
Average	"	0.133
Maximum	"	0.400

Masonry Restoration	UNIT	MAN/ HOURS
04550.10 Refractories		
Flue liners		
Rectangular		
8" x 12"	L.F.	0.133
12" x 12"	"	0.145
12" x 18"	"	0.160
Round		
18" dia.	L.F.	0.190
24" dia.	"	0.229

Metal Fastening

Metal Fastening	UNIT	MAN/ HOURS
05050.10 Structural Welding		
Welding		
Single pass		
1/8"	L.F.	0.040
3/16"	"	0.053
1/4"	"	0.067
05050.95 Metal Lintels		
Lintels, steel		
Plain	Lb.	0.020
Galvanized	"	0.020
05120.10 Structural Steel		
Beams and girders, A-36		
Welded	TON	4.800
Bolted	"	4.364
Columns		
Pipe		
6" dia.	Lb.	0.005

Cold Formed Framing

Cold Formed Framing	UNIT	MAN/ HOURS
05410.10 Metal Framing		
Furring channel, galvanized		
Beams and columns, 3/4"		
12" o.c.	S.F.	0.080
16" o.c.	"	0.073
Walls, 3/4"		
12" o.c.	S.F.	0.040
16" o.c.	"	0.033
24" o.c.	"	0.027
1-1/2"		
12" o.c.	S.F.	0.040
16" o.c.	"	0.033
24" o.c.	"	0.027
Stud, load bearing		
16" o.c.		
16 ga.		
2-1/2"	S.F.	0.036
3-5/8"	"	0.036
4"	"	0.036
6"	"	0.040
18 ga.		
2-1/2"	S.F.	0.036
3-5/8"	"	0.036
4"	"	0.036
6"	"	0.040

Cold Formed Framing	UNIT	MAN/ HOURS
05410.10 Metal Framing *(Cont.)*		
8"	S.F.	0.040
20 ga.		
2-1/2"	S.F.	0.036
3-5/8"	"	0.036
4"	"	0.036
6"	"	0.040
8"	"	0.040
24" o.c.		
16 ga.		
2-1/2"	S.F.	0.031
3-5/8"	"	0.031
4"	"	0.031
6"	"	0.033
8"	"	0.033
18 ga.		
2-1/2"	S.F.	0.031
3-5/8"	"	0.031
4"	"	0.031
6"	"	0.033
8"	"	0.033
20 ga.		
2-1/2"	S.F.	0.031
3-5/8"	"	0.031
4"	"	0.031
6"	"	0.033
8"	"	0.033

Metal Fabrications

Metal Fabrications	UNIT	MAN/ HOURS
05520.10 Railings		
Railing, pipe		
1-1/4" diameter, welded steel		
2-rail		
Primed	L.F.	0.160
Galvanized	"	0.160
3-rail		
Primed	L.F.	0.200
Galvanized	"	0.200
Wall mounted, single rail, welded steel		
Primed	L.F.	0.123
Galvanized	"	0.123
1-1/2" diameter, welded steel		
2-rail		
Primed	L.F.	0.160
Galvanized	"	0.160
3-rail		

Metal Fabrications	UNIT	MAN/ HOURS
05520.10 Railings (Cont.)		
Primed	L.F.	0.200
Galvanized	"	0.200
Wall mounted, single rail, welded steel		
Primed	L.F.	0.123
Galvanized	"	0.123
2" diameter, welded steel		
2-rail		
Primed	L.F.	0.178
Galvanized	"	0.178
3-rail		
Primed	L.F.	0.229
Galvanized	"	0.229
Wall mounted, single rail, welded steel		
Primed	L.F.	0.133
Galvanized	"	0.133

Misc. Fabrications	UNIT	MAN/ HOURS
05700.10 Ornamental Metal		
Railings, square bars, 6" o.c., shaped top rails		
Steel	L.F.	0.400
Aluminum	"	0.400
Bronze	"	0.533
Stainless steel	"	0.533
Laminated metal or wood handrails		
2-1/2" round or oval shape	L.F.	0.400
Aluminum louvers		
Residential use, fixed type, with screen		
8" x 8"	EA.	0.400
12" x 12"	"	0.400
12" x 18"	"	0.400
14" x 24"	"	0.400
18" x 24"	"	0.400
30" x 24"	"	0.444

06 WOOD AND PLASTICS

Fasteners And Adhesives

06050.10 Accessories

Description	UNIT	MAN/HOURS
Column/post base, cast aluminum		
4" x 4"	EA.	0.200
6" x 6"	"	0.200
Bridging, metal, per pair		
12" o.c.	EA.	0.080
16" o.c.	"	0.073
Anchors		
Bolts, threaded two ends, with nuts and washers		
1/2" dia.		
4" long	EA.	0.050
7-1/2" long	"	0.050
3/4" dia.		
7-1/2" long	EA.	0.050
15" long	"	0.050
Framing anchors		
10 gauge	EA.	0.067
Bolts, carriage		
1/4 x 4	EA.	0.080
5/16 x 6	"	0.084
3/8 x 6	"	0.084
1/2 x 6	"	0.084
Joist and beam hangers		
18 ga.		
2 x 4	EA.	0.080
2 x 6	"	0.080
2 x 8	"	0.080
2 x 10	"	0.089
2 x 12	"	0.100
16 ga.		
3 x 6	EA.	0.089
3 x 8	"	0.089
3 x 10	"	0.094
3 x 12	"	0.107
3 x 14	"	0.114
4 x 6	"	0.089
4 x 8	"	0.089
4 x 10	"	0.094
4 x 12	"	0.107
4 x 14	"	0.114
Rafter anchors, 18 ga., 1-1/2" wide		
5-1/4" long	EA.	0.067
10-3/4" long	"	0.067
Shear plates		
2-5/8" dia.	EA.	0.062
4" dia.	"	0.067
Sill anchors		
Embedded in concrete	EA.	0.080
Split rings		
2-1/2" dia.	EA.	0.089
4" dia.	"	0.100
Strap ties, 14 ga., 1-3/8" wide		
12" long	EA.	0.067

Fasteners And Adhesives

06050.10 Accessories *(Cont.)*

Description	UNIT	MAN/HOURS
18" long	EA.	0.073
24" long	"	0.080
36" long	"	0.089
Toothed rings		
2-5/8" dia.	EA.	0.133
4" dia.	"	0.160

Rough Carpentry

06110.10 Blocking

Description	UNIT	MAN/HOURS
Steel construction		
Walls		
2x4	L.F.	0.053
2x6	"	0.062
2x8	"	0.067
2x10	"	0.073
2x12	"	0.080
Ceilings		
2x4	L.F.	0.062
2x6	"	0.073
2x8	"	0.080
2x10	"	0.089
2x12	"	0.100
Wood construction		
Walls		
2x4	L.F.	0.044
2x6	"	0.050
2x8	"	0.053
2x10	"	0.057
2x12	"	0.062
Ceilings		
2x4	L.F.	0.050
2x6	"	0.057
2x8	"	0.062
2x10	"	0.067
2x12	"	0.073

06110.20 Ceiling Framing

Description	UNIT	MAN/HOURS
Ceiling joists		
12" o.c.		
2x4	S.F.	0.019
2x6	"	0.020
2x8	"	0.021
2x10	"	0.022
2x12	"	0.024
16" o.c.		

Rough Carpentry

06110.20 Ceiling Framing (Cont.)

Rough Carpentry	UNIT	MAN/ HOURS
2x4	S.F.	0.015
2x6	"	0.016
2x8	"	0.017
2x10	"	0.017
2x12	"	0.018
24" o.c.		
2x4	S.F.	0.013
2x6	"	0.013
2x8	"	0.014
2x10	"	0.015
2x12	"	0.016
Headers and nailers		
2x4	L.F.	0.026
2x6	"	0.027
2x8	"	0.029
2x10	"	0.031
2x12	"	0.033
Sister joists for ceilings		
2x4	L.F.	0.057
2x6	"	0.067
2x8	"	0.080
2x10	"	0.100
2x12	"	0.133

06110.30 Floor Framing

Rough Carpentry	UNIT	MAN/ HOURS
Floor joists		
12" o.c.		
2x6	S.F.	0.016
2x8	"	0.016
2x10	"	0.017
2x12	"	0.017
2x14	"	0.017
3x6	"	0.017
3x8	"	0.017
3x10	"	0.018
3x12	"	0.019
3x14	"	0.020
4x6	"	0.017
4x8	"	0.017
4x10	"	0.018
4x12	"	0.019
4x14	"	0.020
16" o.c.		
2x6	S.F.	0.013
2x8	"	0.014
2x10	"	0.014
2x12	"	0.014
2x14	"	0.015
3x6	"	0.014
3x8	"	0.014
3x10	"	0.015
3x12	"	0.015
3x14	"	0.016

Rough Carpentry

06110.30 Floor Framing (Cont.)

Rough Carpentry	UNIT	MAN/ HOURS
4x6	S.F.	0.014
4x8	"	0.014
4x10	"	0.015
4x12	"	0.015
4x14	"	0.016
Sister joists for floors		
2x4	L.F.	0.050
2x6	"	0.057
2x8	"	0.067
2x10	"	0.080
2x12	"	0.100
3x6	"	0.080
3x8	"	0.089
3x10	"	0.100
3x12	"	0.114
4x6	"	0.080
4x8	"	0.089
4x10	"	0.100
4x12	"	0.114

06110.40 Furring

Rough Carpentry	UNIT	MAN/ HOURS
Furring, wood strips		
Walls		
On masonry or concrete walls		
1x2 furring		
12" o.c.	S.F.	0.025
16" o.c.	"	0.023
24" o.c.	"	0.021
1x3 furring		
12" o.c.	S.F.	0.025
16" o.c.	"	0.023
24" o.c.	"	0.021
On wood walls		
1x2 furring		
12" o.c.	S.F.	0.018
16" o.c.	"	0.016
24" o.c.	"	0.015
1x3 furring		
12" o.c.	S.F.	0.018
16" o.c.	"	0.016
24" o.c.	"	0.015
Ceilings		
On masonry or concrete ceilings		
1x2 furring		
12" o.c.	S.F.	0.044
16" o.c.	"	0.040
24" o.c.	"	0.036
1x3 furring		
12" o.c.	S.F.	0.044
16" o.c.	"	0.040
24" o.c.	"	0.036
On wood ceilings		
1x2 furring		

Rough Carpentry	UNIT	MAN/ HOURS
06110.40 Furring *(Cont.)*		
12" o.c.	S.F.	0.030
16" o.c.	"	0.027
24" o.c.	"	0.024
1x3		
12" o.c.	S.F.	0.030
16" o.c.	"	0.027
24" o.c.	"	0.024
06110.50 Roof Framing		
Roof framing		
Rafters, gable end		
0-2 pitch (flat to 2-in-12)		
12" o.c.		
2x4	S.F.	0.017
2x6	"	0.017
2x8	"	0.018
2x10	"	0.019
2x12	"	0.020
16" o.c.		
2x6	S.F.	0.014
2x8	"	0.015
2x10	"	0.015
2x12	"	0.016
24" o.c.		
2x6	S.F.	0.012
2x8	"	0.013
2x10	"	0.013
2x12	"	0.013
4-6 pitch (4-in-12 to 6-in-12)		
12" o.c.		
2x4	S.F.	0.017
2x6	"	0.018
2x8	"	0.019
2x10	"	0.020
2x12	"	0.021
16" o.c.		
2x6	S.F.	0.015
2x8	"	0.015
2x10	"	0.016
2x12	"	0.017
24" o.c.		
2x6	S.F.	0.013
2x8	"	0.013
2x10	"	0.014
2x12	"	0.015
8-12 pitch (8-in-12 to 12-in-12)		
12" o.c.		
2x4	S.F.	0.018
2x6	"	0.019
2x8	"	0.020
2x10	"	0.021
2x12	"	0.022
16" o.c.		

Rough Carpentry	UNIT	MAN/ HOURS
06110.50 Roof Framing *(Cont.)*		
2x6	S.F.	0.015
2x8	"	0.016
2x10	"	0.017
2x12	"	0.017
24" o.c.		
2x6	S.F.	0.013
2x8	"	0.013
2x10	"	0.014
2x12	"	0.014
Ridge boards		
2x6	L.F.	0.040
2x8	"	0.044
2x10	"	0.050
2x12	"	0.057
Hip rafters		
2x6	L.F.	0.029
2x8	"	0.030
2x10	"	0.031
2x12	"	0.032
Jack rafters		
4-6 pitch (4-in-12 to 6-in-12)		
16" o.c.		
2x6	S.F.	0.024
2x8	"	0.024
2x10	"	0.026
2x12	"	0.027
24" o.c.		
2x6	S.F.	0.018
2x8	"	0.019
2x10	"	0.020
2x12	"	0.020
8-12 pitch (8-in-12 to 12-in-12)		
16" o.c.		
2x6	S.F.	0.025
2x8	"	0.026
2x10	"	0.027
2x12	"	0.028
24" o.c.		
2x6	S.F.	0.019
2x8	"	0.020
2x10	"	0.020
2x12	"	0.021
Sister rafters		
2x4	L.F.	0.057
2x6	"	0.067
2x8	"	0.080
2x10	"	0.100
2x12	"	0.133
Fascia boards		
2x4	L.F.	0.040
2x6	"	0.040
2x8	"	0.044
2x10	"	0.044

06 WOOD AND PLASTICS

Rough Carpentry

06110.50 Roof Framing (Cont.)	UNIT	MAN/ HOURS
2x12	L.F.	0.050
Cant strips		
Fiber		
3x3	L.F.	0.023
4x4	"	0.024
Wood		
3x3	L.F.	0.024

06110.60 Sleepers	UNIT	MAN/ HOURS
Sleepers, over concrete		
12" o.c.		
1x2	S.F.	0.018
1x3	"	0.019
2x4	"	0.022
2x6	"	0.024
16" o.c.		
1x2	S.F.	0.016
1x3	"	0.016
2x4	"	0.019
2x6	"	0.020

06110.65 Soffits	UNIT	MAN/ HOURS
Soffit framing		
2x3	L.F.	0.057
2x4	"	0.062
2x6	"	0.067
2x8	"	0.073

06110.70 Wall Framing	UNIT	MAN/ HOURS
Framing wall, studs		
12" o.c.		
2x3	S.F.	0.015
2x4	"	0.015
2x6	"	0.016
2x8	"	0.017
16" o.c.		
2x3	S.F.	0.013
2x4	"	0.013
2x6	"	0.013
2x8	"	0.014
24" o.c.		
2x3	S.F.	0.011
2x4	"	0.011
2x6	"	0.011
2x8	"	0.012
Plates, top or bottom		
2x3	L.F.	0.024
2x4	"	0.025
2x6	"	0.027
2x8	"	0.029
Headers, door or window		
2x6		
Single		

Rough Carpentry

06110.70 Wall Framing (Cont.)	UNIT	MAN/ HOURS
3' long	EA.	0.400
6' long	"	0.500
Double		
3' long	EA.	0.444
6' long	"	0.571
2x8		
Single		
4' long	EA.	0.500
8' long	"	0.615
Double		
4' long	EA.	0.571
8' long	"	0.727
2x10		
Single		
5' long	EA.	0.615
10' long	"	0.800
Double		
5' long	EA.	0.667
10' long	"	0.800
2x12		
Single		
6' long	EA.	0.615
12' long	"	0.800
Double		
6' long	EA.	0.727
12' long	"	0.889

06115.10 Floor Sheathing	UNIT	MAN/ HOURS
Sub-flooring, plywood, CDX		
1/2" thick	S.F.	0.010
5/8" thick	"	0.011
3/4" thick	"	0.013
Structural plywood		
1/2" thick	S.F.	0.010
5/8" thick	"	0.011
3/4" thick	"	0.012
Board type subflooring		
1x6		
Minimum	S.F.	0.018
Maximum	"	0.020
1x8		
Minimum	S.F.	0.017
Maximum	"	0.019
1x10		
Minimum	S.F.	0.016
Maximum	"	0.018
Underlayment		
Hardboard, 1/4" tempered	S.F.	0.010
Plywood, CDX		
3/8" thick	S.F.	0.010
1/2" thick	"	0.011
5/8" thick	"	0.011
3/4" thick	"	0.012

06 WOOD AND PLASTICS

Rough Carpentry

06115.20 Roof Sheathing

Rough Carpentry	UNIT	MAN/ HOURS
Sheathing		
Plywood, CDX		
3/8" thick	S.F.	0.010
1/2" thick	"	0.011
5/8" thick	"	0.011
3/4" thick	"	0.012
Structural plywood		
3/8" thick	S.F.	0.010
1/2" thick	"	0.011
5/8" thick	"	0.011
3/4" thick	"	0.012

06115.30 Wall Sheathing

Rough Carpentry	UNIT	MAN/ HOURS
Sheathing		
Plywood, CDX		
3/8" thick	S.F.	0.012
1/2" thick	"	0.012
5/8" thick	"	0.013
3/4" thick	"	0.015
Waferboard		
3/8" thick	S.F.	0.012
1/2" thick	"	0.012
5/8" thick	"	0.013
3/4" thick	"	0.015
Structural plywood		
3/8" thick	S.F.	0.012
1/2" thick	"	0.012
5/8" thick	"	0.013
3/4" thick	"	0.015
Gypsum, 1/2" thick	"	0.012
Asphalt impregnated fiberboard, 1/2" thick	"	0.012

06125.10 Wood Decking

Rough Carpentry	UNIT	MAN/ HOURS
Decking, T&G solid		
Cedar		
3" thick	S.F.	0.020
4" thick	"	0.021
Fir		
3" thick	S.F.	0.020
4" thick	"	0.021
Southern yellow pine		
3" thick	S.F.	0.023
4" thick	"	0.025
White pine		
3" thick	S.F.	0.020
4" thick	"	0.021

06130.10 Heavy Timber

Rough Carpentry	UNIT	MAN/ HOURS
Mill framed structures		
Beams to 20' long		
Douglas fir		
6x8	L.F.	0.080
6x10	"	0.083

06130.10 Heavy Timber *(Cont.)*

Rough Carpentry	UNIT	MAN/ HOURS
6x12	L.F.	0.089
6x14	"	0.092
6x16	"	0.096
8x10	"	0.083
8x12	"	0.089
8x14	"	0.092
8x16	"	0.096
Southern yellow pine		
6x8	L.F.	0.080
6x10	"	0.083
6x12	"	0.089
6x14	"	0.092
6x16	"	0.096
8x10	"	0.083
8x12	"	0.089
8x14	"	0.092
8x16	"	0.096
Columns to 12' high		
Douglas fir		
6x6	L.F.	0.120
8x8	"	0.120
10x10	"	0.133
12x12	"	0.133
Southern yellow pine		
6x6	L.F.	0.120
8x8	"	0.120
10x10	"	0.133
12x12	"	0.133
Posts, treated		
4x4	L.F.	0.032
6x6	"	0.040

06190.20 Wood Trusses

Rough Carpentry	UNIT	MAN/ HOURS
Truss, fink, 2x4 members		
3-in-12 slope		
24' span	EA.	0.686
26' span	"	0.686
28' span	"	0.727
30' span	"	0.727
34' span	"	0.774
38' span	"	0.774
5-in-12 slope		
24' span	EA.	0.706
28' span	"	0.727
30' span	"	0.750
32' span	"	0.750
40' span	"	0.800
Gable, 2x4 members		
5-in-12 slope		
24' span	EA.	0.706
26' span	"	0.706
28' span	"	0.727
30' span	"	0.750

Rough Carpentry	UNIT	MAN/ HOURS
06190.20 Wood Trusses *(Cont.)*		
32' span	EA.	0.750
36' span	"	0.774
40' span	"	0.800
King post type, 2x4 members		
4-in-12 slope		
16' span	EA.	0.649
18' span	"	0.667
24' span	"	0.706
26' span	"	0.706
30' span	"	0.750
34' span	"	0.750
38' span	"	0.774
42' span	"	0.828

Finish Carpentry	UNIT	MAN/ HOURS
06200.10 Finish Carpentry		
Mouldings and trim		
Apron, flat		
9/16 x 2	L.F.	0.040
9/16 x 3-1/2	"	0.042
Base		
Colonial		
7/16 x 2-1/4	L.F.	0.040
7/16 x 3	"	0.040
7/16 x 3-1/4	"	0.040
9/16 x 3	"	0.042
9/16 x 3-1/4	"	0.042
11/16 x 2-1/4	"	0.044
Ranch		
7/16 x 2-1/4	L.F.	0.040
7/16 x 3-1/4	"	0.040
9/16 x 2-1/4	"	0.042
9/16 x 3	"	0.042
9/16 x 3-1/4	"	0.042
Casing		
11/16 x 2-1/2	L.F.	0.036
11/16 x 3-1/2	"	0.038
Chair rail		
9/16 x 2-1/2	L.F.	0.040
9/16 x 3-1/2	"	0.040
Closet pole		
1-1/8" dia.	L.F.	0.053
1-5/8" dia.	"	0.053
Cove		
9/16 x 1-3/4	L.F.	0.040

Finish Carpentry	UNIT	MAN/ HOURS
06200.10 Finish Carpentry *(Cont.)*		
11/16 x 2-3/4	L.F.	0.040
Crown		
9/16 x 1-5/8	L.F.	0.053
9/16 x 2-5/8	"	0.062
11/16 x 3-5/8	"	0.067
11/16 x 4-1/4	"	0.073
11/16 x 5-1/4	"	0.080
Drip cap		
1-1/16 x 1-5/8	L.F.	0.040
Glass bead		
3/8 x 3/8	L.F.	0.050
1/2 x 9/16	"	0.050
5/8 x 5/8	"	0.050
3/4 x 3/4	"	0.050
Half round		
1/2	L.F.	0.032
5/8	"	0.032
3/4	"	0.032
Lattice		
1/4 x 7/8	L.F.	0.032
1/4 x 1-1/8	"	0.032
1/4 x 1-3/8	"	0.032
1/4 x 1-3/4	"	0.032
1/4 x 2	"	0.032
Ogee molding		
5/8 x 3/4	L.F.	0.040
11/16 x 1-1/8	"	0.040
11/16 x 1-3/8	"	0.040
Parting bead		
3/8 x 7/8	L.F.	0.050
Quarter round		
1/4 x 1/4	L.F.	0.032
3/8 x 3/8	"	0.032
1/2 x 1/2	"	0.032
11/16 x 11/16	"	0.035
3/4 x 3/4	"	0.035
1-1/16 x 1-1/16	"	0.036
Railings, balusters		
1-1/8 x 1-1/8	L.F.	0.080
1-1/2 x 1-1/2	"	0.073
Screen moldings		
1/4 x 3/4	L.F.	0.067
5/8 x 5/16	"	0.067
Shoe		
7/16 x 11/16	L.F.	0.032
Sash beads		
1/2 x 3/4	L.F.	0.067
1/2 x 7/8	"	0.067
1/2 x 1-1/8	"	0.073
5/8 x 7/8	"	0.073
Stop		
5/8 x 1-5/8		
Colonial	L.F.	0.050

Finish Carpentry	UNIT	MAN/ HOURS
06200.10 Finish Carpentry *(Cont.)*		
Ranch	L.F.	0.050
Stools		
11/16 x 2-1/4	L.F.	0.089
11/16 x 2-1/2	"	0.089
11/16 x 5-1/4	"	0.100
Exterior trim, casing, select pine, 1x3	"	0.040
Douglas fir		
1x3	L.F.	0.040
1x4	"	0.040
1x6	"	0.044
1x8	"	0.050
Cornices, white pine, #2 or better		
1x2	L.F.	0.040
1x4	"	0.040
1x6	"	0.044
1x8	"	0.047
1x10	"	0.050
1x12	"	0.053
Shelving, pine		
1x8	L.F.	0.062
1x10	"	0.064
1x12	"	0.067
Plywood shelf, 3/4", with edge band, 12" wide	"	0.080
Adjustable shelf, and rod, 12" wide		
3' to 4' long	EA.	0.200
5' to 8' long	"	0.267
Prefinished wood shelves with brackets and supports		
8" wide		
3' long	EA.	0.200
4' long	"	0.200
6' long	"	0.200
10" wide		
3' long	EA.	0.200
4' long	"	0.200
6' long	"	0.200
06220.10 Millwork		
Countertop, laminated plastic		
25" x 7/8" thick		
Minimum	L.F.	0.200
Average	"	0.267
Maximum	"	0.320
25" x 1-1/4" thick		
Minimum	L.F.	0.267
Average	"	0.320
Maximum	"	0.400
Add for cutouts	EA.	0.500
Backsplash, 4" high, 7/8" thick	L.F.	0.160
Plywood, sanded, A-C		
1/4" thick	S.F.	0.027
3/8" thick	"	0.029
1/2" thick	"	0.031
A-D		

Finish Carpentry	UNIT	MAN/ HOURS
06220.10 Millwork *(Cont.)*		
1/4" thick	S.F.	0.027
3/8" thick	"	0.029
1/2" thick	"	0.031
Base cab., 34-1/2" high, 24" deep, hardwood		
Minimum	L.F.	0.320
Average	"	0.400
Maximum	"	0.533
Wall cabinets		
Minimum	L.F.	0.267
Average	"	0.320
Maximum	"	0.400
Oil borne		
Water borne		

Architectural Woodwork	UNIT	MAN/ HOURS
06420.10 Panel Work		
Hardboard, tempered, 1/4" thick		
Natural faced	S.F.	0.020
Plastic faced	"	0.023
Pegboard, natural	"	0.020
Plastic faced	"	0.023
Untempered, 1/4" thick		
Natural faced	S.F.	0.020
Plastic faced	"	0.023
Pegboard, natural	"	0.020
Plastic faced	"	0.023
Plywood unfinished, 1/4" thick		
Birch		
Natural	S.F.	0.027
Select	"	0.027
Knotty pine	"	0.027
Cedar (closet lining)		
Standard boards T&G	S.F.	0.027
Particle board	"	0.027
Plywood, prefinished, 1/4" thick, premium grade		
Birch veneer	S.F.	0.032
Cherry veneer	"	0.032
Chestnut veneer	"	0.032
Lauan veneer	"	0.032
Mahogany veneer	"	0.032
Oak veneer (red)	"	0.032
Pecan veneer	"	0.032
Rosewood veneer	"	0.032
Teak veneer	"	0.032
Walnut veneer	"	0.032

Architectural Woodwork	UNIT	MAN/ HOURS
06430.10 Stairwork		
Risers, 1x8, 42" wide		
White oak	EA.	0.400
Pine	"	0.400
Treads, 1-1/16" x 9-1/2" x 42"		
White oak	EA.	0.500
06440.10 Columns		
Column, hollow, round wood		
12" diameter		
10' high	EA.	0.800
12' high	"	0.857
14' high	"	0.960
16' high	"	1.200
24" diameter		
16' high	EA.	1.200
18' high	"	1.263
20' high	"	1.263
22' high	"	1.333
24' high	"	1.333

Moisture Protection

07100.10 Waterproofing	UNIT	MAN/ HOURS
Membrane waterproofing, elastomeric		
Butyl		
1/32" thick	S.F.	0.032
1/16" thick	"	0.033
Neoprene		
1/32" thick	S.F.	0.032
1/16" thick	"	0.033
Plastic vapor barrier (polyethylene)		
4 mil	S.F.	0.003
6 mil	"	0.003
10 mil	"	0.004
Bituminous membrane, asphalt felt, 15 lb.		
One ply	S.F.	0.020
Two ply	"	0.024
Modified asphalt membrane, fibrous asphalt		
One ply	S.F.	0.033
Two ply	"	0.040
Three ply	"	0.044
Asphalt coated protective board		
1/8" thick	S.F.	0.020
1/4" thick	"	0.020
3/8" thick	"	0.020
1/2" thick	"	0.021
Cement protective board		
3/8" thick	S.F.	0.027
1/2" thick	"	0.027
Fluid applied, neoprene		
50 mil	S.F.	0.027
90 mil	"	0.027
Bentonite waterproofing, panels		
3/16" thick	S.F.	0.020
1/4" thick	"	0.020

07150.10 Dampproofing	UNIT	MAN/ HOURS
Silicone dampproofing, sprayed on		
Concrete surface		
1 coat	S.F.	0.004
2 coats	"	0.006
Concrete block		
1 coat	S.F.	0.005
2 coats	"	0.007
Brick		
1 coat	S.F.	0.006
2 coats	"	0.008

07160.10 Bituminous Dampproofing	UNIT	MAN/ HOURS
Building paper, asphalt felt		
15 lb	S.F.	0.032
30 lb	"	0.033
Asphalt, troweled, cold, primer plus		
1 coat	S.F.	0.027
2 coats	"	0.040
3 coats	"	0.050

Moisture Protection

07160.10 Bituminous Dampproofing *(Cont.)*	UNIT	MAN/ HOURS
Fibrous asphalt, hot troweled, primer plus		
1 coat	S.F.	0.032
2 coats	"	0.044
3 coats	"	0.057
Asphaltic paint dampproofing, per coat		
Brush on	S.F.	0.011
Spray on	"	0.009

07190.10 Vapor Barriers	UNIT	MAN/ HOURS
Vapor barrier, polyethylene		
2 mil	S.F.	0.004
6 mil	"	0.004
8 mil	"	0.004
10 mil	"	0.004

Insulation

07210.10 Batt Insulation	UNIT	MAN/ HOURS
Ceiling, fiberglass, unfaced		
3-1/2" thick, R11	S.F.	0.009
6" thick, R19	"	0.011
9" thick, R30	"	0.012
Suspended ceiling, unfaced		
3-1/2" thick, R11	S.F.	0.009
6" thick, R19	"	0.010
9" thick, R30	"	0.011
Crawl space, unfaced		
3-1/2" thick, R11	S.F.	0.012
6" thick, R19	"	0.013
9" thick, R30	"	0.015
Wall, fiberglass		
Paper backed		
2" thick, R7	S.F.	0.008
3" thick, R8	"	0.009
4" thick, R11	"	0.009
6" thick, R19	"	0.010
Foil backed, 1 side		
2" thick, R7	S.F.	0.008
3" thick, R11	"	0.009
4" thick, R14	"	0.009
6" thick, R21	"	0.010
Foil backed, 2 sides		
2" thick, R7	S.F.	0.009
3" thick, R11	"	0.010
4" thick, R14	"	0.011
6" thick, R21	"	0.011

07 THERMAL AND MOISTURE

Insulation	UNIT	MAN/ HOURS
07210.10 Batt Insulation *(Cont.)*		
Unfaced		
2" thick, R7	S.F.	0.008
3" thick, R9	"	0.009
4" thick, R11	"	0.009
6" thick, R19	"	0.010
Mineral wool batts		
Paper backed		
2" thick, R6	S.F.	0.008
4" thick, R12	"	0.009
6" thick, R19	"	0.010
Fasteners, self adhering, attached to ceiling deck		
2-1/2" long	EA.	0.013
4-1/2" long	"	0.015
Capped, self-locking washers	"	0.008
07210.20 Board Insulation		
Insulation, rigid		
Fiberglass, roof		
0.75" thick, R2.78	S.F.	0.007
1.06" thick, R4.17	"	0.008
1.31" thick, R5.26	"	0.008
1.63" thick, R6.67	"	0.008
2.25" thick, R8.33	"	0.009
Perlite board, roof		
1.00" thick, R2.78	S.F.	0.007
1.50" thick, R4.17	"	0.007
2.00" thick, R5.92	"	0.007
2.50" thick, R6.67	"	0.008
3.00" thick, R8.33	"	0.008
4.00" thick, R10.00	"	0.008
5.25" thick, R14.29	"	0.009
Rigid urethane		
Roof		
1" thick, R6.67	S.F.	0.007
1.20" thick, R8.33	"	0.007
1.50" thick, R11.11	"	0.007
2" thick, R14.29	"	0.007
2.25" thick, R16.67	"	0.008
Wall		
1" thick, R6.67	S.F.	0.008
1.5" thick, R11.11	"	0.009
2" thick, R14.29	"	0.009
Polystyrene		
Roof		
1.0" thick, R4.17	S.F.	0.007
1.5" thick, R6.26	"	0.007
2.0" thick, R8.33	"	0.007
Wall		
1.0" thick, R4.17	S.F.	0.008
1.5" thick, R6.26	"	0.009
2.0" thick, R8.33	"	0.009
Rigid board insulation, deck		
Mineral fiberboard		

Insulation	UNIT	MAN/ HOURS
07210.20 Board Insulation *(Cont.)*		
1" thick, R3.0	S.F.	0.007
2" thick, R5.26	"	0.007
Fiberglass		
1" thick, R4.3	S.F.	0.007
2" thick, R8.5	"	0.007
Polystyrene		
1" thick, R5.4	S.F.	0.007
2" thick, R10.8	"	0.007
Urethane		
.75" thick, R5.4	S.F.	0.007
1" thick, R6.4	"	0.007
1.5" thick, R10.7	"	0.007
2" thick, R14.3	"	0.007
Foamglass		
1" thick, R1.8	S.F.	0.007
2" thick, R5.26	"	0.007
Wood fiber		
1" thick, R3.85	S.F.	0.007
2" thick, R7.7	"	0.007
Particle board		
3/4" thick, R2.08	S.F.	0.007
1" thick, R2.77	"	0.007
2" thick, R5.50	"	0.007
07210.60 Loose Fill Insulation		
Blown-in type		
Fiberglass		
5" thick, R11	S.F.	0.007
6" thick, R13	"	0.008
9" thick, R19	"	0.011
Rockwool, attic application		
6" thick, R13	S.F.	0.008
8" thick, R19	"	0.010
10" thick, R22	"	0.012
12" thick, R26	"	0.013
15" thick, R30	"	0.016
Poured type		
Fiberglass		
1" thick, R4	S.F.	0.005
2" thick, R8	"	0.006
3" thick, R12	"	0.007
4" thick, R16	"	0.008
Mineral wool		
1" thick, R3	S.F.	0.005
2" thick, R6	"	0.006
3" thick, R9	"	0.007
4" thick, R12	"	0.008
Vermiculite or perlite		
2" thick, R4.8	S.F.	0.006
3" thick, R7.2	"	0.007
4" thick, R9.6	"	0.008

Insulation	UNIT	MAN/ HOURS
07210.60 Loose Fill Insulation *(Cont.)*		
Masonry, poured vermiculite or perlite		
4" block	S.F.	0.004
6" block	"	0.005
8" block	"	0.006
10" block	"	0.006
12" block	"	0.007
07210.70 Sprayed Insulation		
Foam, sprayed on		
Polystyrene		
1" thick, R4	S.F.	0.008
2" thick, R8	"	0.011
Urethane		
1" thick, R4	S.F.	0.008
2" thick, R8	"	0.011

Shingles And Tiles	UNIT	MAN/ HOURS
07310.10 Asphalt Shingles		
Standard asphalt shingles, strip shingles		
210 lb/square	SQ.	0.800
235 lb/square	"	0.889
240 lb/square	"	1.000
260 lb/square	"	1.143
300 lb/square	"	1.333
385 lb/square	"	1.600
Roll roofing, mineral surface		
90 lb	SQ.	0.571
110 lb	"	0.667
140 lb	"	0.800
07310.50 Metal Shingles		
Aluminum, .020" thick		
Plain	SQ.	1.600
Colors	"	1.600
Steel, galvanized		
26 ga.		
Plain	SQ.	1.600
Colors	"	1.600
24 ga.		
Plain	SQ.	1.600
Colors	"	1.600
Porcelain enamel, 22 ga.		
Minimum	SQ.	2.000
Average	"	2.000
Maximum	"	2.000

Shingles And Tiles	UNIT	MAN/ HOURS
07310.60 Slate Shingles		
Slate shingles		
Pennsylvania		
Ribbon	SQ.	4.000
Clear	"	4.000
Vermont		
Black	SQ.	4.000
Gray	"	4.000
Green	"	4.000
Red	"	4.000
Replacement shingles		
Small jobs	EA.	0.267
Large jobs	S.F.	0.133
07310.70 Wood Shingles		
Wood shingles, on roofs		
White cedar, #1 shingles		
4" exposure	SQ.	2.667
5" exposure	"	2.000
#2 shingles		
4" exposure	SQ.	2.667
5" exposure	"	2.000
Resquared and rebutted		
4" exposure	SQ.	2.667
5" exposure	"	2.000
On walls		
White cedar, #1 shingles		
4" exposure	SQ.	4.000
5" exposure	"	3.200
6" exposure	"	2.667
#2 shingles		
4" exposure	SQ.	4.000
5" exposure	"	3.200
6" exposure	"	2.667
07310.80 Wood Shakes		
Shakes, hand split, 24" red cedar, on roofs		
5" exposure	SQ.	4.000
7" exposure	"	3.200
9" exposure	"	2.667
On walls		
6" exposure	SQ.	4.000
8" exposure	"	3.200
10" exposure	"	2.667

07 THERMAL AND MOISTURE

Roofing And Siding	UNIT	MAN/ HOURS
07410.10 Manufactured Roofs		
Aluminum roof panels, for steel framing		
Corrugated		
Unpainted finish		
.024"	S.F.	0.020
.030"	"	0.020
Painted finish		
.024"	S.F.	0.020
.030"	"	0.020
Steel roof panels, for structural steel framing		
Corrugated, painted		
18 ga.	S.F.	0.020
20 ga.	"	0.020
07460.10 Metal Siding Panels		
Aluminum siding panels		
Corrugated		
Plain finish		
.024"	S.F.	0.032
.032"	"	0.032
Painted finish		
.024"	S.F.	0.032
.032"	"	0.032
Steel siding panels		
Corrugated		
22 ga.	S.F.	0.053
24 ga.	"	0.053
26 ga.	"	0.053
Box rib		
20 ga.	S.F.	0.053
22 ga.	"	0.053
24 ga.	"	0.053
26 ga.	"	0.053
07460.50 Plastic Siding		
Horizontal vinyl siding, solid		
8" wide		
Standard	S.F.	0.031
Insulated	"	0.031
10" wide		
Standard	S.F.	0.029
Insulated	"	0.029
Vinyl moldings for doors and windows	L.F.	0.032
07460.60 Plywood Siding		
Rough sawn cedar, 3/8" thick	S.F.	0.027
Fir, 3/8" thick	"	0.027
Texture 1-11, 5/8" thick		
Cedar	S.F.	0.029
Fir	"	0.029
Redwood	"	0.029
Southern Yellow Pine	"	0.029

Roofing And Siding	UNIT	MAN/ HOURS
07460.70 Steel Siding		
Ribbed, sheets, galvanized		
22 ga.	S.F.	0.032
24 ga.	"	0.032
26 ga.	"	0.032
28 ga.	"	0.032
Primed		
24 ga.	S.F.	0.032
26 ga.	"	0.032
28 ga.	"	0.032
07460.80 Wood Siding		
Beveled siding, cedar		
A grade		
1/2 x 6	S.F.	0.040
1/2 x 8	"	0.032
3/4 x 10	"	0.027
Clear		
1/2 x 6	S.F.	0.040
1/2 x 8	"	0.032
3/4 x 10	"	0.027
B grade		
1/2 x 6	S.F.	0.040
1/2 x 8	"	0.320
3/4 x 10	"	0.027
Board and batten		
Cedar		
1x6	S.F.	0.040
1x8	"	0.032
1x10	"	0.029
1x12	"	0.026
Pine		
1x6	S.F.	0.040
1x8	"	0.032
1x10	"	0.029
1x12	"	0.026
Redwood		
1x6	S.F.	0.040
1x8	"	0.032
1x10	"	0.029
1x12	"	0.026
Tongue and groove		
Cedar		
1x4	S.F.	0.044
1x6	"	0.042
1x8	"	0.040
1x10	"	0.038
Pine		
1x4	S.F.	0.044
1x6	"	0.042
1x8	"	0.040
1x10	"	0.038
Redwood		
1x4	S.F.	0.044

Roofing And Siding

07460.80 Wood Siding (Cont.)

Roofing And Siding	UNIT	MAN/ HOURS
1x6	S.F.	0.042
1x8	"	0.040
1x10	"	0.038

Membrane Roofing

07510.10 Built-up Asphalt Roofing

Membrane Roofing	UNIT	MAN/ HOURS
Built-up roofing, asphalt felt, including gravel		
2 ply	SQ.	2.000
3 ply	"	2.667
4 ply	"	3.200
Walkway, for built-up roofs		
3' x 3' x		
1/2" thick	S.F.	0.027
3/4" thick	"	0.027
1" thick	"	0.027
Cant strip, 4" x 4"		
Treated wood	L.F.	0.023
Foamglass	"	0.020
Mineral fiber	"	0.020
New gravel for built-up roofing, 400 lb/sq	SQ.	1.600
Roof gravel (ballast)	C.Y.	4.000
Aluminum coating, top surfacing, for built-up roofing	SQ.	1.333
Remove 4-ply built-up roof (includes gravel)	"	4.000
Remove & replace gravel, includes flood coat	"	2.667

07530.10 Single-ply Roofing

Membrane Roofing	UNIT	MAN/ HOURS
Elastic sheet roofing		
Neoprene, 1/16" thick	S.F.	0.010
EPDM rubber		
45 mil	S.F.	0.010
60 mil	"	0.010
PVC		
45 mil	S.F.	0.010
60 mil	"	0.010
Flashing		
Pipe flashing, 90 mil thick		
1" pipe	EA.	0.200
2" pipe	"	0.200
3" pipe	"	0.211
4" pipe	"	0.211
5" pipe	"	0.222
6" pipe	"	0.222
8" pipe	"	0.235
10" pipe	"	0.267
12" pipe	"	0.267

Membrane Roofing

07530.10 Single-ply Roofing (Cont.)

Membrane Roofing	UNIT	MAN/ HOURS
Neoprene flashing, 60 mil thick strip		
6" wide	L.F.	0.067
12" wide	"	0.100
18" wide	"	0.133
24" wide	"	0.200
Adhesives		
Mastic sealer, applied at joints only		
1/4" bead	L.F.	0.004
Fluid applied roofing		
Urethane, 2 part, elastomeric membrane		
1" thick	S.F.	0.013
Vinyl liquid roofing, 2 coats, 2 mils per coat	"	0.011
Silicone roofing, 2 coats sprayed, 16 mil per coat	"	0.013
Inverted roof system		
Insulated membrane with coarse gravel ballast		
3 ply with 2" polystyrene	S.F.	0.013
Ballast, 3/4" through 1-1/2" gravel, 100lb/sf	"	0.800
Walkway for membrane roofs, 1/2" thick	"	0.027

Flashing And Sheet Metal

07610.10 Metal Roofing

Flashing And Sheet Metal	UNIT	MAN/ HOURS
Sheet metal roofing, copper, 16 oz, batten seam	SQ.	5.333
Standing seam	"	5.000
Aluminum roofing, natural finish		
Corrugated, on steel frame		
.0175" thick	SQ.	2.286
.0215" thick	"	2.286
.024" thick	"	2.286
.032" thick	"	2.286
V-beam, on steel frame		
.032" thick	SQ.	2.286
.040" thick	"	2.286
.050" thick	"	2.286
Ridge cap		
.019" thick	L.F.	0.027
Corrugated galvanized steel roofing, on steel frame		
28 ga.	SQ.	2.286
26 ga.	"	2.286
24 ga.	"	2.286
22 ga.	"	2.286
26 ga., factory insulated with 1" polystyrene	"	3.200
Ridge roll		
10" wide	L.F.	0.027
20" wide	"	0.032

Flashing And Sheet Metal	UNIT	MAN/ HOURS
07620.10 Flashing And Trim		
Counter flashing		
Aluminum, .032"	S.F.	0.080
Stainless steel, .015"	"	0.080
Copper		
16 oz.	S.F.	0.080
20 oz.	"	0.080
24 oz.	"	0.080
32 oz.	"	0.080
Valley flashing		
Aluminum, .032"	S.F.	0.050
Stainless steel, .015	"	0.050
Copper		
16 oz.	S.F.	0.050
20 oz.	"	0.067
24 oz.	"	0.050
32 oz.	"	0.050
Base flashing		
Aluminum, .040"	S.F.	0.067
Stainless steel, .018"	"	0.067
Copper		
16 oz.	S.F.	0.067
20 oz.	"	0.050
24 oz.	"	0.067
32 oz.	"	0.067
Waterstop, "T" section, 22 ga.		
1-1/2" x 3"	L.F.	0.040
2" x 2"	"	0.040
4" x 3"	"	0.040
6" x 4"	"	0.040
8" x 4"	"	0.040
Scupper outlets		
10" x 10" x 4"	EA.	0.200
22" x 4" x 4"	"	0.200
8" x 8" x 5"	"	0.200
Flashing and trim, aluminum		
.019" thick	S.F.	0.057
.032" thick	"	0.057
.040" thick	"	0.062
Neoprene sheet flashing, .060" thick	"	0.050
Copper, paper backed		
2 oz.	S.F.	0.080
5 oz.	"	0.080
Drainage boots, roof, cast iron		
2 x 3	L.F.	0.100
3 x 4	"	0.100
4 x 5	"	0.107
4 x 6	"	0.107
5 x 7	"	0.114
Pitch pocket, copper, 16 oz.		
4 x 4	EA.	0.200
6 x 6	"	0.200
8 x 8	"	0.200
8 x 10	"	0.200

Flashing And Sheet Metal	UNIT	MAN/ HOURS
07620.10 Flashing And Trim *(Cont.)*		
8 x 12	EA.	0.200
Reglets, copper 10 oz.	L.F.	0.053
Stainless steel, .020"	"	0.053
Gravel stop		
Aluminum, .032"		
4"	L.F.	0.027
6"	"	0.027
8"	"	0.031
10"	"	0.031
Copper, 16 oz.		
4"	L.F.	0.027
6"	"	0.027
8"	"	0.031
10"	"	0.031
07620.20 Gutters And Downspouts		
Copper gutter and downspout		
Downspouts, 16 oz. copper		
Round		
3" dia.	L.F.	0.053
4" dia.	"	0.053
Rectangular, corrugated		
2" x 3"	L.F.	0.050
3" x 4"	"	0.050
Rectangular, flat surface		
2" x 3"	L.F.	0.053
3" x 4"	"	0.053
Lead-coated copper downspouts		
Round		
3" dia.	L.F.	0.050
4" dia.	"	0.057
Rectangular, corrugated		
2" x 3"	L.F.	0.053
3" x 4"	"	0.053
Rectangular, plain		
2" x 3"	L.F.	0.053
3" x 4"	"	0.053
Gutters, 16 oz. copper		
Half round		
4" wide	L.F.	0.080
5" wide	"	0.089
Type K		
4" wide	L.F.	0.080
5" wide	"	0.089
Lead-coated copper gutters		
Half round		
4" wide	L.F.	0.080
6" wide	"	0.089
Type K		
4" wide	L.F.	0.080
5" wide	"	0.089
Aluminum gutter and downspout		
Downspouts		

Flashing And Sheet Metal

07620.20 Gutters And Downspouts (Cont.)	UNIT	MAN/ HOURS
2" x 3"	L.F.	0.053
3" x 4"	"	0.057
4" x 5"	"	0.062
Round		
3" dia.	L.F.	0.053
4" dia.	"	0.057
Gutters, stock units		
4" wide	L.F.	0.084
5" wide	"	0.089
Galvanized steel gutter and downspout		
Downspouts, round corrugated		
3" dia.	L.F.	0.053
4" dia.	"	0.053
5" dia.	"	0.057
6" dia.	"	0.057
Rectangular		
2" x 3"	L.F.	0.053
3" x 4"	"	0.050
4" x 4"	"	0.050
Gutters, stock units		
5" wide		
Plain	L.F.	0.089
Painted	"	0.089
6" wide		
Plain	L.F.	0.094
Painted	"	0.094

Roofing Specialties

07700.10 Manufactured Specialties	UNIT	MAN/ HOURS
Moisture relief vent		
Aluminum	EA.	0.114
Copper	"	0.114
Smoke vent, 48" x 48"		
Aluminum	EA.	2.000
Galvanized steel	"	2.000
Heat/smoke vent, 48" x 96"		
Aluminum	EA.	2.667
Galvanized steel	"	2.667
Ridge vent strips		
Mill finish	L.F.	0.053
Soffit vents		
Mill finish		
2-1/2" wide	L.F.	0.032

Skylights

07810.10 Plastic Skylights	UNIT	MAN/ HOURS
Single thickness, not including mounting curb		
2' x 4'	EA.	1.000
4' x 4'	"	1.333
5' x 5'	"	2.000
6' x 8'	"	2.667
Double thickness, not including mounting curb		
2' x 4'	EA.	1.000
4' x 4'	"	1.333
5' x 5'	"	2.000
6' x 8'	"	2.667
Metal framed skylights		
Translucent panels, 2-1/2" thick	S.F.	0.080
Continuous vaults, 8' wide		
Single glazed	S.F.	0.100
Double glazed	"	0.114

07820.10 Solar Skylight	UNIT	MAN/ HOURS
Tubular solar skylight, basic kit		
Min.	EA.	2.667
Ave.	"	4.000
Max.	"	8.000
Tubular solar skylight dome, 10" Diameter		
Min.	EA.	0.800
Ave.	"	1.000
Max.	"	1.333
14" Diameter		
Min.	EA.	0.800
Ave.	"	1.000
Max.	"	1.333
Straight extension tube, 10" Diameter X 12" long		
Min.	EA.	0.667
Ave.	"	0.800
Max.	"	1.000
24" long		
Min.	EA.	0.667
Ave.	"	0.800
Max.	"	1.000
36" long		
Min.	EA.	0.667
Ave.	"	0.800
Max.	"	1.000
48" long		
Min.	EA.	0.800
Ave.	"	1.000
Max.	"	1.333
14" Diameter X 12" long		
Min.	EA.	0.667
Ave.	"	0.800
Max.	"	1.000
24" long		
Min.	EA.	0.667
Ave.	"	0.800
Max.	"	1.000

Skylights	UNIT	MAN/ HOURS
07820.10 Solar Skylight *(Cont.)*		
36" long		
Min.	EA.	0.667
Ave.	"	0.800
Max.	"	1.000
90 Degree extension tubes, 10" Diameter		
Min.	EA.	0.444
Ave.	"	0.500
Max.	"	0.571
14" Diameter		
Min.	EA.	0.444
Ave.	"	0.500
Max.	"	0.571
Bottom tube adaptor, 10" Diameter		
Min.	EA.	0.444
Ave.	"	0.500
Max.	"	0.571
14" Diameter		
Min.	EA.	0.444
Ave.	"	0.500
Max.	"	0.571
Top tube adaptor, 10" Diameter		
Min.	EA.	0.444
Ave.	"	0.500
Max.	"	0.571
14" Diameter		
Min.	EA.	0.444
Ave.	"	0.500
Max.	"	0.571
Tube flashing		
Min.	EA.	0.667
Ave.	"	0.800
Max.	"	1.000
Daylight dimmer switch		
Min.	EA.	0.444
Ave.	"	0.500
Max.	"	0.571
Dimmer		
Min.	EA.	0.444
Ave.	"	0.500
Max.	"	0.571

Joint Sealers	UNIT	MAN/ HOURS
07920.10 Caulking		
Caulk exterior, two component		
1/4 x 1/2	L.F.	0.040
3/8 x 1/2	"	0.044
1/2 x 1/2	"	0.050
Caulk interior, single component		
1/4 x 1/2	L.F.	0.038
3/8 x 1/2	"	0.042
1/2 x 1/2	"	0.047

08 DOORS AND WINDOWS

Metal	UNIT	MAN/ HOURS
08110.10 Metal Doors		
Flush hollow metal, std. duty, 20 ga., 1-3/8" thick		
2-6 x 6-8	EA.	0.889
2-8 x 6-8	"	0.889
3-0 x 6-8	"	0.889
1-3/4" thick		
2-6 x 6-8	EA.	0.889
2-8 x 6-8	"	0.889
3-0 x 6-8	"	0.889
2-6 x 7-0	"	0.889
2-8 x 7-0	"	0.889
3-0 x 7-0	"	0.889
Heavy duty, 20 ga., unrated, 1-3/4"		
2-8 x 6-8	EA.	0.889
3-0 x 6-8	"	0.889
2-8 x 7-0	"	0.889
3-0 x 7-0	"	0.889
3-4 x 7-0	"	0.889
18 ga., 1-3/4", unrated door		
2-0 x 7-0	EA.	0.889
2-4 x 7-0	"	0.889
2-6 x 7-0	"	0.889
2-8 x 7-0	"	0.889
3-0 x 7-0	"	0.889
3-4 x 7-0	"	0.889
2", unrated door		
2-0 x 7-0	EA.	1.000
2-4 x 7-0	"	1.000
2-6 x 7-0	"	1.000
2-8 x 7-0	"	1.000
3-0 x 7-0	"	1.000
3-4 x 7-0	"	1.000
Galvanized metal door		
3-0 x 7-0	EA.	1.000
08110.40 Metal Door Frames		
Hollow metal, stock, 18 ga., 4-3/4" x 1-3/4"		
2-0 x 7-0	EA.	1.000
2-4 x 7-0	"	1.000
2-6 x 7-0	"	1.000
2-8 x 7-0	"	1.000
3-0 x 7-0	"	1.000
4-0 x 7-0	"	1.333
5-0 x 7-0	"	1.333
6-0 x 7-0	"	1.333
16 ga., 6-3/4" x 1-3/4"		
2-0 x 7-0	EA.	1.000
2-4 x 7-0	"	1.000
2-6 x 7-0	"	1.000
2-8 x 7-0	"	1.000
3-0 x 7-0	"	1.000
4-0 x 7-0	"	1.333
6-0 x 7-0	"	1.333

Wood And Plastic	UNIT	MAN/ HOURS
08210.10 Wood Doors		
Solid core, 1-3/8" thick		
Birch faced		
2-4 x 7-0	EA.	1.000
2-8 x 7-0	"	1.000
3-0 x 7-0	"	1.000
3-4 x 7-0	"	1.000
2-4 x 6-8	"	1.000
2-6 x 6-8	"	1.000
2-8 x 6-8	"	1.000
3-0 x 6-8	"	1.000
Lauan faced		
2-4 x 6-8	EA.	1.000
2-8 x 6-8	"	1.000
3-0 x 6-8	"	1.000
3-4 x 6-8	"	1.000
Tempered hardboard faced		
2-4 x 7-0	EA.	1.000
2-8 x 7-0	"	1.000
3-0 x 7-0	"	1.000
3-4 x 7-0	"	1.000
Hollow core, 1-3/8" thick		
Birch faced		
2-4 x 7-0	EA.	1.000
2-8 x 7-0	"	1.000
3-0 x 7-0	"	1.000
3-4 x 7-0	"	1.000
Lauan faced		
2-4 x 6-8	EA.	1.000
2-6 x 6-8	"	1.000
2-8 x 6-8	"	1.000
3-0 x 6-8	"	1.000
3-4 x 6-8	"	1.000
Tempered hardboard faced		
2-4 x 7-0	EA.	1.000
2-6 x 7-0	"	1.000
2-8 x 7-0	"	1.000
3-0 x 7-0	"	1.000
3-4 x 7-0	"	1.000
Solid core, 1-3/4" thick		
Birch faced		
2-4 x 7-0	EA.	1.000
2-6 x 7-0	"	1.000
2-8 x 7-0	"	1.000
3-0 x 7-0	"	1.000
3-4 x 7-0	"	1.000
Lauan faced		
2-4 x 7-0	EA.	1.000
2-6 x 7-0	"	1.000
2-8 x 7-0	"	1.000
3-4 x 7-0	"	1.000
3-0 x 7-0	"	1.000
Tempered hardboard faced		
2-4 x 7-0	EA.	1.000

Wood And Plastic	UNIT	MAN/ HOURS
08210.10 Wood Doors *(Cont.)*		
2-6 x 7-0	EA.	1.000
2-8 x 7-0	"	1.000
3-0 x 7-0	"	1.000
3-4 x 7-0	"	1.000
Hollow core, 1-3/4" thick		
Birch faced		
2-4 x 7-0	EA.	1.000
2-6 x 7-0	"	1.000
2-8 x 7-0	"	1.000
3-0 x 7-0	"	1.000
3-4 x 7-0	"	1.000
Lauan faced		
2-4 x 6-8	EA.	1.000
2-6 x 6-8	"	1.000
2-8 x 6-8	"	1.000
3-0 x 6-8	"	1.000
3-4 x 6-8	"	1.000
Tempered hardboard		
2-4 x 7-0	EA.	1.000
2-6 x 7-0	"	1.000
2-8 x 7-0	"	1.000
3-0 x 7-0	"	1.000
3-4 x 7-0	"	1.000
Add-on, louver	"	0.800
Glass	"	0.800
Exterior doors, 3-0 x 7-0 x 2-1/2", solid core		
Carved		
One face	EA.	2.000
Two faces	"	2.000
Closet doors, 1-3/4" thick		
Bi-fold or bi-passing, includes frame and trim		
Paneled		
4-0 x 6-8	EA.	1.333
6-0 x 6-8	"	1.333
Louvered		
4-0 x 6-8	EA.	1.333
6-0 x 6-8	"	1.333
Flush		
4-0 x 6-8	EA.	1.333
6-0 x 6-8	"	1.333
Primed		
4-0 x 6-8	EA.	1.333
6-0 x 6-8	"	1.333
08210.90 Wood Frames		
Frame, interior, pine		
2-6 x 6-8	EA.	1.143
2-8 x 6-8	"	1.143
3-0 x 6-8	"	1.143
5-0 x 6-8	"	1.143
6-0 x 6-8	"	1.143
2-6 x 7-0	"	1.143
2-8 x 7-0	"	1.143

Wood And Plastic	UNIT	MAN/ HOURS
08210.90 Wood Frames *(Cont.)*		
3-0 x 7-0	EA.	1.143
5-0 x 7-0	"	1.600
6-0 x 7-0	"	1.600
Exterior, custom, with threshold, including trim		
Walnut		
3-0 x 7-0	EA.	2.000
6-0 x 7-0	"	2.000
Oak		
3-0 x 7-0	EA.	2.000
6-0 x 7-0	"	2.000
Pine		
2-4 x 7-0	EA.	1.600
2-6 x 7-0	"	1.600
2-8 x 7-0	"	1.600
3-0 x 7-0	"	1.600
3-4 x 7-0	"	1.600
6-0 x 7-0	"	2.667
08300.10 Special Doors		
Metal clad doors, including electric motor		
Light duty		
Minimum	S.F.	0.133
Maximum	"	0.320
Accordion folding, tracks and fittings included		
Vinyl covered, 2 layers	S.F.	0.320
Woven mahogany and vinyl	"	0.320
Economy vinyl	"	0.320
Rigid polyvinyl chloride	"	0.320
Sectional wood overhead, frames not incl.		
Commercial grade, HD, 1-3/4" thick, manual		
8' x 8'	EA.	6.667
10' x 10'	"	7.273
Sectional metal overhead doors, complete		
Residential grade, manual		
9' x 7'	EA.	3.200
16' x 7'	"	4.000
Commercial grade		
8' x 8'	EA.	6.667
10' x 10'	"	7.273
12' x 12'	"	8.000
Sliding glass doors		
Tempered plate glass, 1/4" thick		
6' wide		
Economy grade	EA.	2.667
Premium grade	"	2.667
12' wide		
Economy grade	EA.	4.000
Premium grade	"	4.000
Insulating glass, 5/8" thick		
6' wide		
Economy grade	EA.	2.667
Premium grade	"	2.667
12' wide		

Wood And Plastic	UNIT	MAN/ HOURS
08300.10 Special Doors *(Cont.)*		
Economy grade	EA.	4.000
Premium grade	"	4.000
1" thick		
6' wide		
Economy grade	EA.	2.667
Premium grade	"	2.667
12' wide		
Economy grade	EA.	4.000
Premium grade	"	4.000
Residential storm door		
Minimum	EA.	1.333
Average	"	1.333
Maximum	"	2.000

Storefronts	UNIT	MAN/ HOURS
08410.10 Storefronts		
Storefront, aluminum and glass		
Minimum	S.F.	0.100
Average	"	0.114
Maximum	"	0.133
Entrance doors, premium, closers, panic dev.,etc.		
1/2" thick glass		
3' x 7'	EA.	6.667
3/4" thick glass		
3' x 7'	EA.	6.667

Metal Windows	UNIT	MAN/ HOURS
08510.10 Steel Windows		
Steel windows, primed		
Casements		
Operable		
Minimum	S.F.	0.047
Maximum	"	0.053
Fixed sash	"	0.040
Double hung	"	0.044
Picture window	"	0.044
Projecting sash		

Metal Windows	UNIT	MAN/ HOURS
08510.10 Steel Windows *(Cont.)*		
Minimum	S.F.	0.050
Maximum	"	0.050
Mullions	L.F.	0.040
08520.10 Aluminum Windows		
Jalousie		
3-0 x 4-0	EA.	1.000
3-0 x 5-0	"	1.000
Fixed window		
6 sf to 8 sf	S.F.	0.114
12 sf to 16 sf	"	0.089
Projecting window		
6 sf to 8 sf	S.F.	0.200
12 sf to 16 sf	"	0.133
Horizontal sliding		
6 sf to 8 sf	S.F.	0.100
12 sf to 16 sf	"	0.080
Double hung		
6 sf to 8 sf	S.F.	0.160
10 sf to 12 sf	"	0.133
Storm window, 0.5 cfm, up to		
60 u.i. (united inches)	EA.	0.400
70 u.i.	"	0.400
80 u.i.	"	0.400
90 u.i.	"	0.444
100 u.i.	"	0.444
2.0 cfm, up to		
60 u.i.	EA.	0.400
70 u.i.	"	0.400
80 u.i.	"	0.400
90 u.i.	"	0.444
100 u.i.	"	0.444

Wood And Plastic	UNIT	MAN/ HOURS
08600.10 Wood Windows		
Double hung		
24" x 36"		
Minimum	EA.	0.800
Average	"	1.000
Maximum	"	1.333
24" x 48"		
Minimum	EA.	0.800
Average	"	1.000
Maximum	"	1.333
30" x 48"		

Wood And Plastic	UNIT	MAN/ HOURS
08600.10 Wood Windows *(Cont.)*		
Minimum	EA.	0.889
Average	"	1.143
Maximum	"	1.600
30" x 60"		
Minimum	EA.	0.889
Average	"	1.143
Maximum	"	1.600
Casement		
1 leaf, 22" x 38" high		
Minimum	EA.	0.800
Average	"	1.000
Maximum	"	1.333
2 leaf, 50" x 50" high		
Minimum	EA.	1.000
Average	"	1.333
Maximum	"	2.000
3 leaf, 71" x 62" high		
Minimum	EA.	1.000
Average	"	1.333
Maximum	"	2.000
4 leaf, 95" x 75" high		
Minimum	EA.	1.143
Average	"	1.600
Maximum	"	2.667
5 leaf, 119" x 75" high		
Minimum	EA.	1.143
Average	"	1.600
Maximum	"	2.667
Picture window, fixed glass, 54" x 54" high		
Minimum	EA.	1.000
Average	"	1.143
Maximum	"	1.333
68" x 55" high		
Minimum	EA.	1.000
Average	"	1.143
Maximum	"	1.333
Sliding, 40" x 31" high		
Minimum	EA.	0.800
Average	"	1.000
Maximum	"	1.333
52" x 39" high		
Minimum	EA.	1.000
Average	"	1.143
Maximum	"	1.333
64" x 72" high		
Minimum	EA.	1.000
Average	"	1.333
Maximum	"	1.600
Awning windows		
34" x 21" high		
Minimum	EA.	0.800
Average	"	1.000
Maximum	"	1.333

Wood And Plastic	UNIT	MAN/ HOURS
08600.10 Wood Windows *(Cont.)*		
40" x 21" high		
Minimum	EA.	0.889
Average	"	1.143
Maximum	"	1.600
48" x 27" high		
Minimum	EA.	0.889
Average	"	1.143
Maximum	"	1.600
60" x 36" high		
Minimum	EA.	1.000
Average	"	1.333
Maximum	"	1.600
Window frame, milled		
Minimum	L.F.	0.160
Average	"	0.200
Maximum	"	0.267

Hardware	UNIT	MAN/ HOURS
08710.20 Locksets		
Latchset, heavy duty		
Cylindrical	EA.	0.500
Mortise	"	0.800
Lockset, heavy duty		
Cylindrical	EA.	0.500
Mortise	"	0.800
Preassembled locks and latches, brass		
Latchset, passage or closet latch	EA.	0.667
Lockset		
Privacy (bath or bathroom)	EA.	0.667
Entry lock	"	0.667
Bored locks and latches, satin chrome plated		
Latchset passage or closet latch	EA.	0.667
Lockset		
Privacy (bath or bedroom)	EA.	0.667
Entry lock	"	0.667
08710.30 Closers		
Door closers		
Surface mounted, traditional type, parallel arm		
Standard	EA.	1.000
Heavy duty	"	1.000
Modern type, parallel arm, standard duty	"	1.000
Overhead, concealed, pivot hung, single acting		
Interior	EA.	1.000
Exterior	"	1.000

Hardware	UNIT	MAN/ HOURS
08710.30 Closers *(Cont.)*		
Floor concealed, single acting, offset, pivoted		
Interior	EA.	2.667
Exterior	"	2.667
08710.40 Door Trim		
Panic device		
Mortise	EA.	2.000
Vertical rod	"	2.000
Labeled, rim type	"	2.000
Mortise	"	2.000
Vertical rod	"	2.000
Door plates		
Kick plate, aluminum, 3 beveled edges		
10" x 28"	EA.	0.400
10" x 30"	"	0.400
10" x 34"	"	0.400
10" x 38"	"	0.400
Push plate, 4" x 16"		
Aluminum	EA.	0.160
Bronze	"	0.160
Stainless steel	"	0.160
Armor plate, 40" x 34"	"	0.320
Pull handle, 4" x 16"		
Aluminum	EA.	0.160
Bronze	"	0.160
Stainless steel	"	0.160
Hasp assembly		
3"	EA.	0.133
4-1/2"	"	0.178
6"	"	0.229
08710.60 Weatherstripping		
Weatherstrip, head and jamb, metal strip, neoprene bulb		
Standard duty	L.F.	0.044
Heavy duty	"	0.050
Spring type		
Metal doors	EA.	2.000
Wood doors	"	2.667
Sponge type with adhesive backing	"	0.800
Astragal		
1-3/4" x 13 ga., aluminum	L.F.	0.067
1-3/8" x 5/8", oak	"	0.053
Thresholds		
Bronze	L.F.	0.200
Aluminum		
Plain	L.F.	0.200
Vinyl insert	"	0.200
Aluminum with grit	"	0.200
Steel		
Plain	L.F.	0.200
Interlocking	"	0.667

Glazing	UNIT	MAN/ HOURS
08810.10 Glazing		
Sheet glass, 1/8" thick	S.F.	0.044
Plate glass, bronze or grey, 1/4" thick	"	0.073
Clear	"	0.073
Polished	"	0.073
Plexiglass		
1/8" thick	S.F.	0.073
1/4" thick	"	0.044
Float glass, clear		
3/16" thick	S.F.	0.067
1/4" thick	"	0.073
3/8" thick	"	0.100
Tinted glass, polished plate, twin ground		
3/16" thick	S.F.	0.067
1/4" thick	"	0.073
3/8" thick	"	0.100
Insulated glass, bronze or gray		
1/2" thick	S.F.	0.133
1" thick	"	0.200
Spandrel, polished, 1 side, 1/4" thick	"	0.073
Tempered glass (safety)		
Clear sheet glass		
1/8" thick	S.F.	0.044
3/16" thick	"	0.062
Clear float glass		
1/4" thick	S.F.	0.067
5/16" thick	"	0.080
3/8" thick	"	0.100
1/2" thick	"	0.133
Tinted float glass		
3/16" thick	S.F.	0.062
1/4" thick	"	0.067
3/8" thick	"	0.100
1/2" thick	"	0.133
Laminated glass		
Float safety glass with polyvinyl plastic layer		
1/4", sheet or float		
Two lites, 1/8" thick, clear glass	S.F.	0.067
1/2" thick, float glass		
Two lites, 1/4" thick, clear glass	S.F.	0.133
Tinted glass	"	0.133
Insulating glass, two lites, clear float glass		
1/2" thick	S.F.	0.133
5/8" thick	"	0.160
3/4" thick	"	0.200
Glass seal edge		
3/8" thick	S.F.	0.133
Tinted glass		
1/2" thick	S.F.	0.133
1" thick	"	0.267
Tempered, clear		
1" thick	S.F.	0.267
Wire reinforced	"	0.267
Plate mirror glass		

Glazing	UNIT	MAN/ HOURS
08810.10 Glazing *(Cont.)*		
1/4" thick		
15 sf	S.F.	0.080
Over 15 sf	"	0.073
Door type, 1/4" thick	"	0.080
Transparent, one way vision, 1/4" thick	"	0.080
Sheet mirror glass		
3/16" thick	S.F.	0.080
1/4" thick	"	0.067
Wall tiles, 12" x 12"		
Clear glass	S.F.	0.044
Veined glass	"	0.044
Wire glass, 1/4" thick		
Clear	S.F.	0.267
Hammered	"	0.267
Obscure	"	0.267
Glazing accessories		
Neoprene glazing gaskets		
1/4" glass	L.F.	0.032
3/8" glass	"	0.033
1/2" glass	"	0.035

Glazed Curtain Walls	UNIT	MAN/ HOURS
08910.10 Glazed Curtain Walls		
Curtain wall, aluminum system, framing sections		
2" x 3"		
Jamb	L.F.	0.067
Horizontal	"	0.067
Mullion	"	0.067
2" x 4"		
Jamb	L.F.	0.100
Horizontal	"	0.100
Mullion	"	0.100
3" x 5-1/2"		
Jamb	L.F.	0.100
Horizontal	"	0.100
Mullion	"	0.100
4" corner mullion	"	0.133
Coping sections		
1/8" x 8"	L.F.	0.133
1/8" x 9"	"	0.133
1/8" x 12-1/2"	"	0.160
Sill section		
1/8" x 6"	L.F.	0.080
1/8" x 7"	"	0.080
1/8" x 8-1/2"	"	0.080

Glazed Curtain Walls	UNIT	MAN/ HOURS
08910.10 Glazed Curtain Walls *(Cont.)*		
Column covers, aluminum		
1/8" x 26"	L.F.	0.200
1/8" x 34"	"	0.211
1/8" x 38"	"	0.211
Doors		
Aluminum framed, standard hardware		
Narrow stile		
2-6 x 7-0	EA.	4.000
3-0 x 7-0	"	4.000
3-6 x 7-0	"	4.000
Wide stile		
2-6 x 7-0	EA.	4.000
3-0 x 7-0	"	4.000
3-6 x 7-0	"	4.000
Flush panel doors, to match adjacent wall panels		
2-6 x 7-0	EA.	5.000
3-0 x 7-0	"	5.000
3-6 x 7-0	"	5.000
Window wall system, complete		
Minimum	S.F.	0.080
Average	"	0.089
Maximum	"	0.114

Support Systems

09110.10 Metal Studs

Support Systems	UNIT	MAN/ HOURS
Studs, non load bearing, galvanized		
2-1/2", 20 ga.		
12" o.c.	S.F.	0.017
16" o.c.	"	0.013
25 ga.		
12" o.c.	S.F.	0.017
16" o.c.	"	0.013
24" o.c.	"	0.011
3-5/8", 20 ga.		
12" o.c.	S.F.	0.020
16" o.c.	"	0.016
24" o.c.	"	0.013
25 ga.		
12" o.c.	S.F.	0.020
16" o.c.	"	0.016
24" o.c.	"	0.013
4", 20 ga.		
12" o.c.	S.F.	0.020
16" o.c.	"	0.016
24" o.c.	"	0.013
25 ga.		
12" o.c.	S.F.	0.020
16" o.c.	"	0.016
24" o.c.	"	0.013
6", 20 ga.		
12" o.c.	S.F.	0.025
16" o.c.	"	0.020
24" o.c.	"	0.017
25 ga.		
12" o.c.	S.F.	0.025
16" o.c.	"	0.020
24" o.c.	"	0.017
Load bearing studs, galvanized		
3-5/8", 16 ga.		
12" o.c.	S.F.	0.020
16" o.c.	"	0.016
18 ga.		
12" o.c.	S.F.	0.013
16" o.c.	"	0.016
4", 16 ga.		
12" o.c.	S.F.	0.020
16" o.c.	"	0.016
6", 16 ga.		
12" o.c.	S.F.	0.025
16" o.c.	"	0.020
Furring		
On beams and columns		
7/8" channel	L.F.	0.053
1-1/2" channel	"	0.062
On ceilings		
3/4" furring channels		
12" o.c.	S.F.	0.033
16" o.c.	"	0.032

09110.10 Metal Studs *(Cont.)*

Support Systems	UNIT	MAN/ HOURS
24" o.c.	S.F.	0.029
1-1/2" furring channels		
12" o.c.	S.F.	0.036
16" o.c.	"	0.033
24" o.c.	"	0.031
On walls		
3/4" furring channels		
12" o.c.	S.F.	0.027
16" o.c.	"	0.025
24" o.c.	"	0.024
1-1/2" furring channels		
12" o.c.	S.F.	0.029
16" o.c.	"	0.027
24" o.c.	"	0.025

Lath And Plaster

09205.10 Gypsum Lath

Lath And Plaster	UNIT	MAN/ HOURS
Gypsum lath, 1/2" thick		
Clipped	S.Y.	0.044
Nailed	"	0.050

09205.20 Metal Lath

Lath And Plaster	UNIT	MAN/ HOURS
Diamond expanded, galvanized		
2.5 lb., on walls		
Nailed	S.Y.	0.100
Wired	"	0.114
On ceilings		
Nailed	S.Y.	0.114
Wired	"	0.133
3.4 lb., on walls		
Nailed	S.Y.	0.100
Wired	"	0.114
On ceilings		
Nailed	S.Y.	0.114
Wired	"	0.133
Flat rib		
2.75 lb., on walls		
Nailed	S.Y.	0.100
Wired	"	0.114
On ceilings		
Nailed	S.Y.	0.114
Wired	"	0.133
3.4 lb., on walls		
Nailed	S.Y.	0.100
Wired	"	0.114

Lath And Plaster	UNIT	MAN/ HOURS
09205.20 Metal Lath *(Cont.)*		
On ceilings		
Nailed	S.Y.	0.114
Wired	"	0.133
Stucco lath		
1.8 lb.	S.Y.	0.100
3.6 lb.	"	0.100
Paper backed		
Minimum	S.Y.	0.080
Maximum	"	0.114
09205.60 Plaster Accessories		
Expansion joint, 3/4", 26 ga., galv.	L.F.	0.020
Plaster corner beads, 3/4", galvanized	"	0.023
Casing bead, expanded flange, galvanized	"	0.020
Expanded wing, 1-1/4" wide, galvanized	"	0.020
Joint clips for lath	EA.	0.004
Metal base, galvanized, 2-1/2" high	L.F.	0.027
Stud clips for gypsum lath	EA.	0.004
Sound deadening board, 1/4"	S.F.	0.013
09210.10 Plaster		
Gypsum plaster, trowel finish, 2 coats		
Ceilings	S.Y.	0.250
Walls	"	0.235
3 coats		
Ceilings	S.Y.	0.348
Walls	"	0.308
Vermiculite plaster		
2 coats		
Ceilings	S.Y.	0.381
Walls	"	0.348
3 coats		
Ceilings	S.Y.	0.471
Walls	"	0.421
Keenes cement plaster		
2 coats		
Ceilings	S.Y.	0.308
Walls	"	0.267
3 coats		
Ceilings	S.Y.	0.348
Walls	"	0.308
On columns, add to installation, 50%	"	
Chases, fascia, and soffits, add to installation, 50%	"	
Beams, add to installation, 50%	"	
Patch holes, average size holes		
1 sf to 5 sf		
Minimum	S.F.	0.133
Average	"	0.160
Maximum	"	0.200
Over 5 sf		
Minimum	S.F.	0.080

Lath And Plaster	UNIT	MAN/ HOURS
09210.10 Plaster *(Cont.)*		
Average	S.F.	0.114
Maximum	"	0.133
Patch cracks		
Minimum	S.F.	0.027
Average	"	0.040
Maximum	"	0.080
09220.10 Portland Cement Plaster		
Stucco, portland, gray, 3 coat, 1" thick		
Sand finish	S.Y.	0.348
Trowel finish	"	0.364
White cement		
Sand finish	S.Y.	0.364
Trowel finish	"	0.400
Scratch coat		
For ceramic tile	S.Y.	0.080
For quarry tile	"	0.080
Portland cement plaster		
2 coats, 1/2"	S.Y.	0.160
3 coats, 7/8"	"	0.200
09250.10 Gypsum Board		
Drywall, plasterboard, 3/8" clipped to		
Metal furred ceiling	S.F.	0.009
Columns and beams	"	0.020
Walls	"	0.008
Nailed or screwed to		
Wood framed ceiling	S.F.	0.008
Columns and beams	"	0.018
Walls	"	0.007
1/2", clipped to		
Metal furred ceiling	S.F.	0.009
Columns and beams	"	0.020
Walls	"	0.008
Nailed or screwed to		
Wood framed ceiling	S.F.	0.008
Columns and beams	"	0.018
Walls	"	0.007
5/8", clipped to		
Metal furred ceiling	S.F.	0.010
Columns and beams	"	0.022
Walls	"	0.009
Nailed or screwed to		
Wood framed ceiling	S.F.	0.010
Columns and beams	"	0.022
Walls	"	0.009
Vinyl faced, clipped to metal studs		
1/2"	S.F.	0.010
5/8"	"	0.010
Taping and finishing joints		
Minimum	S.F.	0.005
Average	"	0.007
Maximum	"	0.008

Lath And Plaster

09250.10 Gypsum Board (Cont.)

Lath And Plaster	UNIT	MAN/ HOURS
Casing bead		
Minimum	L.F.	0.023
Average	"	0.027
Maximum	"	0.040
Corner bead		
Minimum	L.F.	0.023
Average	"	0.027
Maximum	"	0.040

Tile

09310.10 Ceramic Tile

Tile	UNIT	MAN/ HOURS
Glazed wall tile, 4-1/4" x 4-1/4"		
Minimum	S.F.	0.057
Average	"	0.067
Maximum	"	0.080
Base, 4-1/4" high		
Minimum	L.F.	0.100
Average	"	0.100
Maximum	"	0.100
Unglazed floor tile		
Portland cem., cushion edge, face mtd		
1" x 1"	S.F.	0.073
2" x 2"	"	0.067
4" x 4"	"	0.067
6" x 6"	"	0.057
12" x 12"	"	0.050
16" x 16"	"	0.044
18" x 18"	"	0.040
Adhesive bed, with white grout		
1" x 1"	S.F.	0.073
2" x 2"	"	0.067
4" x 4"	"	0.067
6" x 6"	"	0.057
12" x 12"	"	0.050
16" x 16"	"	0.044
18" x 18"	"	0.040
Organic adhesive bed, thin set, back mounted		
1" x 1"	S.F.	0.073
2" x 2"	"	0.067
Porcelain floor tile		
1" x 1"	S.F.	0.073
2" x 2"	"	0.070
4" x 4"	"	0.067
6" x 6"	"	0.057
12" x 12"	"	0.050

09310.10 Ceramic Tile (Cont.)

Tile	UNIT	MAN/ HOURS
16" x 16"	S.F.	0.044
18" x 18"	"	0.040
Unglazed wall tile		
Organic adhesive, face mounted cushion edge		
1" x 1"		
Minimum	S.F.	0.067
Average	"	0.073
Maximum	"	0.080
2" x 2"		
Minimum	S.F.	0.062
Average	"	0.067
Maximum	"	0.073
Back mounted		
1" x 1"		
Minimum	S.F.	0.067
Average	"	0.073
Maximum	"	0.080
2" x 2"		
Minimum	S.F.	0.062
Average	"	0.067
Maximum	"	0.073
Conductive floor tile, unglazed square edged		
Portland cement bed		
1 x 1	S.F.	0.100
1-9/16 x 1-9/16	"	0.100
Dry set		
1 x 1	S.F.	0.100
1-9/16 x 1-9/16	"	0.100
Epoxy bed with epoxy joints		
1 x 1	S.F.	0.100
1-9/16 x 1-9/16	"	0.100
Ceramic accessories		
Towel bar, 24" long		
Minimum	EA.	0.320
Average	"	0.400
Maximum	"	0.533
Soap dish		
Minimum	EA.	0.533
Average	"	0.667
Maximum	"	0.800

09330.10 Quarry Tile

Tile	UNIT	MAN/ HOURS
Floor		
4 x 4 x 1/2"	S.F.	0.107
6 x 6 x 1/2"	"	0.100
6 x 6 x 3/4"	"	0.100
12 x 12x 3/4"	"	0.089
16x1 6 x 3/4"	"	0.080
18 x 18 x 3/4"	"	0.067
Medallion		
36" dia.	EA.	2.000
48" dia.	"	2.000
Wall, applied to 3/4" portland cement bed		

Tile	UNIT	MAN/ HOURS
09330.10 Quarry Tile (Cont.)		
4 x 4 x 1/2"	S.F.	0.160
6 x 6 x 3/4"	"	0.133
Cove base		
5 x 6 x 1/2" straight top	L.F.	0.133
6 x 6 x 3/4" round top	"	0.133
Moldings		
2 x 12	L.F.	0.080
4 x 12	"	0.080
Stair treads 6 x 6 x 3/4"	"	0.200
Window sill 6 x 8 x 3/4"	"	0.160
For abrasive surface, add to material, 25%		
09410.10 Terrazzo		
Floors on concrete, 1-3/4" thick, 5/8" topping		
Gray cement	S.F.	0.114
White cement	"	0.114
Sand cushion, 3" thick, 5/8" top, 1/4"		
Gray cement	S.F.	0.133
White cement	"	0.133
Monolithic terrazzo, 3-1/2" base slab, 5/8" topping	"	0.100
Terrazzo wainscot, cast-in-place, 1/2" thick	"	0.200
Base, cast in place, terrazzo cove type, 6" high	L.F.	0.114
Curb, cast in place, 6" wide x 6" high, polished top	"	0.400
Stairs, cast-in-place, topping on concrete or metal		
1-1/2" thick treads, 12" wide	L.F.	0.400
Combined tread and riser	"	1.000
Precast terrazzo, thin set		
Terrazzo tiles, non-slip surface		
9" x 9" x 1" thick	S.F.	0.114
12" x 12"		
1" thick	S.F.	0.107
1-1/2" thick	"	0.114
18" x 18" x 1-1/2" thick	"	0.114
24" x 24" x 1-1/2" thick	"	0.094
Terrazzo wainscot		
12" x 12" x 1" thick	S.F.	0.200
18" x 18" x 1-1/2" thick	"	0.229
Base		
6" high		
Straight	L.F.	0.062
Coved	"	0.062
8" high		
Straight	L.F.	0.067
Coved	"	0.067
Terrazzo curbs		
8" wide x 8" high	L.F.	0.320
6" wide x 6" high	"	0.267
Precast terrazzo stair treads, 12" wide		
1-1/2" thick		
Diamond pattern	L.F.	0.145
Non-slip surface	"	0.145
2" thick		
Diamond pattern	L.F.	0.145

Tile	UNIT	MAN/ HOURS
09410.10 Terrazzo (Cont.)		
Non-slip surface	L.F.	0.160
Stair risers, 1" thick to 6" high		
Straight sections	L.F.	0.080
Cove sections	"	0.080
Combined tread and riser		
Straight sections		
1-1/2" tread, 3/4" riser	L.F.	0.229
3" tread, 1" riser	"	0.229
Curved sections		
2" tread, 1" riser	L.F.	0.267
3" tread, 1" riser	"	0.267
Stair stringers, notched for treads and risers		
1" thick	L.F.	0.200
2" thick	"	0.267
Landings, structural, nonslip		
1-1/2" thick	S.F.	0.133
3" thick	"	0.160
Conductive terrazzo, spark proof industrial floor		
Epoxy terrazzo		
Floor	S.F.	0.050
Base	"	0.067
Polyacrylate		
Floor	S.F.	0.050
Base	"	0.067
Polyester		
Floor	S.F.	0.032
Base	"	0.040
Synthetic latex mastic		
Floor	S.F.	0.050
Base	"	0.067

Acoustical Treatment	UNIT	MAN/ HOURS
09510.10 Ceilings And Walls		
Acoustical panels, suspension system not included		
Fiberglass panels		
5/8" thick		
2' x 2'	S.F.	0.011
2' x 4'	"	0.009
3/4" thick		
2' x 2'	S.F.	0.011
2' x 4'	"	0.009
Glass cloth faced fiberglass panels		
3/4" thick	S.F.	0.013
1" thick	"	0.013
Mineral fiber panels		

09 FINISHES

Acoustical Treatment

09510.10 Ceilings And Walls (Cont.)

Acoustical Treatment	UNIT	MAN/HOURS
5/8" thick		
2' x 2'	S.F.	0.011
2' x 4'	"	0.009
3/4" thick		
2' x 2'	S.F.	0.011
2' x 4'	"	0.009
Wood fiber panels		
1/2" thick		
2' x 2'	S.F.	0.011
2' x 4'	"	0.009
5/8" thick		
2' x 2'	S.F.	0.011
2' x 4'	"	0.009
3/4" thick		
2' x 2'	S.F.	0.011
2' x 4'	"	0.009
2" thick		
2' x 2'	S.F.	0.013
2' x 4'	"	0.010
Air distributing panels		
3/4" thick	S.F.	0.020
5/8" thick	"	0.016
Acoustical tiles, suspension system not included		
Fiberglass tile, 12" x 12"		
5/8" thick	S.F.	0.015
3/4" thick	"	0.018
Glass cloth faced fiberglass tile		
3/4" thick	S.F.	0.018
3" thick	"	0.020
Mineral fiber tile, 12" x 12"		
5/8" thick		
Standard	S.F.	0.016
Vinyl faced	"	0.016
3/4" thick		
Standard	S.F.	0.016
Vinyl faced	"	0.016
Fire rated	"	0.016
Aluminum or mylar faced	"	0.016
Wood fiber tile, 12" x 12"		
1/2" thick	S.F.	0.016
3/4" thick	"	0.016
Metal pan units, 24 ga. steel		
12" x 12"	S.F.	0.032
12" x 24"	"	0.027
Aluminum, .025" thick		
12" x 12"	S.F.	0.032
12" x 24"	"	0.027
Anodized aluminum, 0.25" thick		
12" x 12"	S.F.	0.032
12" x 24"	"	0.027
Stainless steel, 24 ga.		
12" x 12"	S.F.	0.032
12" x 24"	"	0.027

Acoustical Treatment

09510.10 Ceilings And Walls (Cont.)

Acoustical Treatment	UNIT	MAN/HOURS
Metal ceiling systems		
.020" thick panels		
10', 12', and 16' lengths	S.F.	0.023
Custom lengths, 3' to 20'	"	0.023
.025" thick panels		
32 sf, 38 sf, and 52 sf pieces	S.F.	0.027
Custom lengths, 10 sf to 65 sf	"	0.027
Sound absorption walls, with fabric cover		
2-6" x 9' x 3/4"	S.F.	0.027
2' x 9' x 1"	"	0.027
Starter spline	L.F.	0.020
Internal spline	"	0.020
Acoustical treatment		
Barriers for plenums		
Leaded vinyl		
0.48 lb per sf	S.F.	0.038
0.87 lb per sf	"	0.040
Aluminum foil, fiberglass reinforcement		
Minimum	S.F.	0.027
Maximum	"	0.040
Aluminum mesh, paper backed	"	0.027
Fibered cement sheet, 3/16" thick	"	0.029
Sheet lead, 1/64" thick	"	0.020
Sound attenuation blanket		
1" thick	S.F.	0.080
1-1/2" thick	"	0.080
2" thick	"	0.080
3" thick	"	0.089
Ceiling suspension systems		
T bar system		
2' x 4'	S.F.	0.008
2' x 2'	"	0.009
Concealed Z bar suspension system, 12" module	"	0.013

Flooring

09550.10 Wood Flooring

Flooring	UNIT	MAN/HOURS
Wood strip flooring, unfinished		
Fir floor		
C and better		
Vertical grain	S.F.	0.027
Flat grain	"	0.027
Oak floor		
Minimum	S.F.	0.038
Average	"	0.038
Maximum	"	0.038

Flooring

09550.10 Wood Flooring *(Cont.)*

Flooring	UNIT	MAN/HOURS
Maple floor		
25/32" x 2-1/4"		
Minimum	S.F.	0.038
Maximum	"	0.038
33/32" x 3-1/4"		
Minimum	S.F.	0.038
Maximum	"	0.038
Wood block industrial flooring		
Creosoted		
2" thick	S.F.	0.021
2-1/2" thick	"	0.025
3" thick	"	0.027
Parquet, 5/16", white oak		
Finished	S.F.	0.040
Unfinished	"	0.040
Gym floor, 2 ply felt, 25/32" maple, finished, in mastic	"	0.044
Over wood sleepers	"	0.050
Finishing, sand, fill, finish, and wax	"	0.020
Refinish sand, seal, and 2 coats of polyurethane	"	0.027
Clean and wax floors	"	0.004

09630.10 Unit Masonry Flooring

Flooring	UNIT	MAN/HOURS
Clay brick		
9 x 4-1/2 x 3" thick		
Glazed	S.F.	0.067
Unglazed	"	0.067
8 x 4 x 3/4" thick		
Glazed	S.F.	0.070
Unglazed	"	0.070

09660.10 Resilient Tile Flooring

Flooring	UNIT	MAN/HOURS
Solid vinyl tile, 1/8" thick, 12" x 12"		
Marble patterns	S.F.	0.020
Solid colors	"	0.020
Travertine patterns	"	0.020
Conductive resilient flooring, vinyl tile		
1/8" thick, 12" x 12"	S.F.	0.023

09665.10 Resilient Sheet Flooring

Flooring	UNIT	MAN/HOURS
Vinyl sheet flooring		
Minimum	S.F.	0.008
Average	"	0.010
Maximum	"	0.013
Cove, to 6"	L.F.	0.016
Fluid applied resilient flooring		
Polyurethane, poured in place, 3/8" thick	S.F.	0.067
Vinyl sheet goods, backed		
0.070" thick	S.F.	0.010
0.093" thick	"	0.010
0.125" thick	"	0.010
0.250" thick	"	0.010

09678.10 Resilient Base And Accessories

Flooring	UNIT	MAN/HOURS
Wall base, vinyl		
Group 1		
4" high	L.F.	0.027
6" high	"	0.027
Group 2		
4" high	L.F.	0.027
6" high	"	0.027
Group 3		
4" high	L.F.	0.027
6" high	"	0.027
Stair accessories		
Treads, 1/4" x 12", rubber diamond surface		
Marbled	L.F.	0.067
Plain	"	0.067
Grit strip safety tread, 12" wide, colors		
3/16" thick	L.F.	0.067
5/16" thick	"	0.067
Risers, 7" high, 1/8" thick, colors		
Flat	L.F.	0.040
Coved	"	0.040
Nosing, rubber		
3/16" thick, 3" wide		
Black	L.F.	0.040
Colors	"	0.040
6" wide		
Black	L.F.	0.067
Colors	"	0.067

Carpet

09680.10 Floor Leveling

Carpet	UNIT	MAN/HOURS
Repair and level floors to receive new flooring		
Minimum	S.Y.	0.027
Average	"	0.067
Maximum	"	0.080

09682.10 Carpet Padding

Carpet	UNIT	MAN/HOURS
Carpet padding		
Foam rubber, waffle type, 0.3" thick	S.Y.	0.040
Jute padding		
Minimum	S.Y.	0.036
Average	"	0.040
Maximum	"	0.044
Sponge rubber cushion		
Minimum	S.Y.	0.036

Carpet	UNIT	MAN/ HOURS
09682.10 Carpet Padding *(Cont.)*		
Average	S.Y.	0.040
Maximum	"	0.044
Urethane cushion, 3/8" thick		
Minimum	S.Y.	0.036
Average	"	0.040
Maximum	"	0.044
09685.10 Carpet		
Carpet, acrylic		
24 oz., light traffic	S.Y.	0.089
28 oz., medium traffic	"	0.089
Residential		
Nylon		
15 oz., light traffic	S.Y.	0.089
28 oz., medium traffic	"	0.089
Commercial		
Nylon		
28 oz., medium traffic	S.Y.	0.089
35 oz., heavy traffic	"	0.089
Wool		
30 oz., medium traffic	S.Y.	0.089
36 oz., medium traffic	"	0.089
42 oz., heavy traffic	"	0.089
Carpet tile		
Foam backed		
Minimum	S.F.	0.016
Average	"	0.018
Maximum	"	0.020
Tufted loop or shag		
Minimum	S.F.	0.016
Average	"	0.018
Maximum	"	0.020
Clean and vacuum carpet		
Minimum	S.Y.	0.004
Average	"	0.005
Maximum	"	0.008
09700.10 Special Flooring		
Epoxy flooring, marble chips		
Epoxy with colored quartz chips in 1/4" base	S.F.	0.044
Heavy duty epoxy topping, 3/16" thick	"	0.044
Epoxy terrazzo		
1/4" thick chemical resistant	S.F.	0.050

Painting	UNIT	MAN/ HOURS
09905.10 Painting Preparation		
Dropcloths		
Minimum	S.F.	0.001
Average	"	0.001
Maximum	"	0.001
Masking		
Paper and tape		
Minimum	L.F.	0.008
Average	"	0.010
Maximum	"	0.013
Doors		
Minimum	EA.	0.100
Average	"	0.133
Maximum	"	0.178
Windows		
Minimum	EA.	0.100
Average	"	0.133
Maximum	"	0.178
Sanding		
Walls and flat surfaces		
Minimum	S.F.	0.005
Average	"	0.007
Maximum	"	0.008
Doors and windows		
Minimum	EA.	0.133
Average	"	0.200
Maximum	"	0.267
Trim		
Minimum	L.F.	0.010
Average	"	0.013
Maximum	"	0.018
Puttying		
Minimum	S.F.	0.012
Average	"	0.016
Maximum	"	0.020
Water cleaning/preparation		
Washing (General)		
Minimum	S.F.	0.001
Average	"	0.001
Maximum	"	0.001
Mildew eradication		
Minimum	S.F.	0.001
Average	"	0.002
Maximum	"	0.003
Remove loose paint		
Minimum	S.F.	0.002
Average	"	0.003
Maximum	"	0.004
Steam clean		
Minimum	S.F.	0.002
Average	"	0.003
Maximum	"	0.004

Painting	UNIT	MAN/ HOURS
09910.05 Ext. Painting, Sitework		
Benches		
Brush		
First Coat		
Minimum	S.F.	0.008
Average	"	0.010
Maximum	"	0.013
Brickwork		
Brush		
First Coat		
Minimum	S.F.	0.005
Average	"	0.007
Maximum	"	0.010
Second Coat		
Minimum	S.F.	0.004
Average	"	0.005
Maximum	"	0.007
Roller		
First Coat		
Minimum	S.F.	0.004
Average	"	0.005
Maximum	"	0.007
Second Coat		
Minimum	S.F.	0.003
Average	"	0.004
Maximum	"	0.005
Spray		
First Coat		
Minimum	S.F.	0.002
Average	"	0.003
Maximum	"	0.004
Second Coat		
Minimum	S.F.	0.002
Average	"	0.003
Maximum	"	0.003
Concrete Block		
Roller		
First Coat		
Minimum	S.F.	0.004
Average	"	0.005
Maximum	"	0.008
Second Coat		
Minimum	S.F.	0.003
Average	"	0.004
Maximum	"	0.007
Spray		
First Coat		
Minimum	S.F.	0.002
Average	"	0.003
Maximum	"	0.003
Second Coat		
Minimum	S.F.	0.001
Average	"	0.002
Maximum	"	0.003

Painting	UNIT	MAN/ HOURS
09910.05 Ext. Painting, Sitework *(Cont.)*		
Fences, Chain Link		
Brush		
First Coat		
Minimum	S.F.	0.008
Average	"	0.009
Maximum	"	0.010
Second Coat		
Minimum	S.F.	0.005
Average	"	0.006
Maximum	"	0.007
Roller		
First Coat		
Minimum	S.F.	0.006
Average	"	0.007
Maximum	"	0.008
Second Coat		
Minimum	S.F.	0.003
Average	"	0.004
Maximum	"	0.005
Spray		
First Coat		
Minimum	S.F.	0.003
Average	"	0.003
Maximum	"	0.003
Second Coat		
Minimum	S.F.	0.002
Average	"	0.002
Maximum	"	0.003
Fences, Wood or Masonry		
Brush		
First Coat		
Minimum	S.F.	0.008
Average	"	0.010
Maximum	"	0.013
Second Coat		
Minimum	S.F.	0.005
Average	"	0.006
Maximum	"	0.008
Roller		
First Coat		
Minimum	S.F.	0.004
Average	"	0.005
Maximum	"	0.006
Second Coat		
Minimum	S.F.	0.003
Average	"	0.004
Maximum	"	0.005
Spray		
First Coat		
Minimum	S.F.	0.003
Average	"	0.004
Maximum	"	0.005
Second Coat		

Painting	UNIT	MAN/ HOURS
09910.05 Ext. Painting, Sitework *(Cont.)*		
Minimum	S.F.	0.002
Average	"	0.003
Maximum	"	0.003
09910.15 Ext. Painting, Buildings		
Decks, Wood, Stained		
Brush		
First Coat		
Minimum	S.F.	0.004
Average	"	0.004
Maximum	"	0.005
Second Coat		
Minimum	S.F.	0.003
Average	"	0.003
Maximum	"	0.003
Roller		
First Coat		
Minimum	S.F.	0.003
Average	"	0.003
Maximum	"	0.003
Second Coat		
Minimum	S.F.	0.003
Average	"	0.003
Maximum	"	0.003
Spray		
First Coat		
Minimum	S.F.	0.003
Average	"	0.003
Maximum	"	0.003
Second Coat		
Minimum	S.F.	0.002
Average	"	0.002
Maximum	"	0.003
Doors, Wood		
Brush		
First Coat		
Minimum	S.F.	0.012
Average	"	0.016
Maximum	"	0.020
Second Coat		
Minimum	S.F.	0.010
Average	"	0.011
Maximum	"	0.013
Roller		
First Coat		
Minimum	S.F.	0.005
Average	"	0.007
Maximum	"	0.010
Second Coat		
Minimum	S.F.	0.004
Average	"	0.004
Maximum	"	0.007
Spray		

Painting	UNIT	MAN/ HOURS
09910.15 Ext. Painting, Buildings *(Cont.)*		
First Coat		
Minimum	S.F.	0.003
Average	"	0.003
Maximum	"	0.004
Second Coat		
Minimum	S.F.	0.002
Average	"	0.002
Maximum	"	0.003
Gutters and Downspouts		
Brush		
First Coat		
Minimum	L.F.	0.010
Average	"	0.011
Maximum	"	0.013
Second Coat		
Minimum	L.F.	0.007
Average	"	0.008
Maximum	"	0.010
Siding, Wood		
Roller		
First Coat		
Minimum	S.F.	0.003
Average	"	0.003
Maximum	"	0.004
Second Coat		
Minimum	S.F.	0.003
Average	"	0.004
Maximum	"	0.004
Spray		
First Coat		
Minimum	S.F.	0.003
Average	"	0.003
Maximum	"	0.003
Second Coat		
Minimum	S.F.	0.002
Average	"	0.003
Maximum	"	0.004
Stucco		
Roller		
First Coat		
Minimum	S.F.	0.004
Average	"	0.004
Maximum	"	0.005
Second Coat		
Minimum	S.F.	0.003
Average	"	0.003
Maximum	"	0.004
Spray		
First Coat		
Minimum	S.F.	0.003
Average	"	0.003
Maximum	"	0.003
Second Coat		

Painting	UNIT	MAN/ HOURS
09910.15 Ext. Painting, Buildings *(Cont.)*		
Minimum	S.F.	0.002
Average	"	0.002
Maximum	"	0.003
Trim		
Brush		
First Coat		
Minimum	L.F.	0.003
Average	"	0.004
Maximum	"	0.005
Second Coat		
Minimum	L.F.	0.003
Average	"	0.003
Maximum	"	0.005
Walls		
Roller		
First Coat		
Minimum	S.F.	0.003
Average	"	0.003
Maximum	"	0.003
Second Coat		
Minimum	S.F.	0.003
Average	"	0.003
Maximum	"	0.003
Spray		
First Coat		
Minimum	S.F.	0.001
Average	"	0.002
Maximum	"	0.002
Second Coat		
Minimum	S.F.	0.001
Average	"	0.001
Maximum	"	0.002
Windows		
Brush		
First Coat		
Minimum	S.F.	0.013
Average	"	0.016
Maximum	"	0.020
Second Coat		
Minimum	S.F.	0.011
Average	"	0.013
Maximum	"	0.016
09910.25 Ext. Painting, Misc.		
Shakes		
Spray		
First Coat		
Minimum	S.F.	0.003
Average	"	0.004
Maximum	"	0.004
Second Coat		
Minimum	S.F.	0.003
Average	"	0.003

Painting	UNIT	MAN/ HOURS
09910.25 Ext. Painting, Misc. *(Cont.)*		
Maximum	S.F.	0.004
Shingles, Wood		
Roller		
First Coat		
Minimum	S.F.	0.004
Average	"	0.005
Maximum	"	0.006
Second Coat		
Minimum	S.F.	0.003
Average	"	0.003
Maximum	"	0.004
Spray		
First Coat		
Minimum	L.F.	0.003
Average	"	0.003
Maximum	"	0.004
Second Coat		
Minimum	L.F.	0.002
Average	"	0.003
Maximum	"	0.003
Shutters and Louvres		
Brush		
First Coat		
Minimum	EA.	0.160
Average	"	0.200
Maximum	"	0.267
Second Coat		
Minimum	EA.	0.100
Average	"	0.123
Maximum	"	0.160
Spray		
First Coat		
Minimum	EA.	0.053
Average	"	0.064
Maximum	"	0.080
Second Coat		
Minimum	EA.	0.040
Average	"	0.053
Maximum	"	0.064
Stairs, metal		
Brush		
First Coat		
Minimum	S.F.	0.009
Average	"	0.010
Maximum	"	0.011
Second Coat		
Minimum	S.F.	0.005
Average	"	0.006
Maximum	"	0.007
Spray		
First Coat		
Minimum	S.F.	0.004

Painting	UNIT	MAN/ HOURS
09910.25 Ext. Painting, Misc. *(Cont.)*		
Average	S.F.	0.006
Maximum	"	0.006
Second Coat		
Minimum	S.F.	0.003
Average	"	0.004
Maximum	"	0.005
09910.35 Int. Painting, Buildings		
Acoustical Ceiling		
Roller		
First Coat		
Minimum	S.F.	0.005
Average	"	0.007
Maximum	"	0.010
Second Coat		
Minimum	S.F.	0.004
Average	"	0.005
Maximum	"	0.007
Spray		
First Coat		
Minimum	S.F.	0.002
Average	"	0.003
Maximum	"	0.003
Second Coat		
Minimum	S.F.	0.002
Average	"	0.002
Maximum	"	0.002
Cabinets and Casework		
Brush		
First Coat		
Minimum	S.F.	0.008
Average	"	0.009
Maximum	"	0.010
Second Coat		
Minimum	S.F.	0.007
Average	"	0.007
Maximum	"	0.008
Spray		
First Coat		
Minimum	S.F.	0.004
Average	"	0.005
Maximum	"	0.006
Second Coat		
Minimum	S.F.	0.003
Average	"	0.003
Maximum	"	0.004
Ceilings		
Roller		
First Coat		
Minimum	S.F.	0.003
Average	"	0.004
Maximum	"	0.004
Second Coat		

Painting	UNIT	MAN/ HOURS
09910.35 Int. Painting, Buildings *(Cont.)*		
Minimum	S.F.	0.003
Average	"	0.003
Maximum	"	0.003
Spray		
First Coat		
Minimum	S.F.	0.002
Average	"	0.002
Maximum	"	0.003
Second Coat		
Minimum	S.F.	0.002
Average	"	0.002
Maximum	"	0.002
Doors, Wood		
Brush		
First Coat		
Minimum	S.F.	0.011
Average	"	0.015
Maximum	"	0.018
Second Coat		
Minimum	S.F.	0.009
Average	"	0.010
Maximum	"	0.011
Spray		
First Coat		
Minimum	S.F.	0.002
Average	"	0.003
Maximum	"	0.004
Second Coat		
Minimum	S.F.	0.002
Average	"	0.002
Maximum	"	0.003
Walls		
Roller		
First Coat		
Minimum	S.F.	0.003
Average	"	0.003
Maximum	"	0.003
Second Coat		
Minimum	S.F.	0.003
Average	"	0.003
Maximum	"	0.003
Spray		
First Coat		
Minimum	S.F.	0.001
Average	"	0.002
Maximum	"	0.002
Second Coat		
Minimum	S.F.	0.001
Average	"	0.001
Maximum	"	0.002

Painting	UNIT	MAN/ HOURS
09955.10 Wall Covering		
Vinyl wall covering		
Medium duty	S.F.	0.011
Heavy duty	"	0.013
Over pipes and irregular shapes		
Lightweight, 13 oz.	S.F.	0.016
Medium weight, 25 oz.	"	0.018
Heavy weight, 34 oz.	"	0.020
Cork wall covering		
1' x 1' squares		
1/4" thick	S.F.	0.020
1/2" thick	"	0.020
3/4" thick	"	0.020
Wall fabrics		
Natural fabrics, grass cloths		
Minimum	S.F.	0.012
Average	"	0.013
Maximum	"	0.016
Flexible gypsum coated wall fabric, fire resistant	"	0.008
Vinyl corner guards		
3/4" x 3/4" x 8'	EA.	0.100
2-3/4" x 2-3/4" x 4'	"	0.100

Specialties	UNIT	MAN/ HOURS
10110.10 Chalkboards		
Chalkboard, metal frame, 1/4" thick		
48"x60"	EA.	0.800
48"x96"	"	0.889
48"x144"	"	1.000
48"x192"	"	1.143
Liquid chalkboard		
48"x60"	EA.	0.800
48"x96"	"	0.889
48"x144"	"	1.000
48"x192"	"	1.143
Map rail, deluxe	L.F.	0.040
10165.10 Toilet Partitions		
Toilet partition, plastic laminate		
Ceiling mounted	EA.	2.667
Floor mounted	"	2.000
Metal		
Ceiling mounted	EA.	2.667
Floor mounted	"	2.000
Front door and side divider, floor mounted		
Porcelain enameled steel	EA.	2.000
Painted steel	"	2.000
Stainless steel	"	2.000
10185.10 Shower Stalls		
Shower receptors		
Precast, terrazzo		
32" x 32"	EA.	0.667
32" x 48"	"	0.800
Concrete		
32" x 32"	EA.	0.667
48" x 48"	"	0.889
Shower door, trim and hardware		
Economy, 24" wide, chrome, tempered glass	EA.	0.800
Porcelain enameled steel, flush	"	0.800
Baked enameled steel, flush	"	0.800
Aluminum, tempered glass, 48" wide, sliding	"	1.000
Folding	"	1.000
Aluminum and tempered glass, molded plastic		
Complete with receptor and door		
32" x 32"	EA.	2.000
36" x 36"	"	2.000
40" x 40"	"	2.286
10210.10 Vents And Wall Louvers		
Block vent, 8"x16"x4" alum., w/screen, mill finish	EA.	0.267
Standard	"	0.250
Vents w/screen, 4" deep, 8" wide, 5" high		
Modular	EA.	0.250
Aluminum gable louvers	S.F.	0.133
Vent screen aluminum, 4" wide, continuous	L.F.	0.027

Specialties	UNIT	MAN/ HOURS
10225.10 Door Louvers		
Fixed, 1" thick, enameled steel		
8"x8"	EA.	0.100
12"x8"	"	0.100
12"x12"	"	0.114
16"x12"	"	0.123
20"x8"	"	0.114
10290.10 Pest Control		
Termite control		
Under slab spraying		
Minimum	S.F.	0.002
Average	"	0.004
Maximum	"	0.008
10350.10 Flagpoles		
Installed in concrete base		
Fiberglass		
25' high	EA.	5.333
Aluminum		
25' high	EA.	5.333
Bonderized steel		
25' high	EA.	6.154
Freestanding tapered, fiberglass		
30' high	EA.	5.714
Wall mounted, with collar, brushed aluminum finish		
15' long	EA.	4.000
Outrigger, wall, including base		
10' long	EA.	5.333
10450.10 Control		
Access control, 7' high, indoor or outdoor impenetrability		
Remote or card control, type B	EA.	10.667
Free passage, type B	"	10.667
Remote or card control, type AA	"	10.667
Free passage, type AA	"	10.667
10550.10 Postal Specialties		
Single mail chute		
Finished aluminum	L.F.	2.000
Bronze	"	2.000
Single mail chute receiving box		
Finished aluminum	EA.	4.000
Bronze	"	4.000
Receiving box, 36" x 20" x 12"		
Finished aluminum	EA.	6.667
Bronze	"	6.667
Locked receiving mail box		
Finished aluminum	EA.	4.000
Bronze	"	4.000
Residential postal accessories		
Letter slot	EA.	0.400
Rural letter box	"	1.000
Apartment house, keyed, 3.5" x 4.5" x 16"	"	0.267

Specialties	UNIT	MAN/ HOURS
10550.10 Postal Specialties *(Cont.)*		
Ranch style	EA.	0.400
10800.10 Bath Accessories		
Grab bar, 1-1/2" dia., stainless steel, wall mounted		
24" long	EA.	0.400
36" long	"	0.421
42" long	"	0.444
48" long	"	0.471
52" long	"	0.500
1" dia., stainless steel		
12" long	EA.	0.348
18" long	"	0.364
24" long	"	0.400
30" long	"	0.421
36" long	"	0.444
48" long	"	0.471
Medicine cabinet, 16 x 22, baked enamel, lighted	"	0.320
With mirror, lighted	"	0.533
Mirror, 1/4" plate glass, up to 10 sf	S.F.	0.080
Mirror, stainless steel frame		
18"x24"	EA.	0.267
18"x32"	"	0.320
18"x36"	"	0.400
24"x30"	"	0.400
24"x36"	"	0.444
24"x48"	"	0.667
24"x60"	"	0.800
30"x30"	"	0.800
30"x72"	"	1.000
48"x72"	"	1.333
Shower rod, 1" diameter		
Chrome finish over brass	EA.	0.400
Stainless steel	"	0.400
Soap dish, stainless steel, wall mounted	"	0.533
Toilet tissue dispenser, stainless, wall mounted		
Single roll	EA.	0.200
Towel bar, stainless steel		
18" long	EA.	0.320
24" long	"	0.364
30" long	"	0.400
36" long	"	0.444
Toothbrush and tumbler holder	"	0.267

Architectural Equipment

11010.10 Maintenance Equipment

Architectural Equipment	UNIT	MAN/HOURS
Vacuum cleaning system		
3 valves		
1.5 hp	EA.	8.889
2.5 hp	"	11.429
5 valves	"	16.000
7 valves	"	20.000

11400.10 Food Service Equipment

Architectural Equipment	UNIT	MAN/HOURS
Unit kitchens		
30" compact kitchen		
Refrigerator, with range, sink	EA.	4.000
Sink only	"	2.667
Range only	"	2.000
Cabinet for upper wall section	"	1.143
Stainless shield, for rear wall	"	0.320
Side wall	"	0.320
42" compact kitchen		
Refrigerator with range, sink	EA.	4.444
Sink only	"	4.000
Cabinet for upper wall section	"	1.333
Stainless shield, for rear wall	"	0.333
Side wall	"	0.333
54" compact kitchen		
Refrigerator, oven, range, sink	EA.	5.714
Cabinet for upper wall section	"	1.600
Stainless shield, for		
Rear wall	EA.	0.364
Side wall	"	0.364
60" compact kitchen		
Refrigerator, oven, range, sink	EA.	5.714
Cabinet for upper wall section	"	1.600
Stainless shield, for		
Rear wall	EA.	0.364
Side wall	"	0.364
72" compact kitchen		
Refrigerator, oven, range, sink	EA.	6.667
Cabinet for upper wall section	"	1.600
Stainless shield for		
Rear wall	EA.	0.400
Side wall	"	0.400
Bake oven		
Single deck		
Minimum	EA.	1.000
Maximum	"	2.000
Double deck		
Minimum	EA.	1.333
Maximum	"	2.000
Triple deck		
Minimum	EA.	1.333
Maximum	"	2.667
Convection type oven, electric, 40" x 45" x 57"		

11400.10 Food Service Equipment *(Cont.)*

Architectural Equipment	UNIT	MAN/HOURS
Minimum	EA.	1.000
Maximum	"	2.000
Range		
Heavy duty, single oven, open top		
Minimum	EA.	1.000
Maximum	"	2.667

11450.10 Residential Equipment

Architectural Equipment	UNIT	MAN/HOURS
Compactor, 4 to 1 compaction	EA.	2.000
Dishwasher, built-in		
2 cycles	EA.	4.000
4 or more cycles	"	4.000
Disposal		
Garbage disposer	EA.	2.667
Heaters, electric, built-in		
Ceiling type	EA.	2.667
Wall type		
Minimum	EA.	2.000
Maximum	"	2.667
Hood for range, 2-speed, vented		
30" wide	EA.	2.667
42" wide	"	2.667
Ice maker, automatic		
30 lb per day	EA.	1.143
50 lb per day	"	4.000
Folding access stairs, disappearing metal stair		
8' long	EA.	1.143
11' long	"	1.143
12' long	"	1.143
Wood frame, wood stair		
22" x 54" x 8'9" long	EA.	0.800
25" x 54" x 10' long	"	0.800
Ranges electric		
Built-in, 30", 1 oven	EA.	2.667
2 oven	"	2.667
Counter top, 4 burner, standard	"	2.000
With grill	"	2.000
Free standing, 21", 1 oven	"	2.667
30", 1 oven	"	1.600
2 oven	"	1.600
Water softener		
30 grains per gallon	EA.	2.667
70 grains per gallon	"	4.000

11470.10 Darkroom Equipment

Architectural Equipment	UNIT	MAN/HOURS
Dryers		
36" x 25" x 68"	EA.	4.000
48" x 25" x 68"	"	4.000
Processors, film		
Black and white	EA.	4.000
Color negatives	"	4.000

Architectural Equipment	UNIT	MAN/ HOURS
11470.10 Darkroom Equipment *(Cont.)*		
Prints	EA.	4.000
Transparencies	"	4.000
Sinks with cabinet and/or stand		
5" sink with stand		
24" x 48"	EA.	2.000
32" x 64"	"	2.667

12 FURNISHINGS

Interior	UNIT	MAN/ HOURS
12302.10 Casework		
Kitchen base cabinet, standard, 24" deep, 35" high		
12"wide	EA.	0.800
18" wide	"	0.800
24" wide	"	0.889
27" wide	"	0.889
36" wide	"	1.000
48" wide	"	1.000
Drawer base, 24" deep, 35" high		
15"wide	EA.	0.800
18" wide	"	0.800
24" wide	"	0.889
27" wide	"	0.889
30" wide	"	0.889
Sink-ready, base cabinet		
30" wide	EA.	0.889
36" wide	"	0.889
42" wide	"	0.889
60" wide	"	1.000
Corner cabinet, 36" wide	"	1.000
Wall cabinet, 12" deep, 12" high		
30" wide	EA.	0.800
36" wide	"	0.800
15" high		
30" wide	EA.	0.889
36" wide	"	0.889
24" high		
30" wide	EA.	0.889
36" wide	"	0.889
30" high		
12" wide	EA.	1.000
18" wide	"	1.000
24" wide	"	1.000
27" wide	"	1.000
30" wide	"	1.143
36" wide	"	1.143
Corner cabinet, 30" high		
24" wide	EA.	1.333
30" wide	"	1.333
36" wide	"	1.333
Wardrobe	"	2.000
Vanity with top, laminated plastic		
24" wide	EA.	2.000
30" wide	"	2.000
36" wide	"	2.667
48" wide	"	3.200
12390.10 Counter Tops		
Stainless steel, counter top, with backsplash	S.F.	0.200
Acid-proof, kemrock surface	"	0.133

Interior	UNIT	MAN/ HOURS
12500.10 Window Treatment		
Drapery tracks, wall or ceiling mounted		
Basic traverse rod		
50 to 90"	EA.	0.400
84 to 156"	"	0.444
136 to 250"	"	0.444
165 to 312"	"	0.500
Traverse rod with stationary curtain rod		
30 to 50"	EA.	0.400
50 to 90"	"	0.400
84 to 156"	"	0.444
136 to 250"	"	0.500
Double traverse rod		
30 to 50"	EA.	0.400
50 to 84"	"	0.400
84 to 156"	"	0.444
136 to 250"	"	0.500
12510.10 Blinds		
Venetian blinds		
2" slats	S.F.	0.020
1" slats	"	0.020
12690.40 Floor Mats		
Recessed entrance mat, 3/8" thick, aluminum link	S.F.	0.400
Steel, flexible	"	0.400

Construction

	UNIT	MAN/ HOURS
13056.10 Vaults		
Floor safes		
1.0 cf	EA.	0.667
1.3 cf	"	1.000
1.9 cf	"	1.333
5.2 cf	"	1.333
13121.10 Pre-engineered Buildings		
Pre-engineered metal building, 40'x100'		
14' eave height	S.F.	0.032
16' eave height	"	0.037
13152.10 Swimming Pool Equipment		
Diving boards		
14' long		
Aluminum	EA.	4.444
Fiberglass	"	4.444
Lights, underwater		
12 volt, with transformer, 100 watt		
Incandescent	EA.	2.000
Halogen	"	2.000
LED	"	2.000
110 volt		
Minimum	EA.	2.000
Maximum	"	2.000
Ground fault interrupter for 110 volt, each light	"	0.667
Pool cover		
Reinforced polyethylene	S.F.	0.062
Vinyl water tube		
Minimum	S.F.	0.062
Maximum	"	0.062
Slides with water tube		
Minimum	EA.	6.667
Maximum	"	6.667
13200.10 Storage Tanks		
Oil storage tank, underground, single wall, no excv.		
Steel		
500 gals	EA.	3.000
1,000 gals	"	4.000
Fiberglass, double wall		
550 gals	EA.	4.000
1,000 gals	"	4.000
Above ground		
Steel, single wall		
275 gals	EA.	2.400
500 gals	"	4.000
1,000 gals	"	4.800
Fill cap	"	0.800
Vent cap	"	0.800
Level indicator	"	0.800

Hazardous Waste

	UNIT	MAN/ HOURS
13280.10 Asbestos Removal		
Enclosure using wood studs & poly, install & remove	S.F.	0.020
13280.12 Duct Insulation Removal		
Remove duct insulation, duct size		
6" x 12"	L.F.	0.044
x 18"	"	0.062
x 24"	"	0.089
8" x 12"	"	0.067
x 18"	"	0.073
x 24"	"	0.100
12" x 12"	"	0.067
x 18"	"	0.089
x 24"	"	0.114
13280.15 Pipe Insulation Removal		
Removal, asbestos insulation		
2" thick, pipe		
1" to 3" dia.	L.F.	0.067

14 CONVEYING

Elevators	UNIT	MAN/ HOURS
14210.10 Elevators		
Hydraulic, based on a shaft of 3 stops, 3 openings		
50 fpm		
2000 lb	EA.	20.000
2500 lb	"	20.000
3000 lb	"	20.870
Small elevators, 4 to 6 passenger capacity		
Electric, push		
2 stops	EA.	20.000
3 stops	"	21.818
4 stops	"	24.000

Lifts	UNIT	MAN/ HOURS
14410.10 Personnel Lifts		
Residential stair climber, per story	EA.	6.667
14410.20 Wheelchair Lifts		
600 lb, Residential	EA.	8.000

Material Handling	UNIT	MAN/ HOURS
14560.10 Chutes		
Linen chutes, stainless steel, with supports		
18" dia.	L.F.	0.057
Hopper	EA.	0.533
Skylight	"	0.800

Basic Materials	UNIT	MAN/ HOURS
15100.10 Specialties		
Wall penetration		
Concrete wall, 6" thick		
2" dia.	EA.	0.267
4" dia.	"	0.400
8" dia.	"	0.571
12" thick		
2" dia.	EA.	0.364
4" dia.	"	0.571
8" dia.	"	0.889
15120.10 Backflow Preventers		
Backflow preventer, flanged, cast iron, with valves		
3" pipe	EA.	4.000
4" pipe	"	4.444
6" pipe	"	6.667
Threaded		
3/4" pipe	EA.	0.500
2" pipe	"	0.800
Reduced pressure assembly, bronze, threaded		
3/4"	EA.	0.500
1"	"	0.571
1-1/4"	"	0.667
1-1/2"	"	0.800
15140.11 Pipe Hangers, Light		
A band, black iron		
1/2"	EA.	0.057
1"	"	0.059
1-1/4"	"	0.062
1-1/2"	"	0.067
2"	"	0.073
2-1/2"	"	0.080
3"	"	0.089
4"	"	0.100
5"	"	0.107
6"	"	0.114
Copper		
1/2"	EA.	0.057
3/4"	"	0.059
1"	"	0.059
1-1/4"	"	0.062
1-1/2"	"	0.067
2"	"	0.073
2-1/2"	"	0.080
3"	"	0.089
4"	"	0.100
2 hole clips, galvanized		
3/4"	EA.	0.053
1"	"	0.055
1-1/4"	"	0.057
1-1/2"	"	0.059
2"	"	0.062
2-1/2"	"	0.064

Basic Materials	UNIT	MAN/ HOURS
15140.11 Pipe Hangers, Light *(Cont.)*		
3"	EA.	0.067
4"	"	0.073
Perforated strap		
3/4"		
Galvanized, 20 ga.	L.F.	0.040
Copper, 22 ga.	"	0.040
J-Hooks		
1/2"	EA.	0.036
3/4"	"	0.036
1"	"	0.038
1-1/4"	"	0.039
1-1/2"	"	0.040
2"	"	0.040
3"	"	0.042
4"	"	0.042
PVC coated hangers, galvanized, 28 ga.		
1-1/2" x 12"	EA.	0.053
2" x 12"	"	0.057
3" x 12"	"	0.062
4" x 12"	"	0.067
Copper, 30 ga.		
1-1/2" x 12"	EA.	0.053
2" x 12"	"	0.057
3" x 12"	"	0.062
4" x 12"	"	0.067
2" x 24"	"	0.062
3" x 24"	"	0.067
4" x 24"	"	0.073
Wire hook hangers		
Black wire, 1/2" x		
4"	EA.	0.040
6"	"	0.042
3/4" x		
4"	EA.	0.042
6"	"	0.044
1" x		
4"	EA.	0.044
6"	"	0.047
1-1/4" x		
4"	EA.	0.047
6"	"	0.050
1-1/2" x		
6"	EA.	0.053
Copper wire hooks		
1/2" x		
4"	EA.	0.040
6"	"	0.042
3/4" x		
4"	EA.	0.042
6"	"	0.044
1" x		

Basic Materials	UNIT	MAN/ HOURS
15140.11 Pipe Hangers, Light *(Cont.)*		
4"	EA.	0.044
6"	"	0.047
1-1/4" x		
6"	EA.	0.047
1-1/2" x		
6"	EA.	0.053
15240.10 Vibration Control		
Vibration isolator, in-line, stainless connector		
1/2"	EA.	0.444
3/4"	"	0.471
1"	"	0.500
1-1/4"	"	0.533

Insulation	UNIT	MAN/ HOURS
15260.10 Fiberglass Pipe Insulation		
Fiberglass insulation on 1/2" pipe		
1" thick	L.F.	0.027
1-1/2" thick	"	0.033
3/4" pipe		
1" thick	L.F.	0.027
1-1/2" thick	"	0.033
1" pipe		
1" thick	L.F.	0.027
1-1/2" thick	"	0.033
1-1/4" pipe		
1" thick	L.F.	0.033
1-1/2" thick	"	0.036
1-1/2" pipe		
1" thick	L.F.	0.033
1-1/2" thick	"	0.036
2" pipe		
1" thick	L.F.	0.033
1-1/2" thick	"	0.036
2-1/2" pipe		
1" thick	L.F.	0.033
1-1/2" thick	"	0.036
3" pipe		
1" thick	L.F.	0.038
1-1/2" thick	"	0.040
4" pipe		
1" thick	L.F.	0.038
1-1/2" thick	"	0.040

Insulation	UNIT	MAN/ HOURS
15260.60 Exterior Pipe Insulation		
Fiberglass insulation, aluminum jacket		
1/2" pipe		
1" thick	L.F.	0.062
1-1/2" thick	"	0.067
1" pipe		
1" thick	L.F.	0.062
1-1/2" thick	"	0.067
2" pipe		
1" thick	L.F.	0.073
1-1/2" thick	"	0.076
3" pipe		
1" thick	L.F.	0.080
1-1/2" thick	"	0.084
4" pipe		
1" thick	L.F.	0.080
1-1/2" thick	"	0.084
15290.10 Ductwork Insulation		
Fiberglass duct insulation, plain blanket		
1-1/2" thick	S.F.	0.010
2" thick	"	0.013
With vapor barrier		
1-1/2" thick	S.F.	0.010
2" thick	"	0.013
Rigid with vapor barrier		
2" thick	S.F.	0.027
3" thick	"	0.032
4" thick	"	0.040
6" thick	"	0.053
Weatherproof, poly, 3" thick, w/vapor barrier	"	0.080
Urethane board with vapor barrier	"	0.100

Plumbing	UNIT	MAN/ HOURS
15410.05 C.I. Pipe, Above Ground		
No hub pipe		
1-1/2" pipe	L.F.	0.057
2" pipe	"	0.067
3" pipe	"	0.080
4" pipe	"	0.133
6" pipe	"	0.160
No hub fittings, 1-1/2" pipe		
1/4 bend	EA.	0.267
1/8 bend	"	0.267
Sanitary tee	"	0.400
Sanitary cross	"	0.400

Plumbing	UNIT	MAN/ HOURS
15410.05 C.I. Pipe, Above Ground *(Cont.)*		
Wye	EA.	0.400
Tapped tee	"	0.267
P-trap	"	0.267
Tapped cross	"	0.267
2" pipe		
1/4 bend	EA.	0.320
1/8 bend	"	0.320
Sanitary tee	"	0.533
Sanitary cross	"	0.533
Wye	"	0.667
Double wye	"	0.667
2x1-1/2" wye & 1/8 bend	"	0.500
Double wye & 1/8 bend	"	0.667
Test tee less 2" plug	"	0.320
Tapped tee		
2"x2"	EA.	0.320
2"x1-1/2"	"	0.320
P-trap		
2"x2"	EA.	0.320
Tapped cross		
2"x1-1/2"	EA.	0.320
3" pipe		
1/4 bend	EA.	0.400
1/8 bend	"	0.400
Sanitary tee	"	0.500
3"x2" sanitary tee	"	0.500
3"x1-1/2" sanitary tee	"	0.500
Sanitary cross	"	0.667
3x2" sanitary cross	"	0.667
Wye	"	0.667
3x2" wye	"	0.667
Double wye	"	0.667
3x2" double wye	"	0.667
3x2" wye & 1/8 bend	"	0.571
3x1-1/2" wye & 1/8 bend	"	0.571
Double wye & 1/8 bend	"	0.667
3x2" double wye & 1/8 bend	"	0.667
3x2" reducer	"	0.364
Test tee, less 3" plug	"	0.400
3x3" tapped tee	"	0.400
3x2" tapped tee	"	0.400
3x1-1/2" tapped tee	"	0.400
P-trap	"	0.400
3x2" tapped cross	"	0.400
3x1-1/2" tapped cross	"	0.400
Closet flange, 3-1/2" deep	"	0.200
4" pipe		
1/4 bend	EA.	0.400
1/8 bend	"	0.400
Sanitary tee	"	0.667
4x3" sanitary tee	"	0.667
4x2" sanitary tee	"	0.667
Sanitary cross	"	0.800

Plumbing	UNIT	MAN/ HOURS
15410.05 C.I. Pipe, Above Ground *(Cont.)*		
4x3" sanitary cross	EA.	0.800
4x2" sanitary cross	"	0.800
Wye	"	0.667
4x3" wye	"	0.667
4x2" wye	"	0.667
Double wye	"	0.800
4x3" double wye	"	0.800
4x2" double wye	"	0.800
Wye & 1/8 bend	"	0.667
4x3" wye & 1/8 bend	"	0.667
4x2" wye & 1/8 bend	"	0.667
Double wye & 1/8 bend	"	0.800
4x3" double wye & 1/8 bend	"	0.800
4x2" double wye & 1/8 bend	"	0.800
4x3" reducer	"	0.400
4x2" reducer	"	0.400
Test tee, less 4" plug	"	0.400
4x2" tapped tee	"	0.400
4x1-1/2" tapped tee	"	0.400
P-trap	"	0.400
4x2" tapped cross	"	0.400
4x1-1/2" tapped cross	"	0.400
Closet flange		
3" deep	EA.	0.400
8" deep	"	0.400
6" pipe		
1/4 bend	EA.	0.667
1/8 bend	"	0.667
Sanitary tee	"	0.800
6x4" sanitary tee	"	0.800
Wye	"	0.800
6x4" wye	"	0.800
6x3" wye	"	0.800
6x2" wye	"	0.800
Double wye	"	1.000
6x4" double wye	"	1.000
Wye & 1/8 bend	"	0.800
6x4" wye & 1/8 bend	"	0.800
6x3" wye & 1/8 bend	"	0.800
6x2" wye & 1/8 bend	"	0.800
6x4" reducer	"	0.444
6x3" reducer	"	0.444
6x2" reducer	"	0.400
Test tee		
Less 6" plug	EA.	0.500
Plug	"	
P-trap	"	0.500

Plumbing	UNIT	MAN/ HOURS
15410.06 C.I. Pipe, Below Ground		
No hub pipe		
1-1/2" pipe	L.F.	0.040
2" pipe	"	0.044
3" pipe	"	0.050
4" pipe	"	0.067
6" pipe	"	0.073
Fittings, 1-1/2"		
1/4 bend	EA.	0.229
1/8 bend	"	0.229
Wye	"	0.320
Wye & 1/8 bend	"	0.229
P-trap	"	0.229
2"		
1/4 bend	EA.	0.267
1/8 bend	"	0.267
Double wye	"	0.500
Wye & 1/8 bend	"	0.400
Double wye & 1/8 bend	"	0.500
P-trap	"	0.267
3"		
1/4 bend	EA.	0.320
1/8 bend	"	0.320
Wye	"	0.500
3x2" wye	"	0.500
Wye & 1/8 bend	"	0.500
Double wye & 1/8 bend	"	0.500
3x2" double wye & 1/8 bend	"	0.500
3x2" reducer	"	0.320
P-trap	"	0.320
4"		
1/4 bend	EA.	0.320
1/8 bend	"	0.320
Wye	"	0.500
4x3" wye	"	0.500
4x2" wye	"	0.500
Double wye	"	0.667
4x3" double wye	"	0.667
4x2" double wye	"	0.667
Wye & 1/8 bend	"	0.500
4x3" wye & 1/8 bend	"	0.500
4x2" wye & 1/8 bend	"	0.500
Double wye & 1/8 bend	"	0.667
4x3" double wye & 1/8 bend	"	0.667
4x2" double wye & 1/8 bend	"	0.667
4x3" reducer	"	0.320
4x2" reducer	"	0.320
6"		
1/4 bend	EA.	0.500
1/8 bend	"	0.500
Wye & 1/8 bend	"	0.667
6x4" wye & 1/8 bend	"	0.667
6x3" wye & 1/8 bend	"	0.667
6x2" wye & 1/8 bend	"	0.667

Plumbing	UNIT	MAN/ HOURS
15410.06 C.I. Pipe, Below Ground *(Cont.)*		
6x3" reducer	EA.	0.364
P-trap	"	0.400
15410.09 Service Weight Pipe		
Service weight pipe, single hub		
2" x 5'	EA.	0.160
3" x 5'	"	0.170
4" x 5'	"	0.178
5" x 5'	"	0.190
6" x 5'	"	0.200
Double hub		
2" x 5'	EA.	0.200
3" x 5'	"	0.216
4" x 5'	"	0.229
5" x 5'	"	0.250
6" x 5'	"	0.267
Single hub		
2" x 10'	EA.	0.200
3" x 10'	"	0.216
4" x 10'	"	0.229
5" x 10'	"	0.250
6" x 10'	"	0.267
Shorty		
2" x 42"	EA.	0.160
3" x 42"	"	0.170
4" x 42"	"	0.178
5" x 42"	"	0.190
6" x 42"	"	0.200
1/8 bend		
2"	EA.	0.267
3"	"	0.320
4"	"	0.364
5"	"	0.381
6"	"	0.400
1/4 bend		
2"	EA.	0.267
3"	"	0.320
4"	"	0.364
5"	"	0.381
6"	"	0.400
Sweep		
2"	EA.	0.267
3"	"	0.320
4"	"	0.364
5"	"	0.381
6"	"	0.400
Sanitary T		
2"	EA.	0.500
3" x 2"	"	0.533
3"	"	0.571
4" x 2"	"	0.615
4" x 3"	"	0.667
4"	"	0.667

Plumbing	UNIT	MAN/ HOURS
15410.09 Service Weight Pipe *(Cont.)*		
5"	EA.	0.727
6"	"	0.727
Tapped sanitary T		
2" x 1-1/2"	EA.	0.571
2" x 2"	"	0.571
3" x 1-1/2"	"	0.615
3" x 2"	"	0.615
4" x 1-1/2"	"	0.667
4" x 2"	"	0.667
Cleanout, dandy, with brass plug		
2", 1-1/2" plug	EA.	0.571
3", 2" plug	"	0.615
4", 3" plug	"	0.667
5", 4" plug	"	0.727
6", 4" plug	"	0.800
15410.10 Copper Pipe		
Type "K" copper		
1/2"	L.F.	0.025
3/4"	"	0.027
1"	"	0.029
DWV, copper		
1-1/4"	L.F.	0.033
1-1/2"	"	0.036
2"	"	0.040
3"	"	0.044
4"	"	0.050
6"	"	0.057
Refrigeration tubing, copper, sealed		
1/8"	L.F.	0.032
3/16"	"	0.033
1/4"	"	0.035
5/16"	"	0.036
3/8"	"	0.038
1/2"	"	0.040
Type "L" copper		
1/4"	L.F.	0.024
3/8"	"	0.024
1/2"	"	0.025
3/4"	"	0.027
1"	"	0.029
Type "M" copper		
1/2"	L.F.	0.025
3/4"	"	0.027
1"	"	0.029
15410.11 Copper Fittings		
Coupling, with stop		
1/4"	EA.	0.267
3/8"	"	0.320
1/2"	"	0.348
5/8"	"	0.400
3/4"	"	0.444

Plumbing	UNIT	MAN/ HOURS
15410.11 Copper Fittings *(Cont.)*		
1"	EA.	0.471
Reducing coupling		
1/4" x 1/8"	EA.	0.320
3/8" x 1/4"	"	0.348
1/2" x		
3/8"	EA.	0.400
1/4"	"	0.400
1/8"	"	0.400
3/4" x		
3/8"	EA.	0.444
1/2"	"	0.444
1" x		
3/8"	EA.	0.500
1" x 1/2"	"	0.500
1" x 3/4"	"	0.500
Slip coupling		
1/4"	EA.	0.267
1/2"	"	0.320
3/4"	"	0.400
1"	"	0.444
Coupling with drain		
1/2"	EA.	0.400
3/4"	"	0.444
1"	"	0.500
Reducer		
3/8" x 1/4"	EA.	0.320
1/2" x 3/8"	"	0.320
3/4" x		
1/4"	EA.	0.364
3/8"	"	0.364
1/2"	"	0.364
1" x		
1/2"	EA.	0.400
3/4"	"	0.400
Female adapters		
1/4"	EA.	0.320
3/8"	"	0.364
1/2"	"	0.400
3/4"	"	0.444
1"	"	0.444
Increasing female adapters		
1/8" x		
3/8"	EA.	0.320
1/2"	"	0.320
1/4" x 1/2"	"	0.348
3/8" x 1/2"	"	0.364
1/2" X		
3/4"	EA.	0.400
1"	"	0.400
3/4" X		
1"	EA.	0.444
Reducing female adapters		
3/8" x 1/4"	EA.	0.364

Plumbing	UNIT	MAN/ HOURS
15410.11 Copper Fittings *(Cont.)*		
1/2" x		
1/4"	EA.	0.400
3/8"	"	0.400
3/4" x 1/2"	"	0.444
1" x		
1/2"	EA.	0.444
3/4"	"	0.444
Female fitting adapters		
1/2"	EA.	0.400
3/4"	"	0.400
3/4" x 1/2"	"	0.421
1"	"	0.444
Male adapters		
1/4"	EA.	0.364
3/8"	"	0.364
Increasing male adapters		
3/8" x 1/2"	EA.	0.364
1/2" x		
3/4"	EA.	0.400
1"	"	0.400
3/4" x		
1"	EA.	0.421
1-1/4"	"	0.421
1" x 1-1/4"	"	0.444
Reducing male adapters		
1/2" x		
1/4"	EA.	0.400
3/8"	"	0.400
3/4" x 1/2"	"	0.421
1" x		
1/2"	EA.	0.444
3/4"	"	0.444
Fitting x male adapters		
1/2"	EA.	0.400
3/4"	"	0.421
1"	"	0.444
90 ells		
1/8"	EA.	0.320
1/4"	"	0.320
3/8"	"	0.364
1/2"	"	0.400
3/4"	"	0.421
1"	"	0.444
Reducing 90 ell		
3/8" x 1/4"	EA.	0.364
1/2" x		
1/4"	EA.	0.400
3/8"	"	0.400
3/4" x 1/2"	"	0.421
1" x		
1/2"	EA.	0.444
3/4"	"	0.444
Street ells, copper		

Plumbing	UNIT	MAN/ HOURS
15410.11 Copper Fittings *(Cont.)*		
1/4"	EA.	0.320
3/8"	"	0.364
1/2"	"	0.400
3/4"	"	0.421
1"	"	0.444
Female, 90 ell		
1/2"	EA.	0.400
3/4"	"	0.421
1"	"	0.444
Female increasing, 90 ell		
3/8" x 1/2"	EA.	0.364
1/2" x		
3/4"	EA.	0.400
1"	"	0.400
3/4" x 1"	"	0.421
1" x 1-1/4"	"	0.444
Female reducing, 90 ell		
1/2" x 3/8"	EA.	0.400
3/4" x 1/2"	"	0.421
1" x		
1/2"	EA.	0.444
3/4"	"	0.444
Male, 90 ell		
1/4"	EA.	0.320
3/8"	"	0.364
1/2"	"	0.400
3/4"	"	0.421
1"	"	0.444
Male, increasing 90 ell		
1/2" x		
3/4"	EA.	0.400
1"	"	0.400
3/4" x 1"	"	0.421
1" x 1-1/4"	"	0.444
Male, reducing 90 ell		
1/2" x 3/8"	EA.	0.400
3/4" x 1/2"	"	0.421
1" x		
1/2"	EA.	0.444
3/4"	"	0.444
Drop ear ells		
1/2"	EA.	0.400
Female drop ear ells		
1/2"	EA.	0.400
1/2" x 3/8"	"	0.400
3/4"	"	0.421
Female flanged sink ell		
1/2"	EA.	0.400
45 ells		
1/4"	EA.	0.320
3/8"	"	0.364
45 street ell		
1/4"	EA.	0.320

Plumbing	UNIT	MAN/ HOURS
15410.11 Copper Fittings *(Cont.)*		
3/8"	EA.	0.364
1/2"	"	0.400
3/4"	"	0.421
1"	"	0.444
Tee		
1/8"	EA.	0.320
1/4"	"	0.320
3/8"	"	0.364
Caps		
1/4"	EA.	0.320
3/8"	"	0.364
Test caps		
1/2"	EA.	0.400
3/4"	"	0.421
1"	"	0.444
Flush bushing		
1/4" x 1/8"	EA.	0.320
1/2" x		
1/4"	EA.	0.400
3/8"	"	0.400
3/4" x		
3/8"	EA.	0.421
1/2"	"	0.421
1" x		
1/2"	EA.	0.444
3/4"	"	0.444
Female flush bushing		
1/2" x		
1/2" x 1/8"	EA.	0.400
1/4"	"	0.400
Union		
1/4"	EA.	0.320
3/8"	"	0.364
Female		
1/2"	EA.	0.400
3/4"	"	0.421
Male		
1/2"	EA.	0.400
3/4"	"	0.421
1"	"	0.444
45 degree wye		
1/2"	EA.	0.400
3/4"	"	0.421
1"	"	0.444
1" x 3/4" x 3/4"	"	0.444
Twin ells		
1" x 3/4" x 3/4"	EA.	0.444
1" x 1" x 1"	"	0.444
90 union ells, male		
1/2"	EA.	0.400
3/4"	"	0.421
1"	"	0.444
DWV fittings, coupling with stop		

Plumbing	UNIT	MAN/ HOURS
15410.11 Copper Fittings *(Cont.)*		
1-1/4"	EA.	0.471
1-1/2"	"	0.500
1-1/2" x 1-1/4"	"	0.500
2"	"	0.533
2" x 1-1/4"	"	0.533
2" x 1-1/2"	"	0.533
3"	"	0.667
3" x 1-1/2"	"	0.667
3" x 2"	"	0.667
4"	"	0.800
Slip coupling		
1-1/2"	EA.	0.500
2"	"	0.533
3"	"	0.667
90 ells		
1-1/2"	EA.	0.500
1-1/2" x 1-1/4"	"	0.500
2"	"	0.533
2" x 1-1/2"	"	0.533
3"	"	0.667
4"	"	0.800
Street, 90 elbows		
1-1/2"	EA.	0.500
2"	"	0.533
3"	"	0.667
4"	"	0.800
Female, 90 elbows		
1-1/2"	EA.	0.500
2"	"	0.533
Male, 90 elbows		
1-1/2"	EA.	0.500
2"	"	0.533
90 with side inlet		
3" x 3" x 1"	EA.	0.667
3" x 3" x 1-1/2"	"	0.667
3" x 3" x 2"	"	0.667
45 ells		
1-1/4"	EA.	0.471
1-1/2"	"	0.500
2"	"	0.533
3"	"	0.667
4"	"	0.800
Street, 45 ell		
1-1/2"	EA.	0.500
2"	"	0.533
3"	"	0.667
60 ell		
1-1/2"	EA.	0.500
2"	"	0.533
3"	"	0.667
22-1/2 ell		
1-1/2"	EA.	0.500
2"	"	0.533

Plumbing	UNIT	MAN/ HOURS
15410.11 Copper Fittings *(Cont.)*		
3"	EA.	0.667
11-1/4 ell		
1-1/2"	EA.	0.500
2"	"	0.533
3"	"	0.667
Wye		
1-1/4"	EA.	0.471
1-1/2"	"	0.500
2"	"	0.533
2" x 1-1/2" x 1-1/2"	"	0.533
2" x 1-1/2" x 2"	"	0.533
2" x 1-1/2" x 2"	"	0.533
3"	"	0.667
3" x 3" x 1-1/2"	"	0.667
3" x 3" x 2"	"	0.667
4"	"	0.800
4" x 4" x 2"	"	0.800
4" x 4" x 3"	"	0.800
Sanitary tee		
1-1/4"	EA.	0.471
1-1/2"	"	0.500
2"	"	0.533
2" x 1-1/2" x 1-1/2"	"	0.533
2" x 1-1/2" x 2"	"	0.533
2" x 2" x 1-1/2"	"	0.533
3"	"	0.667
3" x 3" x 1-1/2"	"	0.667
3" x 3" x 2"	"	0.667
4"	"	0.800
4" x 4" x 3"	"	0.800
Female sanitary tee		
1-1/2"	EA.	0.500
Long turn tee		
1-1/2"	EA.	0.500
2"	"	0.533
3" x 1-1/2"	"	0.667
Double wye		
1-1/2"	EA.	0.500
2"	"	0.533
2" x 2" x 1-1/2" x 1-1/2"	"	0.533
3"	"	0.667
3" x 3" x 1-1/2" x 1-1/2"	"	0.667
3" x 3" x 2" x 2"	"	0.667
4" x 4" x 1-1/2" x 1-1/2"	"	0.800
Double sanitary tee		
1-1/2"	EA.	0.500
2"	"	0.533
2" x 2" x 1-1/2"	"	0.533
3"	"	0.667
3" x 3" x 1-1/2" x 1-1/2"	"	0.667
3" x 3" x 2" x 2"	"	0.667
4" x 4" x 1-1/2" x 1-1/2"	"	0.800
Long		

Plumbing	UNIT	MAN/ HOURS
15410.11 Copper Fittings *(Cont.)*		
2" x 1-1/2"	EA.	0.533
Twin elbow		
1-1/2"	EA.	0.500
2"	"	0.533
2" x 1-1/2" x 1-1/2"	"	0.533
Spigot adapter, manoff		
1-1/2" x 2"	EA.	0.500
1-1/2" x 3"	"	0.500
2"	"	0.533
2" x 3"	"	0.533
2" x 4"	"	0.533
3"	"	0.667
3" x 4"	"	0.667
4"	"	0.800
No-hub adapters		
1-1/2" x 2"	EA.	0.500
2"	"	0.533
2" x 3"	"	0.533
3"	"	0.667
3" x 4"	"	0.667
4"	"	0.800
Fitting reducers		
1-1/2" x 1-1/4"	EA.	0.500
2" x 1-1/2"	"	0.533
3" x 1-1/2"	"	0.667
3" x 2"	"	0.667
Slip joint (Desanco)		
1-1/4"	EA.	0.471
1-1/2"	"	0.500
1-1/2" x 1-1/4"	"	0.500
Street x slip joint (Desanco)		
1-1/2"	EA.	0.500
1-1/2" x 1-1/4"	"	0.500
Flush bushing		
1-1/2" x 1-1/4"	EA.	0.500
2" x 1-1/2"	"	0.533
3" x 1-1/2"	"	0.667
3" x 2"	"	0.667
Male hex trap bushing		
1-1/4" x 1-1/2"	EA.	0.471
1-1/2"	"	0.500
1-1/2" x 2"	"	0.500
2"	"	0.533
Round trap bushing		
1-1/2"	EA.	0.500
2"	"	0.533
Female adapter		
1-1/4"	EA.	0.471
1-1/2"	"	0.500
1-1/2" x 2"	"	0.500
2"	"	0.533
2" x 1-1/2"	"	0.533
3"	"	0.667

Plumbing	UNIT	MAN/ HOURS
15410.11 Copper Fittings *(Cont.)*		
Fitting x female adapter		
1-1/2"	EA.	0.500
2"	"	0.533
Male adapters		
1-1/4"	EA.	0.471
1-1/4" x 1-1/2"	"	0.471
1-1/2"	"	0.500
1-1/2" x 2"	"	0.500
2"	"	0.533
2" x 1-1/2"	"	0.533
3"	"	0.667
Male x slip joint adapters		
1-1/2" x 1-1/4"	EA.	0.500
Dandy cleanout		
1-1/2"	EA.	0.500
2"	"	0.533
3"	"	0.667
End cleanout, flush pattern		
1-1/2" x 1"	EA.	0.500
2" x 1-1/2"	"	0.533
3" x 2-1/2"	"	0.667
Copper caps		
1-1/2"	EA.	0.500
2"	"	0.533
Closet flanges		
3"	EA.	0.667
4"	"	0.800
Drum traps, with cleanout		
1-1/2" x 3" x 6"	EA.	0.500
P-trap, swivel, with cleanout		
1-1/2"	EA.	0.500
P-trap, solder union		
1-1/2"	EA.	0.500
2"	"	0.533
With cleanout		
1-1/2"	EA.	0.500
2"	"	0.533
2" x 1-1/2"	"	0.533
Swivel joint, with cleanout		
1-1/2" x 1-1/4"	EA.	0.500
1-1/2"	"	0.500
2" x 1-1/2"	"	0.533
Estabrook TY, with inlets		
3", with 1-1/2" inlet	EA.	0.667
Fine thread adapters		
1/2"	EA.	0.400
1/2" x 1/2" IPS	"	0.400
1/2" x 3/4" IPS	"	0.400
1/2" x male	"	0.400
1/2" x female	"	0.400
Copper pipe fittings		
1/2"		
90 deg ell	EA.	0.178
45 deg ell	EA.	0.178
Tee	"	0.229
Cap	"	0.089
Coupling	"	0.178
Union	"	0.200
3/4"		
90 deg ell	EA.	0.200
45 deg ell	"	0.200
Tee	"	0.267
Cap	"	0.094
Coupling	"	0.200
Union	"	0.229
1"		
90 deg ell	EA.	0.267
45 deg ell	"	0.267
Tee	"	0.320
Cap	"	0.133
Coupling	"	0.267
Union	"	0.267
15410.14 Brass I.p.s. Fittings		
Fittings, iron pipe size, 45 deg ell		
1/8"	EA.	0.320
1/4"	"	0.320
3/8"	"	0.364
1/2"	"	0.400
3/4"	"	0.421
1"	"	0.444
90 deg ell		
1/8"	EA.	0.320
1/4"	"	0.320
3/8"	"	0.364
1/2"	"	0.400
3/4"	"	0.421
1"	"	0.444
90 deg ell, reducing		
1/4" x 1/8"	EA.	0.320
3/8" x 1/8"	"	0.364
3/8" x 1/4"	"	0.364
1/2" x 1/4"	"	0.400
1/2" x 3/8"	"	0.400
3/4" x 1/2"	"	0.421
1" x 3/8"	"	0.444
1" x 1/2"	"	0.444
1" x 3/4"	"	0.444
Street ell, 45 deg		
1/2"	EA.	0.400
3/4"	"	0.421
90 deg		
1/8"	EA.	0.320
1/4"	"	0.320
3/8"	"	0.364
1/2"	"	0.400

Plumbing	UNIT	MAN/ HOURS
15410.14 Brass I.p.s. Fittings *(Cont.)*		
3/4"	EA.	0.421
1"	"	0.444
Tee, 1/8"	"	0.320
1/4"	"	0.320
3/8"	"	0.364
1/2"	"	0.400
3/4"	"	0.421
1"	"	0.444
Tee, reducing, 3/8" x		
1/4"	EA.	0.364
1/2"	"	0.364
1/2" x		
1/4"	EA.	0.400
3/8"	"	0.400
3/4"	"	0.400
3/4" x		
1/4"	EA.	0.421
1/2"	"	0.421
1"	"	0.421
1" x		
1/2"	EA.	0.444
3/4"	"	0.444
Tee, reducing		
1/2" x 3/8" x 1/2"	EA.	0.400
3/4" x 1/2" x 1/2"	"	0.421
3/4" x 1/2" x 3/4"	"	0.421
1" x 1/2" x 1/2"	"	0.444
1" x 1/2" x 3/4"	"	0.444
1" x 3/4" x 1/2"	"	0.444
1" x 3/4" x 3/4"	"	0.444
Union		
1/8"	EA.	0.320
1/4"	"	0.320
3/8"	"	0.364
1/2"	"	0.400
3/4"	"	0.421
1"	"	0.444
Brass face bushing		
3/8" x 1/4"	EA.	0.364
1/2" x 3/8"	"	0.400
3/4" x 1/2"	"	0.421
1" x 3/4"	"	0.444
Hex bushing, 1/4" x 1/8"	"	0.320
1/2" x		
1/4"	EA.	0.400
3/8"	"	0.400
5/8" x		
1/8"	EA.	0.400
1/4"	"	0.400
3/4" x		
1/8"	EA.	0.421
1/4"	"	0.421
3/8"	"	0.421

Plumbing	UNIT	MAN/ HOURS
15410.14 Brass I.p.s. Fittings *(Cont.)*		
1/2"	EA.	0.421
1" x		
1/4"	EA.	0.444
3/8"	"	0.444
1/2"	"	0.444
3/4"	"	0.444
Caps		
1/8"	EA.	0.320
1/4"	"	0.320
3/8"	"	0.364
1/2"	"	0.400
3/4"	"	0.421
1"	"	0.444
Couplings		
1/8"	EA.	0.320
1/4"	"	0.320
3/8"	"	0.364
1/2"	"	0.400
3/4"	"	0.421
1"	"	0.444
Couplings, reducing, 1/4" x 1/8"	"	0.320
3/8" x		
1/8"	EA.	0.364
1/4"	"	0.364
1/2" x		
1/8"	EA.	0.400
1/4"	"	0.400
3/8"	"	0.400
3/4" x		
1/4"	EA.	0.421
3/8"	"	0.421
1/2"	"	0.421
1" x		
1/2"	EA.	0.421
3/4"	"	0.421
Square head plug, solid		
1/8"	EA.	0.320
1/4"	"	0.320
3/8"	"	0.364
1/2"	"	0.400
3/4"	"	0.421
Cored		
1/2"	EA.	0.400
3/4"	"	0.421
1"	"	0.444
Countersunk		
1/2"	EA.	0.400
3/4"	"	0.421
Locknut		
3/4"	EA.	0.421
1"	"	0.444
Close standard red nipple, 1/8"	"	0.320
1/8" x		

Plumbing	UNIT	MAN/ HOURS
15410.14 Brass I.p.s. Fittings *(Cont.)*		
1-1/2"	EA.	0.320
2"	"	0.320
2-1/2"	"	0.320
3"	"	0.320
3-1/2"	"	0.320
4"	"	0.320
4-1/2"	"	0.320
5"	"	0.320
5-1/2"	"	0.320
6"	"	0.320
1/4" x close	"	0.320
1/4" x		
1-1/2"	EA.	0.320
2"	"	0.320
2-1/2"	"	0.320
3"	"	0.320
3-1/2"	"	0.320
4"	"	0.320
4-1/2"	"	0.320
5"	"	0.320
5-1/2"	"	0.320
6"	"	0.320
3/8" x close	"	0.364
3/8" x		
1-1/2"	EA.	0.364
2"	"	0.364
2-1/2"	"	0.364
3"	"	0.364
3-1/2"	"	0.364
4"	"	0.364
4-1/2"	"	0.364
5"	"	0.364
5-1/2"	"	0.364
6"	"	0.364
1/2" x close	"	0.400
1/2" x		
1-1/2"	EA.	0.400
2"	"	0.400
2-1/2"	"	0.400
3"	"	0.400
3-1/2"	"	0.400
4"	"	0.400
4-1/2"	"	0.400
5"	"	0.400
5-1/2"	"	0.400
6"	"	0.400
7-1/2"	"	0.400
8"	"	0.400
3/4" x close	"	0.421
3/4" x		
1-1/2"	EA.	0.421
2"	"	0.421
2-1/2"	"	0.421

Plumbing	UNIT	MAN/ HOURS
15410.14 Brass I.p.s. Fittings *(Cont.)*		
3"	EA.	0.421
3-1/2"	"	0.421
4"	"	0.421
4-1/2"	"	0.421
5"	"	0.421
5-1/2"	"	0.421
6"	"	0.421
1" x close	"	0.444
1" x		
2"	EA.	0.444
2-1/2"	"	0.444
3"	"	0.444
3-1/2"	"	0.444
4"	"	0.444
4-1/2"	"	0.444
5"	"	0.444
5-1/2"	"	0.444
6"	"	0.444
15410.15 Brass Fittings		
Compression fittings, union		
3/8"	EA.	0.133
1/2"	"	0.133
5/8"	"	0.133
Union elbow		
3/8"	EA.	0.133
1/2"	"	0.133
5/8"	"	0.133
Union tee		
3/8"	EA.	0.133
1/2"	"	0.133
5/8"	"	0.133
Male connector		
3/8"	EA.	0.133
1/2"	"	0.133
5/8"	"	0.133
Female connector		
3/8"	EA.	0.133
1/2"	"	0.133
5/8"	"	0.133
Brass flare fittings, union		
3/8"	EA.	0.129
1/2"	"	0.129
5/8"	"	0.129
90 deg elbow union		
3/8"	EA.	0.129
1/2"	"	0.129
5/8"	"	0.129
Three way tee		
3/8"	EA.	0.216
1/2"	"	0.216
5/8"	"	0.216
Cross		

Plumbing	UNIT	MAN/ HOURS
15410.15 Brass Fittings *(Cont.)*		
3/8"	EA.	0.286
1/2"	"	0.286
5/8"	"	0.286
Male connector, half union		
3/8"	EA.	0.129
1/2"	"	0.129
5/8"	"	0.129
Female connector, half union		
3/8"	EA.	0.129
1/2"	"	0.129
5/8"	"	0.129
Long forged nut		
3/8"	EA.	0.129
1/2"	"	0.129
5/8"	"	0.129
Short forged nut		
3/8"	EA.	0.129
1/2"	"	0.129
5/8"	"	0.129
Sleeve		
1/8"	EA.	0.160
1/4"	"	0.160
5/16"	"	0.160
3/8"	"	0.160
1/2"	"	0.160
5/8"	"	0.160
Tee		
1/4"	EA.	0.229
5/16"	"	0.229
Male tee		
5/16" x 1/8"	EA.	0.229
Female union		
1/8" x 1/8"	EA.	0.200
1/4" x 3/8"	"	0.200
3/8" x 1/4"	"	0.200
3/8" x 1/2"	"	0.200
5/8" x 1/2"	"	0.229
Male union, 1/4"		
1/4" x 1/4"	EA.	0.200
3/8"	"	0.200
1/2"	"	0.200
5/16" x		
1/8"	EA.	0.200
1/4"	"	0.200
3/8"	"	0.200
3/8" x		
1/8"	EA.	0.200
1/4"	"	0.200
1/2"	"	0.200
5/8" x		
3/8"	EA.	0.229
1/2"	"	0.229
Female elbow, 1/4" x 1/4"	"	0.229

Plumbing	UNIT	MAN/ HOURS
15410.15 Brass Fittings *(Cont.)*		
5/16" x		
1/8"	EA.	0.229
1/4"	"	0.229
3/8" x		
3/8"	EA.	0.229
1/2"	"	0.229
Male elbow, 1/8" x 1/8"	"	0.229
3/16" x 1/4"	"	0.229
1/4" x		
1/8"	EA.	0.229
1/4"	"	0.229
3/8"	"	0.229
5/16" x		
1/8"	EA.	0.229
1/4"	"	0.229
3/8"	"	0.229
3/8" x		
1/8"	EA.	0.229
1/4"	"	0.229
3/8"	"	0.229
1/2"	"	0.229
1/2" x		
1/4"	EA.	0.267
3/8"	"	0.267
1/2"	"	0.267
5/8" x		
3/8"	EA.	0.267
1/2"	"	0.267
3/4"	"	0.267
Union		
1/8"	EA.	0.229
3/16"	"	0.229
1/4"	"	0.229
5/16"	"	0.229
3/8"	"	0.229
Reducing union		
3/8" x 1/4"	EA.	0.267
5/8" x		
3/8"	EA.	0.267
1/2"	"	0.267
15410.17 Chrome Plated Fittings		
Fittings		
90 ell		
3/8"	EA.	0.200
1/2"	"	0.200
45 ell		
3/8"	EA.	0.200
1/2"	"	0.200
Tee		
3/8"	EA.	0.267
1/2"	"	0.267
Coupling		

Plumbing

15410.17 Chrome Plated Fittings (Cont.)

Plumbing	UNIT	MAN/ HOURS
3/8"	EA.	0.200
1/2"	"	0.200
Union		
3/8"	EA.	0.200
1/2"	"	0.200
Tee		
1/2" x 3/8" x 3/8"	EA.	0.267
1/2" x 3/8" x 1/2"	"	0.267

15410.30 PVC/CPVC Pipe

Plumbing	UNIT	MAN/ HOURS
PVC schedule 40		
1/2" pipe	L.F.	0.033
3/4" pipe	"	0.036
1" pipe	"	0.040
1-1/4" pipe	"	0.044
1-1/2" pipe	"	0.050
2" pipe	"	0.057
2-1/2" pipe	"	0.067
3" pipe	"	0.080
4" pipe	"	0.100
6" pipe	"	0.200
Fittings, 1/2"		
90 deg ell	EA.	0.100
45 deg ell	"	0.100
Tee	"	0.114
Reducing insert	"	0.133
Threaded	"	0.100
Male adapter	"	0.133
Female adapter	"	0.100
Union	"	0.160
Cap	"	0.133
Flange	"	0.160
3/4"		
90 deg elbow	EA.	0.133
45 deg elbow	"	0.133
Tee	"	0.160
Reducing insert	"	0.114
Threaded	"	0.133
1"		
90 deg elbow	EA.	0.160
45 deg elbow	"	0.160
Tee	"	0.178
Reducing insert	"	0.160
Threaded	"	0.178
Male adapter	"	0.200
Female adapter	"	0.200
Union	"	0.267
Cap	"	0.160
Flange	"	0.267
1-1/4"		
90 deg elbow	EA.	0.229
45 deg elbow	"	0.229
Tee	"	0.267

Plumbing

15410.30 PVC/CPVC Pipe (Cont.)

Plumbing	UNIT	MAN/ HOURS
Reducing insert	EA.	0.267
Threaded	"	0.267
Male adapter	"	0.267
Female adapter	"	0.267
Union	"	0.320
Cap	"	0.267
Flange	"	0.320
1-1/2"		
90 deg elbow	EA.	0.229
45 deg elbow	"	0.229
Tee	"	0.267
Reducing insert	"	0.267
Threaded	"	0.267
Male adapter	"	0.267
Female adapter	"	0.267
Union	"	0.400
Cap	"	0.267
Flange	"	0.400
2"		
90 deg elbow	EA.	0.267
45 deg elbow	"	0.267
Tee	"	0.320
Reducing insert	"	0.320
Threaded	"	0.320
Male adapter	"	0.320
Female adapter	"	0.320
Union	"	0.500
Cap	"	0.320
Flange	"	0.500
2-1/2"		
90 deg elbow	EA.	0.500
45 deg elbow	"	0.500
Tee	"	0.533
Reducing insert	"	0.533
Threaded	"	0.533
Male adapter	"	0.533
Female adapter	"	0.533
Union	"	0.667
Cap	"	0.500
Flange	"	0.667
3"		
90 deg elbow	EA.	0.667
45 deg elbow	"	0.667
Tee	"	0.727
Reducing insert	"	0.667
Threaded	"	0.667
Male adapter	"	0.667
Female adapter	"	0.667
Union	"	0.800
Cap	"	0.667
Flange	"	0.800
4"		
90 deg elbow	EA.	0.800

Plumbing	UNIT	MAN/HOURS
15410.30 PVC/CPVC Pipe *(Cont.)*		
45 deg elbow	EA.	0.800
Tee	"	0.889
Reducing insert	"	0.800
Threaded	"	0.800
Male adapter	"	0.800
Female adapter	"	0.800
Union	"	1.000
Cap	"	0.800
Flange	"	1.000
PVC schedule 80 pipe		
1-1/2" pipe	L.F.	0.050
2" pipe	"	0.057
3" pipe	"	0.080
4" pipe	"	0.100
Fittings, 1-1/2"		
90 deg elbow	EA.	0.267
45 deg elbow	"	0.267
Tee	"	0.400
Reducing insert	"	0.267
Threaded	"	0.267
Male adapter	"	0.267
Female adapter	"	0.267
Union	"	0.400
Cap	"	0.267
Flange	"	0.400
2"		
90 deg elbow	EA.	0.320
45 deg elbow	"	0.320
Tee	"	0.500
Reducing insert	"	0.320
Threaded	"	0.320
Male adapter	"	0.320
Female adapter	"	0.320
2-1/2"		
90 deg elbow	EA.	0.500
45 deg elbow	"	0.500
Tee	"	0.667
Reducing insert	"	0.500
Threaded	"	0.500
Male adapter	"	0.500
Female adapter	"	0.500
Union	"	0.667
Cap	"	0.500
Flange	"	0.667
3"		
90 deg elbow	EA.	0.667
45 deg elbow	"	0.667
Tee	"	0.800
Reducing insert	"	0.667
Threaded	"	0.667
Male adapter	"	0.667
Female adapter	"	0.667
Union	"	0.800

Plumbing	UNIT	MAN/HOURS
15410.30 PVC/CPVC Pipe *(Cont.)*		
Cap	EA.	0.667
Flange	"	0.800
4"		
90 deg elbow	EA.	0.800
45 deg elbow	"	0.800
Tee	"	1.000
Reducing insert	"	0.800
Threaded	"	0.800
Male adapter	"	0.800
Union	"	1.000
Cap	"	0.800
Flange	"	1.000
CPVC schedule 40		
1/2" pipe	L.F.	0.033
3/4" pipe	"	0.036
1" pipe	"	0.040
1-1/4" pipe	"	0.044
1-1/2" pipe	"	0.050
2" pipe	"	0.057
Fittings, CPVC, schedule 80		
1/2", 90 deg ell	EA.	0.080
Tee	"	0.133
3/4", 90 deg ell	"	0.080
Tee	"	0.133
1", 90 deg ell	"	0.089
Tee	"	0.145
1-1/4", 90 deg ell	"	0.089
Tee	"	0.145
1-1/2", 90 deg ell	"	0.160
Tee	"	0.200
2", 90 deg ell	"	0.160
Tee	"	0.200
15410.33 ABS DWV Pipe		
Schedule 40 ABS		
1-1/2" pipe	L.F.	0.040
2" pipe	"	0.044
3" pipe	"	0.057
4" pipe	"	0.080
6" pipe	"	0.100
Fittings		
1/8 bend		
1-1/2"	EA.	0.160
2"	"	0.200
3"	"	0.267
4"	"	0.320
6"	"	0.400
Tee, sanitary		
1-1/2"	EA.	0.267
2"	"	0.320
3"	"	0.400
4"	"	0.500
6"	"	0.667

Plumbing	UNIT	MAN/ HOURS
15410.33 ABS DWV Pipe *(Cont.)*		
Tee, sanitary reducing		
2 x 1-1/2 x 1-1/2	EA.	0.320
2 x 1-1/2 x 2	"	0.333
2 x 2 x 1-1/2	"	0.364
3 x 3 x 1-1/2	"	0.400
3 x 3 x 2	"	0.444
4 x 4 x 1-1/2	"	0.500
4 x 4 x 2	"	0.571
4 x 4 x 3	"	0.615
6 x 6 x 4	"	0.667
Wye		
1-1/2"	EA.	0.229
2"	"	0.320
3"	"	0.400
4"	"	0.500
6"	"	0.667
Reducer		
2 x 1-1/2	EA.	0.200
3 x 1-1/2	"	0.267
3 x 2	"	0.267
4 x 2	"	0.320
4 x 3	"	0.320
6 x 4	"	0.400
P-trap		
1-1/2"	EA.	0.267
2"	"	0.296
3"	"	0.348
4"	"	0.400
6"	"	0.500
Double sanitary, tee		
1-1/2"	EA.	0.320
2"	"	0.400
3"	"	0.500
4"	"	0.667
Long sweep, 1/4 bend		
1-1/2"	EA.	0.160
2"	"	0.200
3"	"	0.267
4"	"	0.400
Wye, standard		
1-1/2"	EA.	0.267
2"	"	0.320
3"	"	0.400
4"	"	0.500
Wye, reducing		
2 x 1-1/2 x 1-1/2	EA.	0.267
2 x 2 x 1-1/2	"	0.320
4 x 4 x 2	"	0.500
4 x 4 x 3	"	0.533
Double wye		
1-1/2"	EA.	0.320
2"	"	0.400
3"	"	0.500

Plumbing	UNIT	MAN/ HOURS
15410.33 ABS DWV Pipe *(Cont.)*		
4"	EA.	0.667
2 x 2 x 1-1/2 x 1-1/2	"	0.400
3 x 3 x 2 x 2	"	0.500
4 x 4 x 3 x 3	"	0.667
Combination wye and 1/8 bend		
1-1/2"	EA.	0.267
2"	"	0.320
3"	"	0.400
4"	"	0.500
2 x 2 x 1-1/2	"	0.320
3 x 3 x 1-1/2	"	0.400
3 x 3 x 2	"	0.400
4 x 4 x 2	"	0.500
4 x 4 x 3	"	0.500
15410.70 Stainless Steel Pipe		
Stainless steel, schedule 40, threaded		
1/2" pipe	L.F.	0.114
3/4" pipe	"	0.118
1" pipe	"	0.123
Fittings, 1/2"		
90 deg ell	EA.	1.000
45 deg ell	"	1.000
Tee	"	1.333
Cap	"	0.500
Reducer	"	0.667
Union	"	1.000
Flange	"	1.000
3/4"		
90 deg ell	EA.	1.000
45 deg ell	"	1.000
Tee	"	1.333
Cap	"	0.500
Reducer	"	0.667
Union	"	1.000
Flange	"	1.000
1"		
90 deg ell	EA.	1.000
45 deg ell	"	1.000
Tee	"	1.333
Cap	"	0.500
Reducer	"	1.000
Union	"	1.000
Flange	"	1.000
Type 304 tubing		
.035 wall		
1/4"	L.F.	0.044
3/8"	"	0.050
1/2"	"	0.057
5/8"	"	0.067
3/4"	"	0.080
7/8"	"	0.089
1"	"	0.100

Plumbing	UNIT	MAN/ HOURS
15410.70 Stainless Steel Pipe *(Cont.)*		
.049 wall		
1/4"	L.F.	0.047
3/8"	"	0.053
1/2"	"	0.062
5/8"	"	0.073
3/4"	"	0.089
7/8"	"	0.100
1"	"	0.114
.065 wall		
1/4"	L.F.	0.053
3/8"	"	0.067
1/2"	"	0.073
5/8"	"	0.089
3/4"	"	0.114
7/8"	"	0.133
1"	"	0.160
Type 316 tubing		
.035 wall		
1/4"	L.F.	0.044
3/8"	"	0.050
1/2"	"	0.057
5/8"	"	0.067
3/4"	"	0.080
7/8"	"	0.089
1"	"	0.100
.049 wall		
1/4"	L.F.	0.053
3/8"	"	0.067
1/2"	"	0.073
5/8"	"	0.089
3/4"	"	0.114
7/8"	"	0.133
1"	"	0.160
.065 wall		
1/4"	L.F.	0.053
3/8"	"	0.067
1/2"	"	0.073
5/8"	"	0.089
3/4"	"	0.114
7/8"	"	0.133
1"	"	0.160
Fittings, 1/4"		
90 deg elbow	EA.	0.160
Union tee	"	0.267
Union	"	0.267
Male connector	"	0.200
3/8"		
90 deg elbow	EA.	0.200
Union tee	"	0.308
Union	"	0.308
Male connector	"	0.200
1/2"		
90 deg elbow	EA.	0.211

Plumbing	UNIT	MAN/ HOURS
15410.70 Stainless Steel Pipe *(Cont.)*		
Union tee	EA.	0.333
Union	"	0.333
Male connector	"	0.200
5/8"		
90 deg elbow	EA.	0.267
Union tee	"	0.400
Union	"	0.400
Male connector	"	0.267
3/4"		
90 deg elbow	EA.	0.267
Union tee	"	0.400
Union	"	0.400
Male connector	"	0.267
7/8"		
90 deg elbow	EA.	0.286
Union tee	"	0.444
Union	"	0.444
Male connector	"	0.286
1"		
90 deg elbow	EA.	0.364
Union tee	"	0.500
Union	"	0.500
Male connector	"	0.400
Type 316 valves		
Gate valves		
1/4"	EA.	0.267
3/8"	"	0.320
1/2"	"	0.348
3/4"	"	0.400
1"	"	0.533
Globe valves		
1/4"	EA.	0.267
3/8"	"	0.320
1/2"	"	0.348
3/4"	"	0.400
1"	"	0.533
Check valves		
1/4"	EA.	0.267
3/8"	"	0.320
1/2"	"	0.348
3/4"	"	0.400
1"	"	0.533
15410.80 Steel Pipe		
Black steel, extra heavy pipe, threaded		
1/2" pipe	L.F.	0.032
3/4" pipe	"	0.032
1" pipe	"	0.040
Fittings, malleable iron, threaded, 1/2" pipe		
90 deg ell	EA.	0.267
45 deg ell	"	0.267
Tee	"	0.400
Reducing tee	"	0.400

Plumbing	UNIT	MAN/ HOURS
15410.80 Steel Pipe *(Cont.)*		
Cap	EA.	0.160
Coupling	"	0.320
Union	"	0.267
Nipple, 4" long	"	0.267
3/4" pipe		
90 deg ell	EA.	0.267
45 deg ell	"	0.400
Tee	"	0.400
Reducing tee	"	0.267
Cap	"	0.160
Coupling	"	0.267
Union	"	0.267
Nipple, 4" long	"	0.267
1" pipe		
90 deg ell	EA.	0.320
45 deg ell	"	0.320
Tee	"	0.444
Reducing tee	"	0.444
Cap	"	0.160
Coupling	"	0.320
Union	"	0.320
Nipple, 4" long	"	0.320
Cast iron fittings		
1/2" pipe		
90 deg. ell	EA.	0.267
45 deg. ell	"	0.267
Tee	"	0.400
Reducing tee	"	0.400
3/4" pipe		
90 deg. ell	EA.	0.267
45 deg. ell	"	0.267
Tee	"	0.400
Reducing tee	"	0.400
1" pipe		
90 deg. ell	EA.	0.320
45 deg. ell	"	0.320
Tee	"	0.444
Reducing tee	"	0.444
15410.82 Galvanized Steel Pipe		
Galvanized pipe		
1/2" pipe	L.F.	0.080
3/4" pipe	"	0.100
1" pipe	"	0.114
90 degree ell, 150 lb malleable iron, galvanized		
1/2"	EA.	0.160
3/4"	"	0.200
1"	"	0.211
45 degree ell, 150 lb m.i., galv.		
1/2"	EA.	0.160
3/4"	"	0.200
1"	"	0.211
Tees, straight, 150 lb m.i., galv.		

Plumbing	UNIT	MAN/ HOURS
15410.82 Galvanized Steel Pipe *(Cont.)*		
1/2"	EA.	0.200
3/4"	"	0.229
1"	"	0.267
Tees, reducing, out, 150 lb m.i., galv.		
1/2"	EA.	0.200
3/4"	"	0.229
1"	"	0.267
Couplings, straight, 150 lb m.i., galv.		
1/2"	EA.	0.160
3/4"	"	0.178
1"	"	0.200
Couplings, reducing, 150 lb m.i., galv		
1/2"	EA.	0.160
3/4"	"	0.178
1"	"	0.200
Caps, 150 lb m.i., galv.		
1/2"	EA.	0.080
3/4"	"	0.084
1"	"	0.089
Unions, 150 lb m.i., galv.		
1/2"	EA.	0.200
3/4"	"	0.229
1"	"	0.267
Nipples, galvanized steel, 4" long		
1/2"	EA.	0.100
3/4"	"	0.107
1"	"	0.114
90 degree reducing ell, 150 lb m.i., galv.		
3/4" x 1/2"	EA.	0.160
1" x 3/4"	"	0.178
Square head plug (C.I.)		
1/2"	EA.	0.089
3/4"	"	0.100
1"	"	0.107
15430.23 Cleanouts		
Cleanout, wall		
2"	EA.	0.533
3"	"	0.533
4"	"	0.667
6"	"	0.800
Floor		
2"	EA.	0.667
3"	"	0.667
4"	"	0.800
6"	"	1.000
15430.25 Hose Bibbs		
Hose bibb		
1/2"	EA.	0.267
3/4"	"	0.267

Plumbing	UNIT	MAN/ HOURS
15430.60 Valves		
Gate valve, 125 lb, bronze, soldered		
1/2"	EA.	0.200
3/4"	"	0.200
1"	"	0.267
Threaded		
1/4", 125 lb	EA.	0.320
1/2"		
125 lb	EA.	0.320
150 lb	"	0.320
300 lb	"	0.320
3/4"		
125 lb	EA.	0.320
150 lb	"	0.320
300 lb	"	0.320
1"		
125 lb	EA.	0.320
150 lb	"	0.320
300 lb	"	0.400
Check valve, bronze, soldered, 125 lb		
1/2"	EA.	0.200
3/4"	"	0.200
1"	"	0.267
Threaded		
1/2"		
125 lb	EA.	0.267
150 lb	"	0.267
200 lb	"	0.267
3/4"		
125 lb	EA.	0.320
150 lb	"	0.320
200 lb	"	0.320
1"		
125 lb	EA.	0.400
150 lb	"	0.400
200 lb	"	0.400
Vertical check valve, bronze, 125 lb, threaded		
1/2"	EA.	0.320
3/4"	"	0.364
1"	"	0.400
Globe valve, bronze, soldered, 125 lb		
1/2"	EA.	0.229
3/4"	"	0.250
1"	"	0.267
Threaded		
1/2"		
125 lb	EA.	0.267
150 lb	"	0.267
300 lb	"	0.267
3/4"		
125 lb	EA.	0.320
150 lb	"	0.320
300 lb	"	0.320
1"		

Plumbing	UNIT	MAN/ HOURS
15430.60 Valves (Cont.)		
125 lb	EA.	0.400
150 lb	"	0.400
300 lb	"	0.400
Ball valve, bronze, 250 lb, threaded		
1/2"	EA.	0.320
3/4"	"	0.320
1"	"	0.400
Angle valve, bronze, 150 lb, threaded		
1/2"	EA.	0.286
3/4"	"	0.320
1"	"	0.320
Balancing valve, meter connections, circuit setter		
1/2"	EA.	0.320
3/4"	"	0.364
1"	"	0.400
Balancing valve, straight type		
1/2"	EA.	0.320
3/4"	"	0.320
Angle type		
1/2"	EA.	0.320
3/4"	"	0.320
Square head cock, 125 lb, bronze body		
1/2"	EA.	0.267
3/4"	"	0.320
1"	"	0.364
Radiator temp control valve, with control and sensor		
1/2" valve	EA.	0.500
1" valve	"	0.500
Pressure relief valve, 1/2", bronze		
Low pressure	EA.	0.320
High pressure	"	0.320
Pressure and temperature relief valve		
Bronze, 3/4"	EA.	0.320
Cast iron, 3/4"		
High pressure	EA.	0.320
Temperature relief	"	0.320
Pressure & temp relief valve	"	0.320
Pressure reducing valve, bronze, threaded, 250 lb		
1/2"	EA.	0.500
3/4"	"	0.500
1"	"	0.500
Solar water temperature regulating valve		
3/4"	EA.	0.667
Tempering valve, threaded		
3/4"	EA.	0.267
1"	"	0.320
Thermostatic mixing valve, threaded		
1/2"	EA.	0.286
3/4"	"	0.320
1"	"	0.348
Sweat connection		
1/2"	EA.	0.286
3/4"	"	0.320

Plumbing	UNIT	MAN/ HOURS
15430.60 Valves *(Cont.)*		
Mixing valve, sweat connection		
1/2"	EA.	0.286
3/4"	"	0.320
Liquid level gauge, aluminum body		
3/4"	EA.	0.320
4125 psi, PVC body		
3/4"	EA.	0.320
150 psi, crs body		
3/4"	EA.	0.320
1"	"	0.320
175 psi, bronze body, 1/2"	"	0.286
15430.65 Vacuum Breakers		
Vacuum breaker, atmospheric, threaded connection		
3/4"	EA.	0.320
1"	"	0.320
Anti-siphon, brass		
3/4"	EA.	0.320
1"	"	0.320
15430.68 Strainers		
Strainer, Y pattern, 125 psi, cast iron body, threaded		
3/4"	EA.	0.286
1"	"	0.320
250 psi, brass body, threaded		
3/4"	EA.	0.320
1"	"	0.320
Cast iron body, threaded		
3/4"	EA.	0.320
1"	"	0.320
15430.70 Drains, Roof & Floor		
Floor drain, cast iron, with cast iron top		
2"	EA.	0.667
3"	"	0.667
4"	"	0.667
6"	"	0.800
Roof drain, cast iron		
2"	EA.	0.667
3"	"	0.667
4"	"	0.667
5"	"	0.800
6"	"	0.800
15430.80 Traps		
Bucket trap, threaded		
3/4"	EA.	0.500
1"	"	0.533
Inverted bucket steam trap, threaded		
3/4"	EA.	0.500
1"	"	0.500
With stainless interior		
1/2"	EA.	0.500

Plumbing	UNIT	MAN/ HOURS
15430.80 Traps *(Cont.)*		
3/4"	EA.	0.500
1"	"	0.500
Brass interior		
3/4"	EA.	0.500
1"	"	0.533
Cast steel body, threaded, high temperature		
3/4"	EA.	0.500
Float trap, 15 psi		
3/4"	EA.	0.500
Float and thermostatic trap, 15 psi		
3/4"	EA.	0.500
Steam trap, cast iron body, threaded, 125 psi		
3/4"	EA.	0.500
Thermostatic trap, low pressure, angle type, 25 psi		
1/2"	EA.	0.500
3/4"	"	0.500
1"	"	0.533
50 psi		
1/2"	EA.	0.500
3/4"	"	0.500
1"	"	0.533
Cast iron body, threaded, 125 psi		
3/4"	EA.	0.500
1"	"	0.571

Plumbing Fixtures	UNIT	MAN/ HOURS
15440.10 Baths		
Bath tub, 5' long		
Minimum	EA.	2.667
Average	"	4.000
Maximum	"	8.000
6' long		
Minimum	EA.	2.667
Average	"	4.000
Maximum	"	8.000
Square tub, whirlpool, 4'x4'		
Minimum	EA.	4.000
Average	"	8.000
Maximum	"	10.000
5'x5'		
Minimum	EA.	4.000
Average	"	8.000
Maximum	"	10.000
6'x6'		
Minimum	EA.	4.000

Plumbing Fixtures	UNIT	MAN/ HOURS
15440.10 Baths (Cont.)		
Average	EA.	8.000
Maximum	"	10.000
For trim and rough-in		
Minimum	EA.	2.667
Average	"	4.000
Maximum	"	8.000
15440.12 Disposals & Accessories		
Continuous feed		
Minimum	EA.	1.600
Average	"	2.000
Maximum	"	2.667
Batch feed, 1/2 hp		
Minimum	EA.	1.600
Average	"	2.000
Maximum	"	2.667
Hot water dispenser		
Minimum	EA.	1.600
Average	"	2.000
Maximum	"	2.667
Epoxy finish faucet	"	1.600
Lock stop assembly	"	1.000
Mounting gasket	"	0.667
Tailpipe gasket	"	0.667
Stopper assembly	"	0.800
Switch assembly, on/off	"	1.333
Tailpipe gasket washer	"	0.400
Stop gasket	"	0.444
Tailpipe flange	"	0.400
Tailpipe	"	0.500
15440.15 Faucets		
Kitchen		
Minimum	EA.	1.333
Average	"	1.600
Maximum	"	2.000
Bath		
Minimum	EA.	1.333
Average	"	1.600
Maximum	"	2.000
Lavatory, domestic		
Minimum	EA.	1.333
Average	"	1.600
Maximum	"	2.000
Hospital, patient rooms		
Minimum	EA.	2.000
Average	"	2.667
Maximum	"	4.000
Washroom		
Minimum	EA.	1.333
Average	"	1.600
Maximum	"	2.000
Handicapped		

Plumbing Fixtures	UNIT	MAN/ HOURS
15440.15 Faucets (Cont.)		
Minimum	EA.	1.600
Average	"	2.000
Maximum	"	2.667
Shower		
Minimum	EA.	1.333
Average	"	1.600
Maximum	"	2.000
For trim and rough-in		
Minimum	EA.	1.600
Average	"	2.000
Maximum	"	4.000
15440.18 Hydrants		
Wall hydrant		
8" thick	EA.	1.333
12" thick	"	1.600
18" thick	"	1.778
24" thick	"	2.000
15440.20 Lavatories		
Lavatory, counter top, porcelain enamel on cast iron		
Minimum	EA.	1.600
Average	"	2.000
Maximum	"	2.667
Wall hung, china		
Minimum	EA.	1.600
Average	"	2.000
Maximum	"	2.667
Handicapped		
Minimum	EA.	2.000
Average	"	2.667
Maximum	"	4.000
For trim and rough-in		
Minimum	EA.	2.000
Average	"	2.667
Maximum	"	4.000
15440.30 Showers		
Shower, fiberglass, 36"x34"x84"		
Minimum	EA.	5.714
Average	"	8.000
Maximum	"	8.000
Steel, 1 piece, 36"x36"		
Minimum	EA.	5.714
Average	"	8.000
Maximum	"	8.000
Receptor, molded stone, 36"x36"		
Minimum	EA.	2.667
Average	"	4.000
Maximum	"	6.667
For trim and rough-in		
Minimum	EA.	3.636
Average	"	4.444

Plumbing Fixtures	UNIT	MAN/ HOURS
15440.30 Showers *(Cont.)*		
Maximum	EA.	8.000
15440.40 Sinks		
Service sink, 24"x29"		
Minimum	EA.	2.000
Average	"	2.667
Maximum	"	4.000
Kitchen sink, single, stainless steel, single bowl		
Minimum	EA.	1.600
Average	"	2.000
Maximum	"	2.667
Double bowl		
Minimum	EA.	2.000
Average	"	2.667
Maximum	"	4.000
Porcelain enamel, cast iron, single bowl		
Minimum	EA.	1.600
Average	"	2.000
Maximum	"	2.667
Double bowl		
Minimum	EA.	2.000
Average	"	2.667
Maximum	"	4.000
Mop sink, 24"x36"x10"		
Minimum	EA.	1.600
Average	"	2.000
Maximum	"	2.667
Washing machine box		
Minimum	EA.	2.000
Average	"	2.667
Maximum	"	4.000
For trim and rough-in		
Minimum	EA.	2.667
Average	"	4.000
Maximum	"	5.333
15440.50 Urinals		
Wall mounted		
Minimum	EA.	2.000
Average	"	2.667
Maximum	"	4.000
For trim and rough-in		
Minimum	EA.	2.000
Average	"	4.000
Maximum	"	5.333
15440.60 Water Closets		
Water closet flush tank, floor mounted		
Minimum	EA.	2.000
Average	"	2.667
Maximum	"	4.000
Handicapped		
Minimum	EA.	2.667

Plumbing Fixtures	UNIT	MAN/ HOURS
15440.60 Water Closets *(Cont.)*		
Average	EA.	4.000
Maximum	"	8.000
Bowl, with flush valve, floor mounted		
Minimum	EA.	2.000
Average	"	2.667
Maximum	"	4.000
Wall mounted		
Minimum	EA.	2.000
Average	"	2.667
Maximum	"	4.000
For trim and rough-in		
Minimum	EA.	2.000
Average	"	2.667
Maximum	"	4.000
15440.70 Water Heaters		
Water heater, electric		
6 gal	EA.	1.333
10 gal	"	1.333
15 gal	"	1.333
20 gal	"	1.600
30 gal	"	1.600
40 gal	"	1.600
52 gal	"	2.000
66 gal	"	2.000
80 gal	"	2.000
100 gal	"	2.667
120 gal	"	2.667
Oil fired		
20 gal	EA.	4.000
50 gal	"	5.714
15440.95 Fixture Carriers		
Lavatory, wall carrier		
Minimum	EA.	0.800
Average	"	1.000
Maximum	"	1.333
Sink, industrial, wall carrier		
Minimum	EA.	0.800
Average	"	1.000
Maximum	"	1.333
Toilets, water closets, wall carrier		
Minimum	EA.	0.800
Average	"	1.000
Maximum	"	1.333
Floor support		
Minimum	EA.	0.667
Average	"	0.800
Maximum	"	1.000
Urinals, wall carrier		
Minimum	EA.	0.800
Average	"	1.000
Maximum	"	1.333

Plumbing Fixtures

15450.40 Storage Tanks	UNIT	MAN/ HOURS
Hot water storage tank, cement lined		
10 gallon	EA.	2.667
70 gallon	"	4.000

Heating & Ventilating

15555.10 Boilers	UNIT	MAN/ HOURS
Cast iron, gas fired, hot water		
115 mbh	EA.	20.000
175 mbh	"	21.818
235 mbh	"	24.000
Steam		
115 mbh	EA.	20.000
175 mbh	"	21.818
235 mbh	"	24.000
Electric, hot water		
115 mbh	EA.	12.000
175 mbh	"	12.000
235 mbh	"	12.000
Steam		
115 mbh	EA.	12.000
175 mbh	"	12.000
235 mbh	"	12.000
Oil fired, hot water		
115 mbh	EA.	16.000
175 mbh	"	18.462
235 mbh	"	21.818
Steam		
115 mbh	EA.	16.000
175 mbh	"	18.462
235 mbh	"	21.818

15610.10 Furnaces	UNIT	MAN/ HOURS
Electric, hot air		
40 mbh	EA.	4.000
60 mbh	"	4.211
80 mbh	"	4.444
100 mbh	"	4.706
125 mbh	"	4.848
160 mbh	"	5.000
200 mbh	"	5.161
Gas fired hot air		
40 mbh	EA.	4.000
60 mbh	"	4.211
80 mbh	"	4.444
100 mbh	"	4.706

Heating & Ventilating

15610.10 Furnaces (Cont.)	UNIT	MAN/ HOURS
125 mbh	EA.	4.848
160 mbh	"	5.000
200 mbh	"	5.161
Oil fired hot air		
40 mbh	EA.	4.000
60 mbh	"	4.211
80 mbh	"	4.444
100 mbh	"	4.706
125 mbh	"	4.848
160 mbh	"	5.000
200 mbh	"	5.161

Refrigeration

15670.10 Condensing Units	UNIT	MAN/ HOURS
Air cooled condenser, single circuit		
3 ton	EA.	1.333
5 ton	"	1.333
With low ambient dampers		
3 ton	EA.	2.000
5 ton	"	2.000

15780.20 Rooftop Units	UNIT	MAN/ HOURS
Packaged, single zone rooftop unit, with roof curb		
2 ton	EA.	8.000
3 ton	"	8.000
4 ton	"	10.000
5 ton	"	13.333

15830.10 Radiation Units	UNIT	MAN/ HOURS
Baseboard radiation unit		
1.7 mbh/lf	L.F.	0.320
2.1 mbh/lf	"	0.400
Enclosure only		
Two tier	L.F.	0.133
Three tier	"	0.133
Copper element only, 3/4" dia.		
Two tier	L.F.	0.200
Three tier	"	0.267
Fin-tube, 16 ga, sloping cover, 1-1/4" steel		
One tier	L.F.	0.267
Two tier	"	0.320
2" steel		
Two tier	L.F.	0.320
Three tier	"	0.400
1-1/4" copper		

Refrigeration

15830.10 Radiation Units *(Cont.)*

Refrigeration	UNIT	MAN/ HOURS
Two tier	L.F.	0.267
18 ga flat cover, 1-1/4" steel		
One tier	L.F.	0.267
Two tier	"	0.320
Three tier	"	0.400
2" steel		
One tier	L.F.	0.267
Two tier	"	0.320
Three tier	"	0.400
1-1/4" copper		
One tier	L.F.	0.267
Two tier	"	0.320
Three tier	"	0.400

15830.20 Fan Coil Units

Description	UNIT	MAN/ HOURS
Fan coil unit, 2 pipe, complete		
200 cfm ceiling hung	EA.	2.667

15830.70 Unit Heaters

Description	UNIT	MAN/ HOURS
Steam unit heater, horizontal		
12,500 btuh, 200 cfm	EA.	1.333
17,000 btuh, 300 cfm	"	1.333
Gas unit heater, horizontal		
27,400 btuh	EA.	3.200
38,000 btuh	"	3.200
Hot water unit heater, horizontal		
12,500 btuh, 200 cfm	EA.	1.333
17,000 btuh, 300 cfm	"	1.333
25,000 btuh, 500 cfm	"	1.333
30,000 btuh, 700 cfm	"	1.333
Cabinet unit heaters, ceiling, exposed, hot water		
200 cfm	EA.	2.667
300 cfm	"	3.200
400 cfm	"	3.810

Air Handling

15855.10 Air Handling Units

Air Handling	UNIT	MAN/ HOURS
Air handling unit, medium pressure, single zone		
1500 cfm	EA.	5.000
3000 cfm	"	8.889
Rooftop air handling units		
4950 cfm	EA.	8.889
7370 cfm	"	11.429

Air Handling

15870.20 Exhaust Fans

Air Handling	UNIT	MAN/ HOURS
Belt drive roof exhaust fans		
640 cfm, 2618 fpm	EA.	1.000
940 cfm, 2604 fpm	"	1.000

Air Distribution

15890.10 Metal Ductwork

Air Distribution	UNIT	MAN/ HOURS
Rectangular duct		
Galvanized steel		
Minimum	Lb.	0.073
Average	"	0.089
Maximum	"	0.133
Aluminum		
Minimum	Lb.	0.160
Average	"	0.200
Maximum	"	0.267
Fittings		
Minimum	EA.	0.267
Average	"	0.400
Maximum	"	0.800

15890.30 Flexible Ductwork

Description	UNIT	MAN/ HOURS
Flexible duct, 1.25" fiberglass		
5" dia.	L.F.	0.040
6" dia.	"	0.044
7" dia.	"	0.047
8" dia.	"	0.050
10" dia.	"	0.057
12" dia.	"	0.062
Flexible duct connector, 3" wide fabric	"	0.133

15910.10 Dampers

Description	UNIT	MAN/ HOURS
Horizontal parallel aluminum backdraft damper		
12" x 12"	EA.	0.200
16" x 16"	"	0.229
20" x 20"	"	0.286
24" x 24"	"	0.400

15940.10 Diffusers

Description	UNIT	MAN/ HOURS
Ceiling diffusers, round, baked enamel finish		
6" dia.	EA.	0.267
8" dia.	"	0.333
10" dia.	"	0.333
12" dia.	"	0.333

Air Distribution	UNIT	MAN/ HOURS
15940.10 Diffusers *(Cont.)*		
Rectangular		
6x6"	EA.	0.267
9x9"	"	0.400
12x12"	"	0.400
15x15"	"	0.400
18x18"	"	0.400
15940.40 Registers And Grilles		
Lay in flush mounted, perforated face, return		
6x6/24x24	EA.	0.320
8x8/24x24	"	0.320
9x9/24x24	"	0.320
10x10/24x24	"	0.320
12x12/24x24	"	0.320
Rectangular, ceiling return, single deflection		
10x10	EA.	0.400
12x12	"	0.400
14x14	"	0.400
16x8	"	0.400
16x16	"	0.400
18x8	"	0.400
20x20	"	0.400
24x12	"	0.400
24x18	"	0.400
36x24	"	0.444
36x30	"	0.444
Wall, return air register		
12x12	EA.	0.200
16x16	"	0.200
18x18	"	0.200
20x20	"	0.200
24x24	"	0.200
Ceiling, return air grille		
6x6	EA.	0.267
8x8	"	0.320
10x10	"	0.320
Ceiling, exhaust grille, aluminum egg crate		
6x6	EA.	0.267
8x8	"	0.320
10x10	"	0.320
12x12	"	0.400
14x14	"	0.400
16x16	"	0.400
18x18	"	0.400

Basic Materials

16050.30 Bus Duct

Basic Materials	UNIT	MAN/ HOURS
Bus duct, 100a, plug-in		
10', 600v	EA.	2.759
With ground	"	4.211
10', 277/480v	"	2.759
With ground	"	4.211
Circuit breakers, with enclosure		
1 pole		
15a-60a	EA.	1.000
70a-100a	"	1.250
2 pole		
15a-60a	EA.	1.100
70a-100a	"	1.301
Circuit breaker, adapter cubicle		
225a	EA.	1.509
400a	"	1.600
Fusible switches, 240v, 3 phase		
30a	EA.	1.000
60a	"	1.250
100a	"	1.509
200a	"	2.105

16110.20 Conduit Specialties

Basic Materials	UNIT	MAN/ HOURS
Rod beam clamp, 1/2"	EA.	0.050
Hanger rod		
3/8"	L.F.	0.040
1/2"	"	0.050
All thread rod		
1/4"	L.F.	0.030
3/8"	"	0.040
1/2"	"	0.050
5/8"	"	0.080
Hanger channel, 1-1/2"		
No holes	EA.	0.030
Holes	"	0.030
Channel strap		
1/2"	EA.	0.050
3/4"	"	0.050
Conduit penetrations, roof and wall, 8" thick		
1/2"	EA.	0.615
3/4"	"	0.615
1"	"	0.800
Threaded rod couplings		
1/4"	EA.	0.050
3/8"	"	0.050
1/2"	"	0.050
5/8"	"	0.050
3/4"	"	0.050
Hex nuts		
1/4"	EA.	0.050
3/8"	"	0.050
1/2"	"	0.050
5/8"	"	0.050
3/4"	"	0.050

16110.20 Conduit Specialties *(Cont.)*

Basic Materials	UNIT	MAN/ HOURS
Square nuts		
1/4"	EA.	0.050
3/8"	"	0.050
3/8"	"	0.050
5/8"	"	0.050
3/4"	"	0.050

16110.21 Aluminum Conduit

Basic Materials	UNIT	MAN/ HOURS
Aluminum conduit		
1/2"	L.F.	0.030
3/4"	"	0.040
1"	"	0.050
90 deg. elbow		
1/2"	EA.	0.190
3/4"	"	0.250
1"	"	0.308
Coupling		
1/2"	EA.	0.050
3/4"	"	0.059
1"	"	0.080

16110.22 EMT Conduit

Basic Materials	UNIT	MAN/ HOURS
EMT conduit		
1/2"	L.F.	0.030
3/4"	"	0.040
1"	"	0.050
90 deg. elbow		
1/2"	EA.	0.089
3/4"	"	0.100
1"	"	0.107
Connector, steel compression		
1/2"	EA.	0.089
3/4"	"	0.089
1"	"	0.089
Coupling, steel, compression		
1/2"	EA.	0.059
3/4"	"	0.059
1"	"	0.059
1 hole strap, steel		
1/2"	EA.	0.040
3/4"	"	0.040
1"	"	0.040
Connector, steel set screw		
1/2"	EA.	0.070
3/4"	"	0.070
1"	"	0.070
Insulated throat		
1/2"	EA.	0.070
3/4"	"	0.070
1"	"	0.070
Connector, die cast set screw		
1/2"	EA.	0.059
3/4"	"	0.059

Basic Materials	UNIT	MAN/ HOURS
16110.22 EMT Conduit *(Cont.)*		
1"	EA.	0.059
Insulated throat		
1/2"	EA.	0.059
3/4"	"	0.059
1"	"	0.059
Coupling, steel set screw		
1/2"	EA.	0.040
3/4"	"	0.040
1"	"	0.040
Diecast set screw		
1/2"	EA.	0.040
3/4"	"	0.040
1"	"	0.040
1 hole malleable straps		
1/2"	EA.	0.040
3/4"	"	0.040
1"	"	0.040
EMT to rigid compression coupling		
1/2"	EA.	0.100
3/4"	"	0.100
1"	"	0.150
Set screw couplings		
1/2"	EA.	0.100
3/4"	"	0.100
1"	"	0.145
Set screw offset connectors		
1/2"	EA.	0.100
3/4"	"	0.100
1"	"	0.145
Compression offset connectors		
1/2"	EA.	0.100
3/4"	"	0.100
1"	"	0.145
Type "LB" set screw condulets		
1/2"	EA.	0.229
3/4"	"	0.296
1"	"	0.381
Type "T" set screw condulets		
1/2"	EA.	0.296
3/4"	"	0.400
1"	"	0.444
Type "C" set screw condulets		
1/2"	EA.	0.250
3/4"	"	0.296
1"	"	0.381
Type "LL" set screw condulets		
1/2"	EA.	0.250
3/4"	"	0.296
1"	"	0.381
Type "LR" set screw condulets		
1/2"	EA.	0.250
3/4"	"	0.296
1"	"	0.381

Basic Materials	UNIT	MAN/ HOURS
16110.22 EMT Conduit *(Cont.)*		
Type "LB" compression condulets		
1/2"	EA.	0.296
3/4"	"	0.500
1"	"	0.500
Type "T" compression condulets		
1/2"	EA.	0.400
3/4"	"	0.444
1"	"	0.615
Condulet covers		
1/2"	EA.	0.123
3/4"	"	0.123
1"	"	0.123
Clamp type entrance caps		
1/2"	EA.	0.250
3/4"	"	0.296
1"	"	0.400
Slip fitter type entrance caps		
1/2"	EA.	0.250
3/4"	"	0.296
1"	"	0.400
16110.23 Flexible Conduit		
Flexible conduit, steel		
3/8"	L.F.	0.030
1/2	"	0.030
3/4"	"	0.040
1"	"	0.040
Flexible conduit, liquid tight		
3/8"	L.F.	0.030
1/2"	"	0.030
3/4"	"	0.040
1"	"	0.040
Connector, straight		
3/8"	EA.	0.080
1/2"	"	0.080
3/4"	"	0.089
1"	"	0.100
Straight insulated throat connectors		
3/8"	EA.	0.123
1/2"	"	0.123
3/4"	"	0.145
1"	"	0.145
90 deg connectors		
3/8"	EA.	0.148
1/2"	"	0.148
3/4"	"	0.170
1"	"	0.182
90 degree insulated throat connectors		
3/8"	EA.	0.145
1/2"	"	0.145
3/4"	"	0.170
1"	"	0.178
Flexible aluminum conduit		

Basic Materials	UNIT	MAN/ HOURS
16110.23 Flexible Conduit *(Cont.)*		
3/8"	L.F.	0.030
1/2"	"	0.030
3/4"	"	0.040
1"	"	0.040
Connector, straight		
3/8"	EA.	0.100
1/2"	"	0.100
3/4"	"	0.107
1"	"	0.123
Straight insulated throat connectors		
3/8"	EA.	0.089
1/2"	"	0.089
3/4"	"	0.089
1"	"	0.100
90 deg connectors		
3/8"	EA.	0.145
1/2"	"	0.145
3/4"	"	0.145
1"	"	0.170
90 deg insulated throat connectors		
3/8"	EA.	0.145
1/2"	"	0.145
3/4"	"	0.145
1"	"	0.170
16110.24 Galvanized Conduit		
Galvanized rigid steel conduit		
1/2"	L.F.	0.040
3/4"	"	0.050
1"	"	0.059
1-1/4"	"	0.080
1-1/2"	"	0.089
2"	"	0.100
90 degree ell		
1/2"	EA.	0.250
3/4"	"	0.308
1"	"	0.381
1-1/4"	"	0.444
1-1/2"	"	0.500
2"	"	0.533
Couplings, with set screws		
1/2"	EA.	0.050
3/4"	"	0.059
1"	"	0.080
1-1/4"	"	0.100
1-1/2"	"	0.123
2"	"	0.145
Split couplings		
1/2"	EA.	0.190
3/4"	"	0.250
1"	"	0.276
1-1/4"	"	0.308
1-1/2"	"	0.381

Basic Materials	UNIT	MAN/ HOURS
16110.24 Galvanized Conduit *(Cont.)*		
2"	EA.	0.571
Erickson couplings		
1/2"	EA.	0.444
3/4"	"	0.500
1"	"	0.615
1-1/4"	"	0.889
1-1/2"	"	1.000
2"	"	1.333
Seal fittings		
1/2"	EA.	0.667
3/4"	"	0.800
1"	"	1.000
1-1/4"	"	1.143
1-1/2"	"	1.333
2"	"	1.600
Entrance fitting, (weather head), threaded		
1/2"	EA.	0.444
3/4"	"	0.500
1"	"	0.571
1-1/4"	"	0.727
1-1/2"	"	0.800
2"	"	0.889
Locknuts		
1/2"	EA.	0.050
3/4"	"	0.050
1"	"	0.050
1-1/4"	"	0.050
1-1/2"	"	0.059
2"	"	0.059
Plastic conduit bushings		
1/2"	EA.	0.123
3/4"	"	0.145
1"	"	0.190
1-1/4"	"	0.222
1-1/2"	"	0.250
2"	"	0.308
Conduit bushings, steel		
1/2"	EA.	0.123
3/4"	"	0.145
1"	"	0.190
1-1/4"	"	0.222
1-1/2"	"	0.250
2"	"	0.308
Pipe cap		
1/2"	EA.	0.050
3/4"	"	0.050
1"	"	0.050
1-1/4"	"	0.080
1-1/2"	"	0.080
2"	"	0.080
GRS elbows, 36" radius		
2"	EA.	0.667
42" radius		

Basic Materials	UNIT	MAN/ HOURS
16110.24 Galvanized Conduit *(Cont.)*		
2"	EA.	0.808
48" radius		
2"	EA.	0.930
Threaded couplings		
1/2"	EA.	0.050
3/4"	"	0.059
1"	"	0.080
1-1/4"	"	0.089
1-1/2"	"	0.100
2"	"	0.107
Threadless couplings		
1/2"	EA.	0.100
3/4"	"	0.123
1"	"	0.145
1-1/4"	"	0.190
1-1/2"	"	0.250
2"	"	0.308
Threadless connectors		
1/2"	EA.	0.100
3/4"	"	0.123
1"	"	0.145
1-1/4"	"	0.190
1-1/2"	"	0.250
2"	"	0.308
Setscrew connectors		
1/2"	EA.	0.080
3/4"	"	0.089
1"	"	0.100
1-1/4"	"	0.123
1-1/2"	"	0.145
2"	"	0.190
Clamp type entrance caps		
1/2"	EA.	0.308
3/4"	"	0.381
1"	"	0.444
1-1/4"	"	0.500
1-1/2"	"	0.615
2"	"	0.727
"LB" condulets		
1/2"	EA.	0.308
3/4"	"	0.381
1"	"	0.444
1-1/4"	"	0.500
1-1/2"	"	0.615
2"	"	0.727
"T" condulets		
1/2"	EA.	0.381
3/4"	"	0.444
1"	"	0.500
1-1/4"	"	0.571
1-1/2"	"	0.615
2"	"	0.727
"X" condulets		

Basic Materials	UNIT	MAN/ HOURS
16110.24 Galvanized Conduit *(Cont.)*		
1/2"	EA.	0.444
3/4"	"	0.500
1"	"	0.571
1-1/4"	"	0.615
1-1/2"	"	0.667
2"	"	0.879
Blank steel condulet covers		
1/2"	EA.	0.100
3/4"	"	0.100
1"	"	0.100
1-1/4"	"	0.123
1-1/2"	"	0.123
2"	"	0.123
Solid condulet gaskets		
1/2"	EA.	0.050
3/4"	"	0.050
1"	"	0.050
1-1/4"	"	0.080
1-1/2"	"	0.080
2"	"	0.080
One-hole malleable straps		
1/2"	EA.	0.040
3/4"	"	0.040
1"	"	0.040
1-1/4"	"	0.050
1-1/2"	"	0.050
2"	"	0.050
One-hole steel straps		
1/2"	EA.	0.040
3/4"	"	0.040
1"	"	0.040
1-1/4"	"	0.050
1-1/2"	"	0.050
2"	"	0.050
Grounding locknuts		
1/2"	EA.	0.080
3/4"	"	0.080
1"	"	0.080
1-1/4"	"	0.089
1-1/2"	"	0.089
2"	"	0.089
Insulated grounding metal bushings		
1/2"	EA.	0.190
3/4"	"	0.222
1"	"	0.250
1-1/4"	"	0.308
1-1/2"	"	0.381
2"	"	0.444

Basic Materials	UNIT	MAN/ HOURS
16110.25 Plastic Conduit		
PVC conduit, schedule 40		
1/2"	L.F.	0.030
3/4"	"	0.030
1"	"	0.040
1-1/4"	"	0.040
1-1/2"	"	0.050
2"	"	0.050
Couplings		
1/2"	EA.	0.050
3/4"	"	0.050
1"	"	0.050
1-1/4"	"	0.059
1-1/2"	"	0.059
2"	"	0.059
90 degree elbows		
1/2"	EA.	0.100
3/4"	"	0.123
1"	"	0.123
1-1/4"	"	0.145
1-1/2"	"	0.190
2"	"	0.222
Terminal adapters		
1/2"	EA.	0.100
3/4"	"	0.100
1"	"	0.100
1-1/4"	"	0.160
1-1/2"	"	0.160
2"	"	0.160
End bells		
1"	EA.	0.100
1-1/4"	"	0.160
1-1/2"	"	0.160
2"	"	0.160
LB conduit body		
1/2"	EA.	0.190
3/4"	"	0.190
1	"	0.190
1-1/4"	"	0.308
1-1/2"	"	0.308
2"	"	0.308
"EB" and "DB" duct, 90 degree elbows		
1-1/2"	EA.	0.145
2"	"	0.229
45 degree elbows		
1-1/2"	EA.	0.229
2"	"	0.229
Couplings		
1-1/2"	EA.	0.059
2"	"	0.059
Bell ends		
1-1/2"	EA.	0.160
2"	"	0.160
Female adapters, 1-1/2"	"	0.200

Basic Materials	UNIT	MAN/ HOURS
16110.25 Plastic Conduit *(Cont.)*		
5 degree couplings		
1-1/2"	EA.	0.070
2"	"	0.070
45 degree elbows		
1/2"	EA.	0.123
3/4"	"	0.145
1"	"	0.145
1-1/4"	"	0.182
1-1/2"	"	0.229
2"	"	0.267
Female adapters		
1/2"	EA.	0.123
3/4"	"	0.123
1"	"	0.123
1-1/4"	"	0.200
1-1/2"	"	0.200
2"	"	0.200
Expansion couplings		
1/2"	EA.	0.123
3/4"	"	0.123
1"	"	0.145
1-1/4"	"	0.200
1-1/2"	"	0.200
2"	"	0.200
Type "T" condulets		
1/2"	EA.	0.296
3/4"	"	0.296
1"	"	0.296
1-1/4"	"	0.500
1-1/2"	"	0.500
2"	"	0.500
16110.27 Plastic Coated Conduit		
Rigid steel conduit, plastic coated		
1/2"	L.F.	0.050
3/4"	"	0.059
1"	"	0.080
1-1/4"	"	0.100
1-1/2"	"	0.123
2"	"	0.145
90 degree elbows		
1/2"	EA.	0.308
3/4"	"	0.381
1"	"	0.444
1-1/4"	"	0.500
1-1/2"	"	0.615
2"	"	0.800
Couplings		
1/2"	EA.	0.059
3/4"	"	0.080
1"	"	0.089
1-1/4"	"	0.107
1-1/2"	"	0.123

Basic Materials	UNIT	MAN/ HOURS
16110.27 Plastic Coated Conduit *(Cont.)*		
2"	EA.	0.145
1 hole conduit straps		
3/4"	EA.	0.050
1"	"	0.050
1-1/4"	"	0.059
1-1/2"	"	0.059
2"	"	0.059
"L.B." condulets with covers		
1/2"	EA.	0.500
3/4"	"	0.500
1"	"	0.615
1-1/4"	"	0.727
1-1/2"	"	0.879
2"	"	1.000
"T" condulets with covers		
1/2"	EA.	0.571
3/4"	"	0.615
1"	"	0.667
1-1/4"	"	0.808
1-1/2"	"	0.941
2"	"	1.053
16110.28 Steel Conduit		
Intermediate metal conduit (IMC)		
1/2"	L.F.	0.030
3/4"	"	0.040
1"	"	0.050
1-1/4"	"	0.059
1-1/2"	"	0.080
2"	"	0.089
90 degree ell		
1/2"	EA.	0.250
3/4"	"	0.308
1"	"	0.381
1-1/4"	"	0.444
1-1/2"	"	0.500
2"	"	0.571
Couplings		
1/2"	EA.	0.050
3/4"	"	0.059
1"	"	0.080
1-1/4"	"	0.089
1-1/2"	"	0.100
2"	"	0.107
16110.35 Surface Mounted Raceway		
Single Raceway		
3/4" x 17/32" Conduit	L.F.	0.040
Mounting Strap	EA.	0.053
Connector	"	0.053
Elbow		
45 degree	EA.	0.050
90 degree	"	0.050

Basic Materials	UNIT	MAN/ HOURS
16110.35 Surface Mounted Raceway *(Cont.)*		
internal	EA.	0.050
external	"	0.050
Switch	"	0.400
Utility Box	"	0.400
Receptacle	"	0.400
3/4" x 21/32" Conduit	L.F.	0.040
Mounting Strap	EA.	0.053
Connector	"	0.053
Elbow		
45 degree	EA.	0.050
90 degree	"	0.050
internal	"	0.050
external	"	0.050
Switch	"	0.400
Utility Box	"	0.400
Receptacle	"	0.400
16120.41 Aluminum Conductors		
Type XHHW, stranded aluminum, 600v		
#8	L.F.	0.005
#6	"	0.006
#4	"	0.008
#2	"	0.009
1/0	"	0.011
2/0	"	0.012
3/0	"	0.014
4/0	"	0.015
THW, stranded		
#8	L.F.	0.005
#6	"	0.006
#4	"	0.008
#3	"	0.009
#1	"	0.010
1/0	"	0.011
2/0	"	0.012
3/0	"	0.012
4/0	"	0.015
XLP, stranded		
#6	L.F.	0.005
#4	"	0.008
#2	"	0.009
#1	"	0.010
1/0	"	0.011
2/0	"	0.012
3/0	"	0.014
4/0	"	0.015
Bare stranded aluminum wire		
#4	L.F.	0.008
#2	"	0.009
1/0	"	0.011
2/0	"	0.012
3/0	"	0.014
4/0	"	0.015

Basic Materials	UNIT	MAN/ HOURS
16120.41 Aluminum Conductors *(Cont.)*		
Triplex XLP cable		
#4	L.F.	0.015
#2	"	0.020
1/0	"	0.030
4/0	"	0.048
Aluminum quadruplex XLP cable		
#4	L.F.	0.018
#2	"	0.023
1/0	"	0.032
2/0	"	0.042
4/0	"	0.064
Triplexed URD-XLP cable		
#6	L.F.	0.011
#4	"	0.014
#2	"	0.018
1/0	"	0.028
2/0	"	0.033
3/0	"	0.040
4/0	"	0.047
Type S.E.U. cable		
#8/3	L.F.	0.025
#6/3	"	0.028
#4/3	"	0.035
#2/3	"	0.038
#1/3	"	0.040
1/0-3	"	0.042
2/0-3	"	0.044
3/0-3	"	0.052
4/0-3	"	0.057
Type S.E.R. cable with ground		
#8/3	L.F.	0.028
#6/3	"	0.035
#4/3	"	0.038
#2/3	"	0.040
#1/3	"	0.044
1/0-3	"	0.050
2/0-3	"	0.055
3/0-3	"	0.059
4/0-3	"	0.067
#6/4	"	0.038
#4/4	"	0.044
#2/4	"	0.044
#1/4	"	0.050
1/0-4	"	0.052
2/0-4	"	0.057
3/0-4	"	0.064
4/0-4	"	0.076

Basic Materials	UNIT	MAN/ HOURS
16120.43 Copper Conductors		
Copper conductors, type THW, solid		
#14	L.F.	0.004
#12	"	0.005
#10	"	0.006
Stranded		
#14	L.F.	0.004
#12	"	0.005
#10	"	0.006
#8	"	0.008
#6	"	0.009
#4	"	0.010
#3	"	0.010
#2	"	0.012
#1	"	0.014
1/0	"	0.016
2/0	"	0.020
3/0	"	0.025
4/0	"	0.028
THHN-THWN, solid		
#14	L.F.	0.004
#12	"	0.005
#10	"	0.006
Stranded		
#14	L.F.	0.004
#12	"	0.005
#10	"	0.006
#8	"	0.008
#6	"	0.009
#4	"	0.010
#2	"	0.012
#1	"	0.014
1/0	"	0.016
2/0	"	0.020
3/0	"	0.025
4/0	"	0.028
XHHW		
#14	L.F.	0.004
#10	"	0.006
#8	"	0.008
#6	"	0.009
#4	"	0.009
#2	"	0.011
#1	"	0.014
1/0	"	0.016
2/0	"	0.019
3/0	"	0.025
XLP, 600v		
#12	L.F.	0.005
#10	"	0.006
#8	"	0.008
#6	"	0.009
#4	"	0.010
#3	"	0.011

Basic Materials	UNIT	MAN/ HOURS
16120.43 Copper Conductors *(Cont.)*		
#2	L.F.	0.012
#1	"	0.014
1/0	"	0.016
2/0	"	0.020
3/0	"	0.026
4/0	"	0.028
Bare solid wire		
#14	L.F.	0.004
#12	"	0.005
#10	"	0.006
#8	"	0.008
#6	"	0.009
#4	"	0.010
#2	"	0.012
Bare stranded wire		
#8	L.F.	0.008
#6	"	0.010
#4	"	0.010
#2	"	0.011
#1	"	0.014
1/0	"	0.018
2/0	"	0.020
3/0	"	0.025
4/0	"	0.028
Type "BX" solid armored cable		
#14/2	L.F.	0.025
#14/3	"	0.028
#14/4	"	0.031
#12/2	"	0.028
#12/3	"	0.031
#12/4	"	0.035
#10/2	"	0.031
#10/3	"	0.035
#10/4	"	0.040
#8/2	"	0.035
#8/3	"	0.040
Steel type, metal clad cable, solid, with ground		
#14/2	L.F.	0.018
#14/3	"	0.020
#14/4	"	0.023
#12/2	"	0.020
#12/3	"	0.025
#12/4	"	0.030
#10/2	"	0.023
#10/3	"	0.028
#10/4	"	0.033
Metal clad cable, stranded, with ground		
#8/2	L.F.	0.028
#8/3	"	0.035
#8/4	"	0.042
#6/2	"	0.030
#6/3	"	0.038
#6/4	"	0.044

Basic Materials	UNIT	MAN/ HOURS
16120.43 Copper Conductors *(Cont.)*		
#4/2	L.F.	0.040
#4/3	"	0.044
#4/4	"	0.055
#3/3	"	0.050
#3/4	"	0.059
#2/3	"	0.057
#2/4	"	0.067
#1/3	"	0.076
#1/4	"	0.084
16120.45 Flat Conductor Cable		
Flat conductor cable, with shield, 3 conductor		
#12 awg	L.F.	0.059
#10 awg	"	0.059
4 conductor		
#12 awg	L.F.	0.080
#10 awg	"	0.080
Transition boxes		
#12 awg	L.F.	0.089
#10 awg	"	0.089
Flat conductor cable communication, with shield		
10 conductor	L.F.	0.059
16 conductor	"	0.070
24 conductor	"	0.100
Power and communication heads, duplex receptacle	EA.	0.800
Double duplex receptacle	"	0.952
Telephone	"	0.800
Receptacle and telephone	"	0.952
Blank cover	"	0.145
Transition boxes		
Surface	EA.	0.727
Flush	"	1.000
Flat conductor cable fittings		
End caps	EA.	0.145
Insulators	"	0.296
Splice connectors	"	0.444
Tap connectors	"	0.444
Cable connectors	"	0.444
Terminal blocks	"	0.615
Tape	"	
16120.47 Sheathed Cable		
Non-metallic sheathed cable		
Type NM cable with ground		
#14/2	L.F.	0.015
#12/2	"	0.016
#10/2	"	0.018
#8/2	"	0.020
#6/2	"	0.025
#14/3	"	0.026
#12/3	"	0.027
#10/3	"	0.027
#8/3	"	0.028

Basic Materials	UNIT	MAN/ HOURS
16120.47 Sheathed Cable *(Cont.)*		
#6/3	L.F.	0.028
#4/3	"	0.032
#2/3	"	0.035
Type U.F. cable with ground		
#14/2	L.F.	0.016
#12/2	"	0.019
#10/2	"	0.020
#8/2	"	0.023
#6/2	"	0.027
#14/3	"	0.020
#12/3	"	0.022
#10/3	"	0.025
#8/3	"	0.028
#6/3	"	0.032
Type S.F.U. cable, 3 conductor		
#8	L.F.	0.028
#6	"	0.031
#3	"	0.040
#2	"	0.044
#1	"	0.050
#1/0	"	0.055
#2/0	"	0.064
#3/0	"	0.070
#4/0	"	0.076
Type SER cable, 4 conductor		
#6	L.F.	0.036
#4	"	0.039
#3	"	0.044
#2	"	0.048
#1	"	0.055
#1/0	"	0.064
#2/0	"	0.067
#3/0	"	0.076
#4/0	"	0.084
Flexible cord, type STO cord		
#18/2	L.F.	0.004
#18/3	"	0.005
#18/4	"	0.006
#16/2	"	0.004
#16/3	"	0.004
#16/4	"	0.005
#14/2	"	0.005
#14/3	"	0.006
#14/4	"	0.007
#12/2	"	0.006
#12/3	"	0.007
#12/4	"	0.008
#10/2	"	0.007
#10/3	"	0.008
#10/4	"	0.009
#8/2	"	0.008
#8/3	"	0.009
#8/4	"	0.010

Basic Materials	UNIT	MAN/ HOURS
16130.10 Floor Boxes		
Aluminum round	EA.	0.727
1 gang	"	0.800
2 gang	"	0.952
3 gang	"	1.000
Steel plate single recept	"	0.145
Duplex receptacle	"	0.182
Twist lock receptacle	"	0.182
Plug, 3/4"	"	0.145
1" plug	"	0.145
Carpet flange	"	0.145
Adjustable bronze plates for round cast boxes		
1/2" plug	EA.	0.145
3/4" plug	"	0.145
1" plug	"	0.145
Combination plug	"	0.200
Duplex receptacle plug	"	0.200
Adjustable aluminum plates for round cast boxes		
1/2" plug	EA.	0.145
3/4" plug	"	0.145
1" plug	"	0.145
Combination plug	"	0.200
Duplex receptacle plug	"	0.200
Adjustable bronze plates for gang type boxes		
1/2" plug	EA.	0.145
3/4" plug	"	0.145
1" plug	"	0.145
Carpet plate		
1 gang	EA.	0.145
2 gang	"	0.145
3 gang	"	0.200
Adjustable aluminum plates for gang type boxes		
1/2" plug	EA.	0.145
3/4" plug	"	0.145
1" plug	"	0.145
Duplex recept	"	0.200
Carpet plate		
1 gang	EA.	0.145
2 gang	"	0.145
3 gang	"	0.200
4 gang carpet plate	"	0.571
Telephone	"	0.500
Floor box nozzles, horizontal		
Duplex recept	EA.	0.533
Single recept	"	0.533
Double duplex recept	"	0.727
Vertical with duplex recept	"	0.615
Double duplex recept	"	0.727

Basic Materials	UNIT	MAN/ HOURS
16130.40 Boxes		
Round cast box, type SEH		
1/2"	EA.	0.348
3/4"	"	0.421
SEHC		
1/2"	EA.	0.348
3/4"	"	0.421
SEHL		
1/2"	EA.	0.348
3/4"	"	0.444
SEHT		
1/2"	EA.	0.421
3/4"	"	0.500
SEHX		
1/2"	EA.	0.500
3/4"	"	0.615
Blank cover	"	0.145
1/2", hub cover	"	0.145
Cover with gasket	"	0.178
Rectangle, type FS boxes		
1/2"	EA.	0.348
3/4"	"	0.400
1"	"	0.500
FSA		
1/2"	EA.	0.348
3/4"	"	0.400
FSC		
1/2"	EA.	0.348
3/4"	"	0.421
1"	"	0.500
FSL		
1/2"	EA.	0.348
3/4"	"	0.400
FSR		
1/2"	EA.	0.348
3/4"	"	0.400
FSS		
1/2"	EA.	0.348
3/4"	"	0.400
FSLA		
1/2"	EA.	0.348
3/4"	"	0.400
FSCA		
1/2"	EA.	0.348
3/4"	"	0.400
FSCC		
1/2"	EA.	0.400
3/4"	"	0.500
FSCT		
1/2"	EA.	0.400
3/4"	"	0.500
1"	"	0.571
FST		
1/2"	EA.	0.500

Basic Materials	UNIT	MAN/ HOURS
16130.40 Boxes *(Cont.)*		
3/4"	EA.	0.571
FSX		
1/2"	EA.	0.615
3/4"	"	0.727
FSCD boxes		
1/2"	EA.	0.615
3/4"	"	0.727
Rectangle, type FS, 2 gang boxes		
1/2"	EA.	0.348
3/4"	"	0.400
1"	"	0.500
FSC, 2 gang boxes		
1/2"	EA.	0.348
3/4"	"	0.400
1"	"	0.500
FSS, 2 gang boxes		
3/4"	EA.	0.400
FS, tandem boxes		
1/2"	EA.	0.400
3/4"	"	0.444
FSC, tandem boxes		
1/2"	EA.	0.400
3/4"	"	0.444
FS, three gang boxes		
3/4"	EA.	0.444
1"	"	0.500
FSS, three gang boxes, 3/4"	"	0.500
Weatherproof cast aluminum boxes, 1 gang, 3 outlets		
1/2"	EA.	0.400
3/4"	"	0.500
2 gang, 3 outlets		
1/2"	EA.	0.500
3/4"	"	0.533
1 gang, 4 outlets		
1/2"	EA.	0.615
3/4"	"	0.727
2 gang, 4 outlets		
1/2"	EA.	0.615
3/4"	"	0.727
1 gang, 5 outlets		
1/2"	EA.	0.727
3/4"	"	0.800
2 gang, 5 outlets		
1/2"	EA.	0.727
3/4"	"	0.800
2 gang, 6 outlets		
1/2"	EA.	0.851
3/4"	"	0.899
2 gang, 7 outlets		
1/2"	EA.	1.000
3/4"	"	1.096
Weatherproof and type FS box covers, blank, 1 gang	"	0.145
Tumbler switch, 1 gang	"	0.145

Basic Materials	UNIT	MAN/ HOURS
16130.40 Boxes (Cont.)		
1 gang, single recept	EA.	0.145
Duplex recept	"	0.145
Despard	"	0.145
Red pilot light	"	0.145
SW and		
Single recept	EA.	0.200
Duplex recept	"	0.200
2 gang		
Blank	EA.	0.182
Tumbler switch	"	0.182
Single recept	"	0.182
Duplex recept	"	0.182
3 gang		
Blank	EA.	0.200
Tumbler switch	"	0.200
4 gang		
Tumbler switch	EA.	0.250
Box covers		
Surface	EA.	0.200
Sealing	"	0.200
Dome	"	0.200
1/2" nipple	"	0.200
3/4" nipple	"	0.200
16130.60 Pull And Junction Boxes		
4"		
Octagon box	EA.	0.114
Box extension	"	0.059
Plaster ring	"	0.059
Cover blank	"	0.059
Square box	"	0.114
Box extension	"	0.059
Plaster ring	"	0.059
Cover blank	"	0.059
4-11/16"		
Square box	EA.	0.114
Box extension	"	0.059
Plaster ring	"	0.059
Cover blank	"	0.059
Switch and device boxes		
2 gang	EA.	0.114
3 gang	"	0.114
4 gang	"	0.160
Device covers		
2 gang	EA.	0.059
3 gang	"	0.059
4 gang	"	0.059
Handy box	"	0.114
Extension	"	0.059
Switch cover	"	0.059
Switch box with knockout	"	0.145
Weatherproof cover, spring type	"	0.080
Cover plate, dryer receptacle 1 gang plastic	"	0.100

Basic Materials	UNIT	MAN/ HOURS
16130.60 Pull And Junction Boxes (Cont.)		
For 4" receptacle, 2 gang	EA.	0.100
Duplex receptacle cover plate, plastic	"	0.059
4", vertical bracket box, 1-1/2" with		
RMX clamps	EA.	0.145
BX clamps	"	0.145
4", octagon device cover		
1 switch	EA.	0.059
1 duplex recept	"	0.059
4", octagon swivel hanger box, 1/2" hub	"	0.059
3/4" hub	"	0.059
4" octagon adjustable bar hangers		
18-1/2"	EA.	0.050
26-1/2"	"	0.050
With clip		
18-1/2"	EA.	0.050
26-1/2"	"	0.050
4", square face bracket boxes, 1-1/2"		
RMX	EA.	0.145
BX	"	0.145
4" square to round plaster rings	"	0.059
2 gang device plaster rings	"	0.059
Surface covers		
1 gang switch	EA.	0.059
2 gang switch	"	0.059
1 single recept	"	0.059
1 20a twist lock recept	"	0.059
1 30a twist lock recept	"	0.059
1 duplex recept	"	0.059
2 duplex recept	"	0.059
Switch and duplex recept	"	0.059
4-11/16" square to round plaster rings	"	0.059
2 gang device plaster rings	"	0.059
Surface covers		
1 gang switch	EA.	0.059
2 gang switch	"	0.059
1 single recept	"	0.059
1 20a twist lock recept	"	0.059
1 30a twist lock recept	"	0.059
1 duplex recept	"	0.059
2 duplex recept	"	0.059
Switch and duplex recept	"	0.059
4" plastic round boxes, ground straps		
Box only	EA.	0.145
Box w/clamps	"	0.200
Box w/16" bar	"	0.229
Box w/24" bar	"	0.250
4" plastic round box covers		
Blank cover	EA.	0.059
Plaster ring	"	0.059
4" plastic square boxes		
Box only	EA.	0.145
Box w/clamps	"	0.200
Box w/hanger	"	0.250

Basic Materials	UNIT	MAN/ HOURS
16130.60 Pull And Junction Boxes *(Cont.)*		
Box w/nails and clamp	EA.	0.250
4" plastic square box covers		
Blank cover	EA.	0.059
1 gang ring	"	0.059
2 gang ring	"	0.059
Round ring	"	0.059
16130.65 Pull Boxes And Cabinets		
Galvanized pull boxes, screw cover		
4x4x4	EA.	0.190
4x6x4	"	0.190
6x6x4	"	0.190
6x8x4	"	0.190
8x8x4	"	0.250
16130.80 Receptacles		
Contractor grade duplex receptacles, 15a 120v		
Duplex	EA.	0.200
125 volt, 20a, duplex, standard grade	"	0.200
Ground fault interrupter type	"	0.296
250 volt, 20a, 2 pole, single, ground type	"	0.200
120/208v, 4 pole, single receptacle, twist lock		
20a	EA.	0.348
50a	"	0.348
125/250v, 3 pole, flush receptacle		
30a	EA.	0.296
50a	"	0.296
60a	"	0.348
277v, 20a, 2 pole, grounding type, twist lock	"	0.200
Dryer receptacle, 250v, 30a/50a, 3 wire	"	0.296
Clock receptacle, 2 pole, grounding type	"	0.200
125v, 20a single recept. grounding type		
Standard grade	EA.	0.200
125/250v, 3 pole, 3 wire surface recepts		
30a	EA.	0.296
50a	"	0.296
60a	"	0.348
Cord set, 3 wire, 6' cord		
30a	EA.	0.296
50a	"	0.296
125/250v, 3 pole, 3 wire cap		
30a	EA.	0.400
50a	"	0.400
60a	"	0.444
16199.10 Utility Poles & Fittings		
Wood pole, creosoted		
25'	EA.	2.353
30'	"	2.963
Treated, wood preservative, 6"x6"		
8'	EA.	0.500
10'	"	0.800
12'	"	0.889

Basic Materials	UNIT	MAN/ HOURS
16199.10 Utility Poles & Fittings *(Cont.)*		
14'	EA.	1.333
16'	"	1.600
18'	"	2.000
20'	"	2.000
Aluminum, brushed, no base		
8'	EA.	2.000
10'	"	2.667
15'	"	2.759
20'	"	3.200
Steel, no base		
10'	EA.	2.500
15'	"	2.963
20'	"	3.810
Concrete, no base		
13'	EA.	5.517
16'	"	7.273
18'	"	8.791
Lightning arrester		
3kv	EA.	1.000
10kv	"	1.600
16350.10 Circuit Breakers		
Molded case, 240v, 15-60a, bolt-on		
1 pole	EA.	0.250
2 pole	"	0.348
70-100a, 2 pole	"	0.533
15-60a, 3 pole	"	0.400
70-100a, 3 pole	"	0.615
Load center circuit breakers, 240v		
1 pole, 10-60a	EA.	0.250
2 pole		
10-60a	EA.	0.400
70-100a	"	0.667
110-150a	"	0.727
Load center, G.F.I. breakers, 240v		
1 pole, 15-30a	EA.	0.296
Tandem breakers, 240v		
1 pole, 15-30a	EA.	0.400
2 pole, 15-30a	"	0.533
16365.10 Fuses		
Fuse, one-time, 250v		
30a	EA.	0.050
60a	"	0.050
100a	"	0.050
200a	"	0.050
16395.10 Grounding		
Ground rods, copper clad, 1/2" x		
6'	EA.	0.667
8'	"	0.727
10'	"	1.000
5/8" x		

Basic Materials	UNIT	MAN/ HOURS
16395.10 Grounding *(Cont.)*		
5'	EA.	0.615
6'	"	0.727
8'	"	1.000
10'	"	1.250
3/4" x		
8'	EA.	0.727
10'	"	0.800
Ground rod clamp		
5/8"	EA.	0.123
3/4"	"	0.123
Coupling, on threaded rods, 3/4"	"	0.050
Ground receptacles	"	0.250
Bus bar, copper, 2" x 1/4"	L.F.	0.145
Copper braid, 1" x 1/8", for door ground	EA.	0.100
Brazed connection for		
#6 wire	EA.	0.500
#2 wire	"	0.800
#2/0 wire	"	1.000
#4/0 wire	"	1.143
Ground rod couplings		
1/2"	EA.	0.100
5/8"	"	0.100
Ground rod, driving stud		
1/2"	EA.	0.100
5/8"	"	0.100
3/4"	"	0.100
Ground rod clamps, #8-2 to		
1" pipe	EA.	0.200
2" pipe	"	0.250
#4-4/0 to		
1" pipe	EA.	0.200
2" pipe	"	0.250

Service And Distribution	UNIT	MAN/ HOURS
16430.20 Metering		
Outdoor wp meter sockets, 1 gang, 240v, 1 phase		
Includes sealing ring, 100a	EA.	1.509
150a	"	1.778
200a	"	2.000
Die cast hubs, 1-1/4"	"	0.320
1-1/2"	"	0.320
2"	"	0.320

Service And Distribution	UNIT	MAN/ HOURS
16470.10 Panelboards		
Indoor load center, 1 phase 240v main lug only		
30a - 2 spaces	EA.	2.000
100a - 8 spaces	"	2.424
150a - 16 spaces	"	2.963
200a - 24 spaces	"	3.478
200a - 42 spaces	"	4.000
Main circuit breaker		
100a - 8 spaces	EA.	2.424
100a - 16 spaces	"	2.759
150a - 16 spaces	"	2.963
150a - 24 spaces	"	3.200
200a - 24 spaces	"	3.478
200a - 42 spaces	"	3.636
120/208v, flush, 3 ph., 4 wire, main only		
100a		
12 circuits	EA.	5.096
20 circuits	"	6.299
30 circuits	"	7.018
225a		
30 circuits	EA.	7.767
42 circuits	"	9.524
16490.10 Switches		
Photo electric switches		
1000 watt		
105-135v	EA.	0.727
Dimmer switch and switch plate		
600w	EA.	0.308
1000w	"	0.348
Dimmer switch incandescent		
1500w	EA.	0.702
2000w	"	0.748
Fluorescent		
12 lamps	EA.	0.500
Time clocks with skip, 40a, 120v		
SPST	EA.	0.748
Contractor grade wall switch 15a, 120v		
Single pole	EA.	0.160
Three way	"	0.200
Four way	"	0.267
Specification grade toggle switches, 20a, 120-277v		
Single pole	EA.	0.200
Double pole	"	0.296
3 way	"	0.250
4 way	"	0.296
30a, 120-277v		
Single pole	EA.	0.200
Double pole	"	0.296
3 way	"	0.250
Combination switch and pilot light, single pole	"	0.296
3 way	"	0.348
Combination switch and receptacle, single pole	"	0.296
3 way	"	0.296

Service And Distribution

16490.10 Switches (Cont.)

Service And Distribution	UNIT	MAN/ HOURS
Switch plates, plastic ivory		
1 gang	EA.	0.080
2 gang	"	0.100
3 gang	"	0.119
4 gang	"	0.145
5 gang	"	0.160
6 gang	"	0.182
Stainless steel		
1 gang	EA.	0.080
2 gang	"	0.100
3 gang	"	0.123
4 gang	"	0.145
5 gang	"	0.160
6 gang	"	0.182
Brass		
1 gang	EA.	0.080
2 gang	"	0.100
3 gang	"	0.123
4 gang	"	0.145
5 gang	"	0.160
6 gang	"	0.182

Lighting

16510.05 Interior Lighting

Lighting	UNIT	MAN/ HOURS
Recessed fluorescent fixtures, 2'x2'		
2 lamp	EA.	0.727
4 lamp	"	0.727
1'x4'		
2 lamp	EA.	0.667
3 lamp	"	0.667
2 lamp w/flange	"	0.727
3 lamp w/flange	"	0.727
2'x4'		
2 lamp	EA.	0.727
3 lamp	"	0.727
4 lamp	"	0.727
2 lamp w/flange	"	1.000
3 lamp w/flange	"	1.000
4 lamp w/flange	"	1.000
Surface mounted incandescent fixtures		
40w	EA.	0.667
75w	"	0.667
100w	"	0.667
150w	"	0.667
Pendant		

Lighting

16510.05 Interior Lighting (Cont.)

Lighting	UNIT	MAN/ HOURS
40w	EA.	0.800
75w	"	0.800
100w	"	0.800
150w	"	0.800
Recessed incandescent fixtures		
40w	EA.	1.509
75w	"	1.509
100w	"	1.509
150w	"	1.509
Light track single circuit		
2'	EA.	0.500
4'	"	0.500
8'	"	1.000
12'	"	1.509
Fittings and accessories		
Dead end	EA.	0.145
Starter kit	"	0.250
Conduit feed	"	0.145
Straight connector	"	0.145
Center feed	"	0.145
L-connector	"	0.145
T-connector	"	0.145
X-connector	"	0.200
Cord and plug	"	0.100
Rigid corner	"	0.145
Flex connector	"	0.145
2 way connector	"	0.200
Spacer clip	"	0.050
Grid box	"	0.145
T-bar clip	"	0.050
Utility hook	"	0.145
Fixtures, square		
R-20	EA.	0.145
R-30	"	0.145
40w flood	"	0.145
40w spot	"	0.145
100w flood	"	0.145
100w spot	"	0.145
Mini spot	"	0.145
Mini flood	"	0.145
Quartz, 500w	"	0.145
R-20 sphere	"	0.145
R-30 sphere	"	0.145
R-20 cylinder	"	0.145
R-30 cylinder	"	0.145
R-40 cylinder	"	0.145
R-30 wall wash	"	0.145
R-40 wall wash	"	0.145

Lighting

Lighting	UNIT	MAN/ HOURS
16510.10 Lighting Industrial		
Surface mounted fluorescent, wrap around lens		
1 lamp	EA.	0.800
2 lamps	"	0.889
Wall mounted fluorescent		
2-20w lamps	EA.	0.500
2-30w lamps	"	0.500
2-40w lamps	"	0.667
Strip fluorescent		
4'		
1 lamp	EA.	0.667
2 lamps	"	0.667
8'		
1 lamp	EA.	0.727
2 lamps	"	0.889
Compact fluorescent		
2-7w	EA.	1.000
2-13w	"	1.333
16670.10 Lightning Protection		
Lightning protection		
Copper point, nickel plated, 12'		
1/2" dia.	EA.	1.000
5/8" dia.	"	1.000
16720.50 Security Systems		
Sensors		
Balanced magnetic door switch, surface mounted	EA.	0.500
With remote test	"	1.000
Flush mounted	"	1.860
Mounted bracket	"	0.348
Mounted bracket spacer	"	0.348
Photoelectric sensor, for fence		
6 beam	EA.	2.759
9 beam	"	4.255
Photoelectric sensor, 12 volt dc		
500' range	EA.	1.600
800' range	"	2.000
Vibration sensor, 30 max per zone	"	0.500
Audio sensor, 30 max per zone	"	0.500
Inertia sensor		
Outdoor	EA.	0.727
Indoor	"	0.500
Monitor panel, with access/secure tone, standard	"	1.739
High security	"	2.000
Emergency power indicator	"	0.500
Monitor rack with 115v power supply		
1 zone	EA.	1.000
10 zone	"	2.500
Monitor cabinet, wall mounted		
1 zone	EA.	1.000
5 zone	"	1.600
10 zone	"	1.739
Audible alarm	"	0.500

Lighting	UNIT	MAN/ HOURS
16720.50 Security Systems *(Cont.)*		
Audible alarm control	EA.	0.348
16750.20 Signaling Systems		
Contractor grade doorbell chime kit		
Chime	EA.	1.000
Doorbutton	"	0.320
Transformer	"	0.500
16770.30 Sound Systems		
Power amplifiers	EA.	3.478
Tuner	"	1.455
Equalizer	"	1.600
Mixer	"	2.222
Cassette Player	"	2.162
Record player	"	1.905
Equipment rack	"	1.290
Speaker		
Wall	EA.	4.000
Paging	"	0.800
Column	"	0.533
Single	"	0.615
Double	"	4.444
Volume control	"	0.533
Plug-in	"	0.800
Desk	"	0.400
Outlet	"	0.400
Stand	"	0.296
Console	"	8.000
Power supply	"	1.290
16780.50 Television Systems		
TV outlet, self terminating, w/cover plate	EA.	0.308
Thru splitter	"	1.600
End of line	"	1.333
In line splitter multitap		
4 way	EA.	1.818
2 way	"	1.702
Equipment cabinet	"	1.600
Antenna		
Broad band uhf	EA.	3.478
Lightning arrester	"	0.727
TV cable	L.F.	0.005
Coaxial cable rg	"	0.005
Cable drill, with replacement tip	EA.	0.500
Cable blocks for in-line taps	"	0.727
In-line taps ptu-series 36 tv system	"	1.143
Control receptacles	"	0.449
Coupler	"	2.424
Head end equipment	"	6.667
TV camera	"	1.667
TV power bracket	"	0.800
TV monitor	"	1.455
Video recorder	"	2.105

Lighting

16780.50 Television Systems *(Cont.)*

Item	UNIT	MAN/ HOURS
Console	EA.	8.502
Selector switch	"	1.379
TV controller	"	1.404

Resistance Heating

16850.10 Electric Heating

Item	UNIT	MAN/ HOURS
Baseboard heater		
2', 375w	EA.	1.000
3', 500w	"	1.000
4', 750w	"	1.143
5', 935w	"	1.333
6', 1125w	"	1.600
7', 1310w	"	1.818
8', 1500w	"	2.000
9', 1680w	"	2.222
10', 1875w	"	2.286
Unit heater, wall mounted		
750w	EA.	1.600
1500w	"	1.667
2000w	"	1.739
2500w	"	1.818
3000w	"	2.000
4000w	"	2.286
Thermostat		
Integral	EA.	0.500
Line voltage	"	0.500
Electric heater connection	"	0.250
Fittings		
Inside corner	EA.	0.400
Outside corner	"	0.400
Receptacle section	"	0.400
Blank section	"	0.400
Infrared heaters		
600w	EA.	1.000
Radiant ceiling heater panels		
500w	EA.	1.000
750w	"	1.000

Controls

16910.40 Control Cable

Item	UNIT	MAN/ HOURS
Control cable, 600v, #14 THWN, PVC jacket		
2 wire	L.F.	0.008
4 wire	"	0.010
Audio cables, shielded, #24 gauge		
3 conductor	L.F.	0.004
4 conductor	"	0.006
5 conductor	"	0.007
6 conductor	"	0.009
#22 gauge		
3 conductor	L.F.	0.004
4 conductor	"	0.006
#20 gauge		
3 conductor	L.F.	0.004
10 conductor	"	0.015
#18 gauge		
3 conductor	L.F.	0.004
4 conductor	"	0.006
Computer cables shielded, #24 gauge		
1 pair	L.F.	0.004
2 pair	"	0.004
3 pair	"	0.006
4 pair	"	0.007
Coaxial cables		
RG 6/u	L.F.	0.006
RG 6a/u	"	0.006
RG 8/u	"	0.006
RG 8a/u	"	0.006
MATV and CCTV camera cables		
1 conductor	L.F.	0.004
2 conductor	"	0.005
4 conductor	"	0.006
Fire alarm cables, #22 gauge		
6 conductor	L.F.	0.010
9 conductor	"	0.015
12 conductor	"	0.016
#18 gauge		
2 conductor	L.F.	0.005
4 conductor	"	0.007
#16 gauge		
2 conductor	L.F.	0.007
4 conductor	"	0.008
#14 gauge		
2 conductor	L.F.	0.008
#12 gauge		
2 conductor	L.F.	0.010
Plastic jacketed thermostat cable		
2 conductor	L.F.	0.004
3 conductor	"	0.005

BNi® Building News

Supporting Construction Reference Data

This section contains information, text, charts and tables on various aspects of construction. The intent is to provide the user with a better understanding of unfamiliar areas in order to be able to estimate better. This information includes actual takeoff data for some areas and also selected explanations of common construction materials, methods and common practices.

TYPICAL BUILDING COST BROKEN DOWN BY CSI FORMAT

(Commercial Construction)

Division	New Construction	Remodeling Construction
1. General Requirements	6 to 8%	Up to 30%
2. Sitework	4 to 6%	
3. Concrete	15 to 20%	
4. Masonry	8 to 12%	
5. Metals	5 to 7%	
6. Wood and Plastics	1 to 5 %	
7. Thermal and Moisture Protection	4 to 6%	
8. Doors and Windows	5 to 7%	Up to 30% (Divisions 8–9)
9. Finishes	8 to 12 %	
10. Specialties	6 to 10% (Divisions 10–14)	
11. Architectural Equipment		
12. Furnishings		
13. Special Construction		
14. Conveying Systems		
15. Mechanical	15 to 25%	Up to 30% (Divisions 15–16)
16. Electrical	8 to 12%	
TOTAL COST	100%	

CONVERSION FACTORS

Change	To	Multiply By
ATMOSPHERES	POUNDS PER SQUARE INCH	14.696
ATMOSPHERES	INCHES OF MERCURY	29.92
ATMOSPHERES	FEET OF WATER	34
BARRELS, OIL	GALLONS, OF OIL	42
BARRELS, CEMENT	POUNDS OF CEMENT	376
BOGS OR SACKS, CEMENT	POUNDS OF CEMENT	94
BTU/MIN	FOOT-POUNDS/SEC	12.96
BTU/MIN	HORSE-POWER	0.02356
BTU/MIN	KILOWATTS	0.01757
BTU/MM	WATTS	17.57
CENTIMETERS	INCHES	0.3937
CENTIMETERS OF MERCURY	ATMOSPHERES	0.01316
CENTIMETERS OF MERCURY	FEET OF WATER	0.4461
CUBIC INCHES	CUBIC FEET	0.00058
CUBIC FEET	CUBIC INCHES	1728
CUBIC FEET	CUBIC YARDS	0.03703
CUBIC YARDS	CUBIC FEET	27
CUBIC INCHES	GALLONS	0.00433
CUBIC FEET	GALLONS	7.48
FEET	INCHES	12
FEET	YARDS	0.3333
YARDS	FEET	3
FEET OF WATER	ATMOSPHERES	0.02950
FEET OF WATER	INCHES OF MERCURY	0.8826
GALLONS	CUBIC INCHES	231
GALLONS	CUBIC FEET	0.1337
GALLONS	POUNDS OF WATER	8.33
GALLONS	QUARTS	4
GALLONS PER MIN	CUBIC FEET SEC	0.002228
GALLONS PER MIN	CUBIC FEET HOUR	8.0208
GALLONS WATER PER MIN	TONS WATER/24 HOURS	6.0086
HORSE-POWER	FOOT-LBS./SEC	550
INCHES	CENTIMETERS	2.540
INCHES	FEET	0.0833
INCHES	MILLIMETERS	25.4
INCHES OF WATER	POUNDS PER SQ. INCH	0.0361
INCHES OF WATER	INCHES OF MERCURY	0.0735
INCHES OF WATER	OUNCES PER SQUARE INCH	0.578
INCHES OF WATER	OUNCES PER SQUARE FOOT	5.2
INCHES OF MERCURY	INCHES OF WATER	13.6
INCHES OF MERCURY	FEET OF WATER	1.1333
INCHES OF MERCURY	POUNDS PER SQUARE INCH	0.4914
KILOMETERS	MILES	0.6214
METERS	INCHES	39.37
MILES	FEET	5280
MILLIMETERS	CENTIMETERS	0.1
MILLIMETERS	INCHES	0.03937
OUNCES (FLUID)	CUBIC INCHES	1.805
OUNCES	POUNDS	0.0625
POUNDS	OUNCES	16
POUNDS PER SQUARE INCH	INCHES OF WATER	27.72
POUNDS PER SQUARE INCH	FEET OF WATER	2.310
POUNDS PER SQUARE INCH	INCHES OF MERCURY	2.04
POUNDS PER SQUARE INCH	ATMOSPHERES	0.0681
QUARTS	CUBIC INCHES	67.20
SQUARE INCHES	SQUARE FEET	0.00694
SQUARE FEET	SQUARE INCHES	144
SQUARE FEET	SQUARE YARDS	0.11111
SQUARE YARDS	SQUARE FEET	9
SQUARE MILES	ACRES	640
SHORT TONS	POUNDS	2000
SHORT TONS	LONG TONS	0.89285
TONS OF WATER/24 HOURS	GALLONS PER MINUTE	0.16643
YARDS	FEET	3
YARDS	CENTIMETERS	91.44
YARDS	INCHES	36

CONVERSION CALCULATIONS

Commercial Measure

16 grams = 1 ounce
16 ounces = 1 pound
2,000 pounds = 1 ton

Long Measure

12 inches = 1 foot
3 feet = 1 yard
16 ½ feet = 1 rod
40 rods = 1 furlong
8 furlongs (5,280 ft.) = 1 mile
3 miles = 1 league

Square Measure

144 square inches = 1 square foot
9 square feet = 1 square yard
30 ¼ square yards = 1 square rod
160 square rods = 1 acre
4,840 square yards = 1 acre
640 acres = 1 square mile
36 square miles = 1 township

Surveyor's Measure

7.92 inches = 1 link
25 links = 1 rod
4 rods (66 ft.) = 1 chain
10 chains = 1 furlong
8 furlongs = 1 mile
1 square mile = 1 section

Cubic Measure

1728 cubic inches = 1 cubic foot
27 cubic feet = 1 cubic yard
128 cubic feet = 1 cord (wood/stone)
231 cubic inches = 1 U.S. gallon
7.48 U.S. Gallons = 1 cubic foot
2150.4 cubic inches = 1 U.S. bushel

Liquid Measure

4 fluid ounces = 1 gill
4 gills = 1 pint
2 pints = 1 quart
4 quarts = 1 gallon
9 gallons = 1 firkin
31 ½ gallons = 1 barrel
2 barrels = 1 hogshead

Dry Measure

2 pints = 1 quart
8 quarts = 1 peck
4 pecks = 1 bushel
2150.42 cubic inches = 1 bushel

SQUARE

$A=a^2$

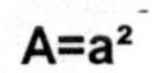

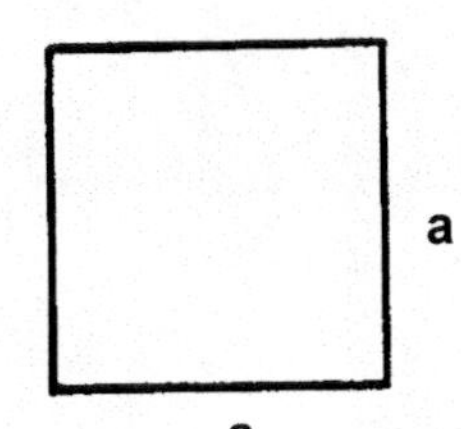

RECTANGLE

$A=bh$

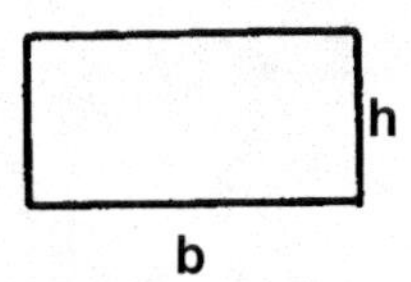

TRIANGLE

$A=\frac{1}{2}bh$

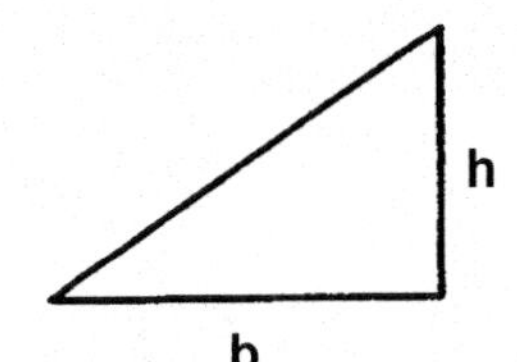

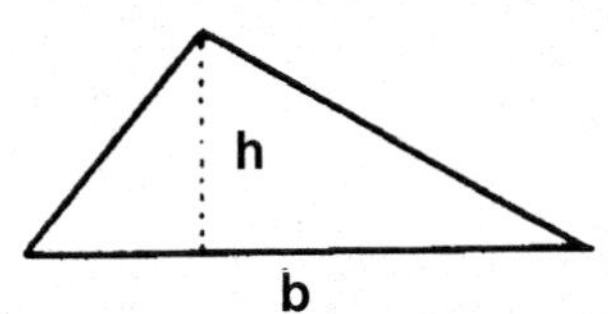

PARALLELOGRAM

$A=bh=ah\ \sin\theta$

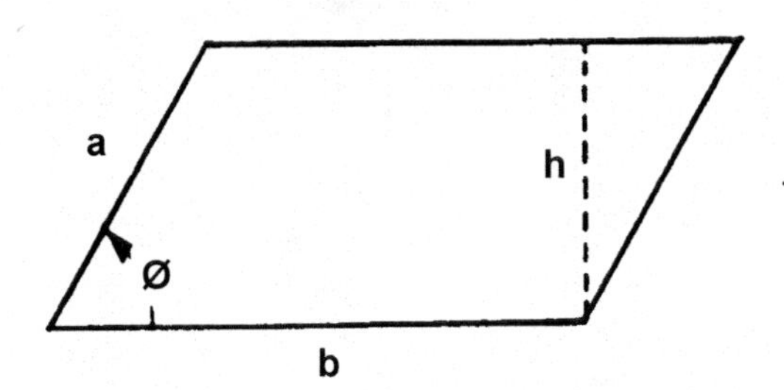

TRAPEZOID

$A=\left(\frac{a+b}{2}\right)h$

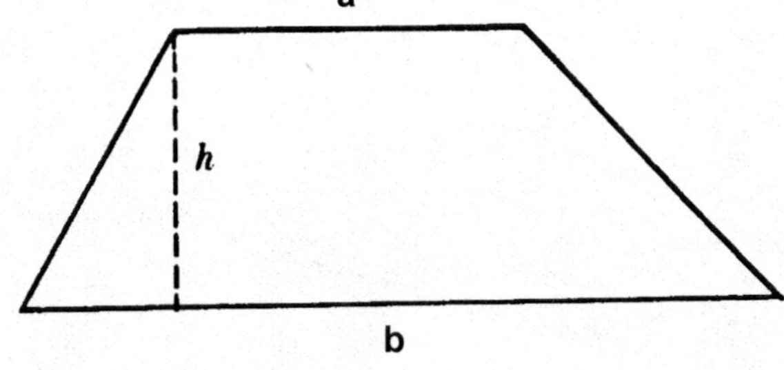

CIRCLE

$A=\pi r^2=\frac{\pi d^2}{4}$

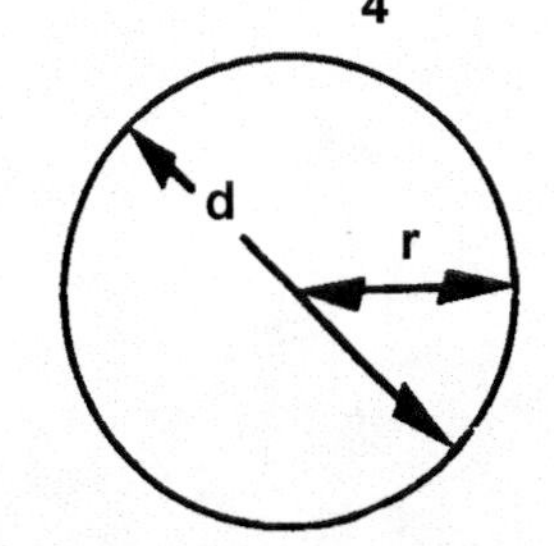

Circumference$=C=2\pi r=\pi d$

ELLIPSE

$A=\pi ab$

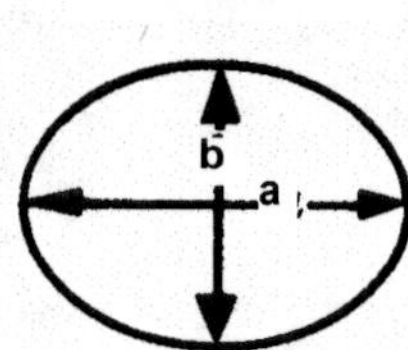

PARABOLA

$A=\frac{2}{3}bh$

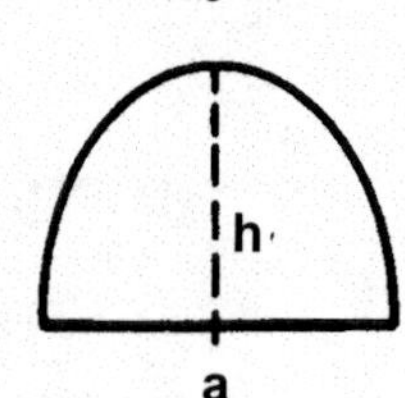

VOLUMES

CUBE

$V=a^3$

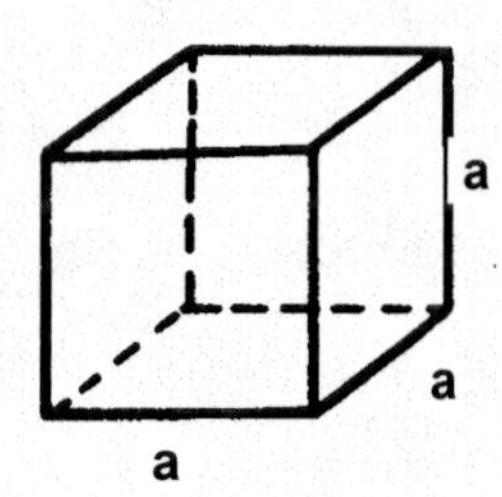

CYLINDER

$V=\pi r^3 h$

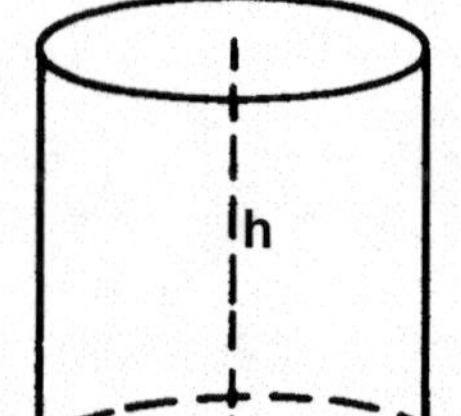

PYRAMID

$V=\frac{1}{3}(\text{Base})\ h$

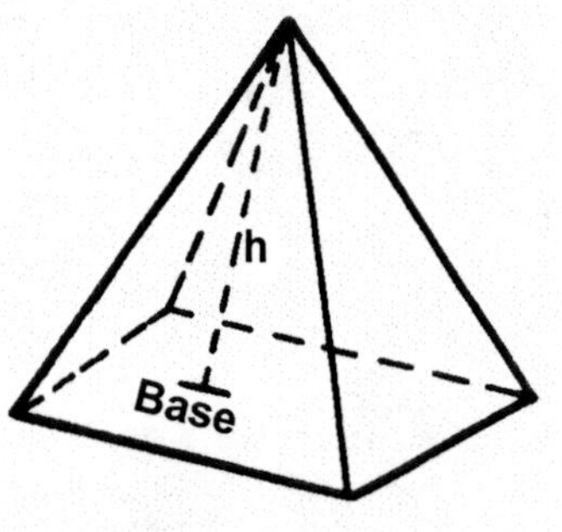

CONE

$V=\frac{1}{3}\pi r^2\ h$

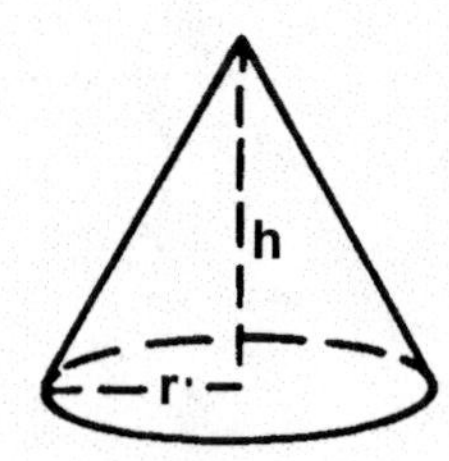

SPHERE

$V=\frac{4}{3}\pi r^3=\frac{1}{6}\pi d^3$

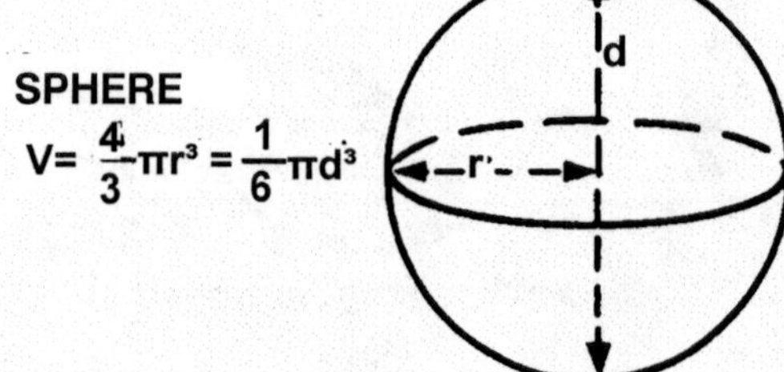

WEDGE

$V=\frac{1}{2}abc$

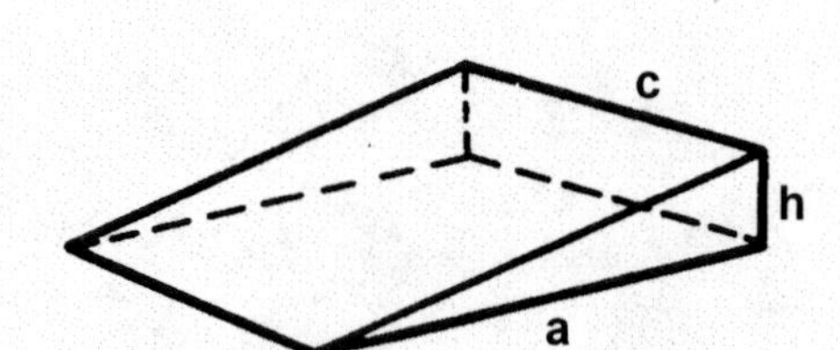

CONVERSION FACTORS
ENGLISH TO SI (SYSTEM INTERNATIONAL)

To Convert from	To	Multiply by
LENGTH		
Inches	Millimetres	25.4[a]
Feet	Metres	0.3048[a]
Yards	Metres	0.9144[a]
Miles (statute)	Kilometres	1.609
AREA		
Square inches	Square millimetres	645.2
Square feet	Square metres	0.0929
Square yards	Square metres	0.8361
VOLUME		
Cubic inches	Cubic millimetres	16.387
Cubic feet	Cubic metres	0.02832
Cubic yards	Cubic metres	0.7646
Gallons (U.S. liquid) [b]	Cubic metres[c]	0.003785
Gallons (Canadian liquid) [b]	Cubic metres[c]	0.004546
Ounces (U.S. liquid) [b]	Millilitres [c,d]	29.57
Quarts (U.S. liquid) [b]	Litres [c,d]	0.9464
Gallons (U.S. liquid) [b]	Litres [c]	3.785
FORCE		
Kilograms force	Newtons	9.807
Pounds force	Newtons	4.448
Pounds force	Kilograms force [d]	0.4536
Kips	Newtons	4448
Kips	Kilograms force [d]	453.6
PRESSURE, STRESS, STRENGTH (FORCE PER UNIT AREA)		
Kilograms force per sq. centimetre	Megapascals	0.09807
Pounds force per square inch (psi)	Megapascals	6895
Kips per square inch	Megapascals	6.895
Pounds force per square inch (psi)	Kilograms force per square centimetre [d]	0.07031
Pounds force per square foot	Pascals	47.88
Pounds force per square foot	Kilograms force per square metre [d]	4.882
SENDING MOMENT OR TORQUE		
Inch-pounds force	Metre-kilog. force [d]	0.01152
Inch-pounds force	Newton-metres	0.1130
Foot-pounds force	Metre-kilog. force [d]	0.1383
Foot-pounds force	Newton-metres	1.356
Metre-kilograms force	Newton-metres	9.807
MASS		
Ounce (avoirdupois)	Grams	28.35
Pounds (avoirdupois)	Kilograms	0.4536
Tons (metric)	Kilograms	1000[a]
Tons, short (2000 pounds)	Kilograms	907.2
Tons, short (2000 pounds)	Megagrams [e]	0.9072
MASS PER UNIT VOLUME		
Pounds mass per cubic foot	Kilog. per cubic metre	16.02
Pounds mass per cubic yard	Kilog. per cubic metre	0.5933
Pds. mass per gallon (U.S. liquid) [b]	Kilog. per cubic metre	119.8
Pds. mass p/gal. (Canadian liquid) [b]	Kilog. per cubic metre	99.78
TEMPERATURE		
Degrees Fahrenheit	Degrees Celsius	tK = (1F - 32)/1.8
Degrees Fahrenheit	Degrees Kelvin	tK = (1F + 459.67)/1.8
Degree Celsius	Degree Kelvin	tK = 1C + 273.15

[a] The factor given is exact.
[b] One U.S. gallon equals 0.8327 Canadian gallon.
[c] 1 litre = 1000 millilitres = 10,000 cubic centimetres = 1 cubic decimetre = 0.001 cubic metre.
[d] Metric but not SI unit.
[e] Called "tonne" in England. Called "metric ton" in other metric systems.

TRENCH BRACING

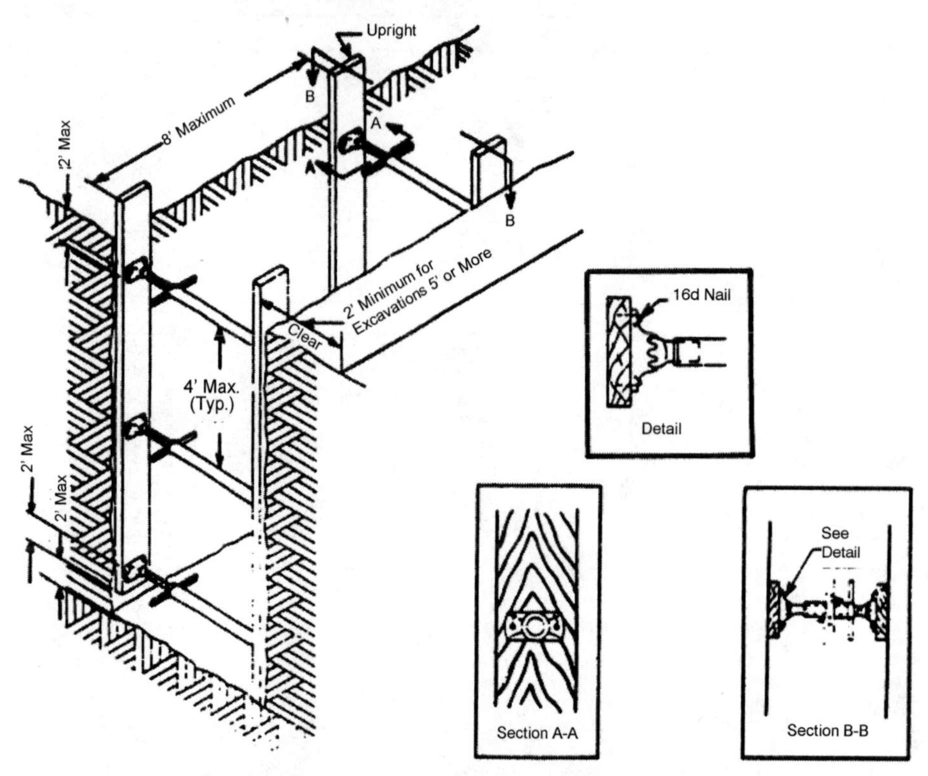

CLOSED VERTICAL SHEETING

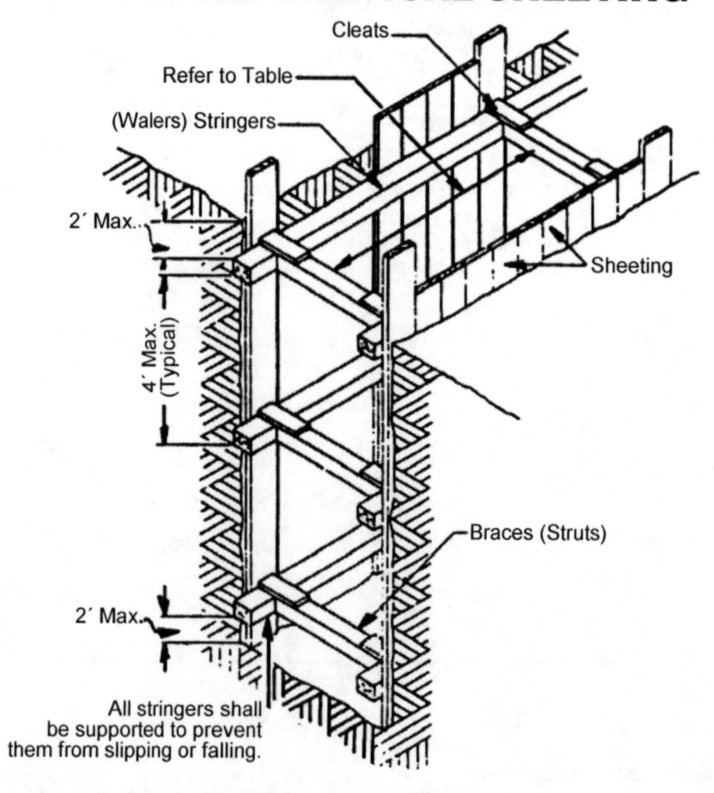

Soil Identification. For most purposes, soils can usually be identified visually and by texture, as described in the chart that follows. For design purposes, however, soils must be formally identified and their performance characteristics determined in a laboratory by skilled soil mechanics.

Classification	Identifying Characteristics
Gravel	Rounded or water-worn pebbles or bulk rock grains. No cohesion or plasticity. Gritty, granular and crunchy underfoot.
Sand	Granular, gritty, loose grains, passing a No. 4 sieve and between .002 and .079 inches in diameter. Individual grains readily seen and felt. No plasticity or cohesion. When dry, it cannot be molded but will crumble when touched. The coarse grains are rounded; the fine grains are visible and angular.
Silt	Fine, barely visible grains passing a No. 200 sieve and between .0002 and .002 inches in diameter. Little or no plasticity and no cohesion. A dried cast is easily crushed. Is permeable and movement of water through the voids occurs easily and is visible. Feels gritty when bitten and will not form a thread.
Clay	Invisible particles under .0002 inches in diameter. Cohesive and highly plastic when moist. Will form a long, thin, flexible thread when rolled between the hands. Does not feel gritty when bitten. Will form hard lumps or clods when dry which resist crushing. Impermeable, with no apparent movement of water through voids.
Muck and Organic Silt	Thoroughly decomposed organic material often mixed with other soils of mineral origin. Usually black with fibrous remains. Odorous when dried and burnt. Found as deposits in swamps, peat bogs and muskeg flats.
Peat	Partly decayed plant material. Mostly organic. Highly fibrous with visible plant remains. Spongy and easily identified.

PILES AND PILE DRIVING

General. A pile is a column driven or jetted into the ground which derives its supporting capabilities from end-bearing on the underlying strata, skin friction between the pile surface and the soil, or from a combination of end-bearing and skin friction.

Piles can be divided into two major classes: **Sheet piles** and **load-bearing piles**. Sheet piling is used primarily to restrain lateral forces as in trench sheeting and bulkheads, or to resist the flow of water as in cofferdams. It is prefabricated and is available in steel, wood or concrete. Load-bearing piles are used primarily to transmit loads through soil formations of low bearing values to formations that are capable of supporting the designed loads. If the load is supported predominantly by the action of soil friction on the surface of the pile, it is called a **friction pile**. If the load is transmitted to the soil primarily through the lower tip, it is called an **end-bearing pile**.

There are several load-bearing pile types, which can be classified according to the material from which they are fabricated:

- Timber (Treated and untreated)
- Concrete (Precast and cast in place)
- Steel (H-Section and steel pipe)
- Composite (A combination of two or more materials)

Some of the additional uses of piling are to: eliminate or control settlement of structures, support bridge piers and abutments and protect them from scour, anchor structures against uplift or overturning, and for numerous marine structures such as docks, wharves, fenders, anchorages, piers, trestles and jetties.

Timber Piles. Timber piles, treated or untreated, are the piles most commonly used throughout the world, primarily because they are readily available, economical, easily handled, can be easily cut off to any desired length after driving and can be easily removed if necessary. On the other hand, they have some serious disadvantages which include: difficulty in securing straight piles of long length, problems in driving them into hard formations and difficulty in splicing to increase their length. They are generally not suitable for use as end-bearing piles under heavy load and they are subject to decay and insect attack. Timber piles are resilient and particularly adaptable for use in waterfront structures such as wharves, docks and piers for anchorages since they will bend or give under load or impact where other materials may break. The ease with which they can be worked and their economy makes them popular for trestle construction and for temporary structures such as falsework or centering. Where timber piles can be driven and cut off below the permanent groundwater level, they will last indefinitely; but above this level in the soil, a timber pile will rot or will be attacked by insects and eventually destroyed. In sea water, marine borers and fungus will act to deteriorate timber piles. Treatment of timber piles increases their life but does not protect them indefinitely.

Concrete Piles. Concrete piles are of two general types, precast and cast-in-place. The advantages in the use of concrete piles are that they can be fabricated to meet the most exacting conditions of design, can be cast in any desired shape or length, possess high strength and have excellent resistance to chemical and biological attack. Certain disadvantages are encountered in the use of precast piles, such as:

(a) Their heavy weight and bulk (which introduces problems in handling and driving).
(b) Problems with hair cracks which often develop in the concrete as a result of shrinkage after curing (which may expose the steel reinforcement to deterioration).
(c) Difficulty encountered in cut-off or splicing.
(d) Susceptibility to damage or breakage in handling and driving.
(e) They are more expensive to fabricate, transport and drive.

Precast piles are fabricated in casting yards. Centrifugally spun piles (or piles with square or octagonal cross-sections) are cast in horizontal forms, while round piles are usually cast in vertical forms. With the exception of relatively short lengths, precast piles must be reinforced to provide the designed column strengths and to resist damage or breakage while being transported or driven.

Precast piles can be tapered or have parallel sides. The reinforcement can be of deformed bars or be prestressed or poststressed with high strength steel tendons. Prestressing or prestressing eliminates the problem of open shrinkage cracks in the concrete. Otherwise, the pile must be protected by coating it with a bituminous or plastic material to prevent ultimate deterioration of the reinforcement. Proper curing of the precast concrete in piles is essential.

Cast-in-place pile types are numerous and vary according to the manufacturer of the shell or inventor of the method. In general, they can be classified into two groups: shell-less types and the shell types. The shell-less type is constructed by driving a steel shell into the ground and filling it with concrete as the shell is pulled from the ground. The shell type is constructed by driving a steel shell into the ground and filling it in place with concrete. Some of the advantages of cast-in-place concrete piles are: lightweight shells are handled and driven easily, lengths of the shell may be increased or decreased easily, shells may be transported in short lengths and quickly assembled, the problem of breakage is eliminated and a driven shell may be inspected for shell damage or an uncased hole for "pinching off." Among the disadvantages are problems encountered in the proper centering of the reinforcement cages, in placing and consolidating the concrete without displacement of the reinforcement steel or segregation of the concrete, and shell damage or "pinching-off" of uncased holes.

Shell type piles are fabricated of heavy gage metal or are fluted, corrugated or spirally reinforced with heavy wire to make them strong enough to be driven without a mandrel.

Other thin-shell types are driven with a collapsible steel mandrel or core inside the casing. In addition to making the driving of a long thin shell possible, the mandrel prevents or minimizes damage to the shell from tearing, buckling, collapsing or from hard objects encountered in driving.

Some shell type piles are fabricated of heavy gauge metal with enlargement at the lower end to increase the end bearing.

These enlargements are formed by withdrawing the casing two to three feet after placing concrete in the lower end of the shell. This wet concrete is then struck by a blow of the pile hammer on a core in the casing and the enlargement is formed. As the shell is withdrawn, the core is used to consolidate the concrete after each batch is placed in the shell. The procedure results in completely filling the hole left by the withdrawal of the shell.

Steel Piles. A steel pile is any pile fabricated entirely of steel. They are usually formed of rolled steel H sections, but heavy steel pipe or box piles (fabricated from sections of steel sheet piles welded together) are also used. The advantages of steel piles are that they are readily available, have a thin uniform section and high strength, will take hard driving, will develop high load-bearing values, are easily cut off or extended, are easily adapted to the structure they are to support, and breakage is eliminated. Some disadvantages are: they will rust and deteriorate unless protected from the elements; acid, soils or water will result in corrosion of the pile; and greater lengths may be required than for other types of piles to achieve the same bearing value unless bearing on rock strata. Pipe pile can either be driven open-end or closed-end and can be unfilled, sand filled or concrete filled. After open-end pipe piles are driven, the material from inside can be removed by an earth auger, air or water jets, or other means, inspected, and then filled with concrete. Concrete filled pipe piles are subject to corrosion on the outside surface only.

Composite Piles. Any pile that is fabricated of two or more materials is called a composite pile. There are three general classes of composite piles: wood with concrete, steel with concrete, and wood with steel. Composite piles are usually used for a special purpose or for reasons of economy.

Where a permanent ground-water table exists and a composite pile is to be used, it will generally be of concrete and wood. The wood portion is driven to below the water table level and the concrete upper portion eliminates problems of decay and insect infestation above the water table. Composite piles of steel and concrete are used where high bearing loads are desired or where driving in hard or rocky soils is expected. Composite wood and steel piles are relatively uncommon.

It is important that the pile design provides for a permanent joint between the two materials used, so constructed that the parts do not separate or shift out of axial alignment during driving operations.

Sheet Piles. Sheet piles are made from the same basic materials as other piling: wood, steel and concrete. They are ordinarily designed so as to interlock along the edges of adjacent piles.

Sheet piles are used where support of a vertical wall of earth is required, such as trench walls, bulkheads, waterfront structures or cofferdams. Wood sheet piling is generally used in temporary installations, but is seldom used where water-tightness is required or hard driving expected. Concrete sheet piling has the capability of resisting much larger lateral loads than wood sheet piling, but considerable difficulty is experienced in securing water-tight joints. The type referred to as "fishmouth" type is designed to permit jetting out the joint and filling with grout, but a seal is not always effected unless the adjacent piles are wedged tightly together. Concrete sheet piling has the advantage that it is the most permanent of all types of sheet piling.

Steel sheet piling is manufactured with a tension-type interlock along its edges. Several different shapes are available to permit versatility in its use. It has the advantages that it can take hard driving, has reasonably water-tight joints and can be easily cut, patched, lengthened or reinforced. It can also be easily extracted and reused. Its principal disadvantage is its vulnerability to corrosion.

Types of Pile Driving Hammers. A pile-driving hammer is used to drive load-bearing or sheet piles. The commonly used types are: drop, single-acting, double-acting, differential acting and diesel hammers. The most recent development is a type of hammer that utilizes high-frequency sound and a dead load as the principal sources of driving energy.

Drop Hammers. These hammers employ the principle of lifting a heavy weight by a cable and releasing it to fall on top of the pile. This type of hammer is rapidly disappearing from use, primarily because other types of pile driving hammers are more efficient. Its disadvantages are that it has a slow rate of driving (four to eight blows per minute), that there is some risk of damaging the pile from excessive impact, that damage may occur in adjacent structures from heavy vibration and that it cannot be used directly for driving piles under water. Drop hammers have the advantages of simplicity of operation, ability to vary the energy by changing the height of fall and they represent a small investment in equipment.

Single-Acting Hammers. These hammers can be operated either on steam or compressed air. The driving energy is provided by a free-falling weight (called a ram) which is raised after each stroke by the action of steam or air on a piston. They are manufactured as either open or closed types. Single-acting hammers are best suited for jobs where dense or elastic soil materials must be penetrated or where long heavy timber or precast concrete piles must be driven. The closed type can be used for underwater pile driving. Its advantages include: faster driving (50 blows or more per minute), reduction in skin friction as a result of more frequent blows, lower velocity of the ram which transmits a greater proportion of its energy to the pile and minimizes piles damage during driving, and it has underwater driving capability. Some of its disadvantages are: requires higher investment in equipment (i.e. steam boiler, air compressor, etc.), higher maintenance costs, greater set-up and moving time required, and a larger operating crew.

Double-Acting Hammers. These hammers are similar to the single-acting hammers except that steam or compressed air is used both to lift the ram and to impart energy to the falling ram. While the action is approximately twice as fast as the single-acting hammer (100 blows per minute or more), the ram is much lighter and operates at a greater velocity, thereby making it particularly useful in high production driving of light or medium-weight piles of moderate lengths in granular soils. The hammer is nearly always fully encased by a steel housing which also permits direct driving of piles under water.

Some of its advantages are: faster driving rate, less static skin friction develops between blows, has underwater driving capability and piles can be driven more easily without leads.

Among its disadvantages are: it is less suitable for driving heavy piles in high-friction soils and the more complicated mechanism results in higher maintenance costs.

Differential-Acting Hammers. This type of hammer is, in effect, a modified double-acting hammer with the actuating mechanism having two different diameters. A large-diameter piston operates in an upper cylinder to accelerate the ram on the downstroke and a small-diameter piston operates in a lower cylinder to raise the ram. The additional energy added to the falling ram is the difference in areas of the two pistons multiplied by the unit pressure of the steam or air used. This hammer is a short-stroke, fast-acting hammer with a cycle rate approximately that of the double-acting hammer.

Its advantages are that it has the speed and characteristics of the double-acting hammer with a ram weight comparable to the single-acting type, and it uses 25 to 35 percent less steam or air. It is also more suitable for driving heavy piles under more difficult driving conditions than the double-acting hammer. It is available in the open or closed-type cases, the latter permitting direct underwater pile driving. Its principal disadvantage is higher maintenance costs.

Diesel Hammers. This hammer is a self-contained driving unit which does not require an auxiliary steam boiler or air compressor. It consists essentially of a ram operating as a piston in a cylinder. When the ram is lifted and allowed to fall in the cylinder, diesel fuel is injected in the compression space between the ram and an anvil placed on top of the pile. The continued downstroke of the ram compresses the air and fuel to ignition heat and the resultant explosion drives the pile downward and the ram upward to start another cycle. This hammer is capable of driving at a rate of from 80 to 100 blows per minute. Its advantages are that it has a low equipment investment cost, is easily moved, requires a small crew, has a high driving rate, does not require a steam boiler or air compressor and can be used with or without leads for most work. Its disadvantages are that it is not self-starting (the ram must be mechanically lifted to start the action) and it does not deliver a uniform blow. The latter disadvantage arises from the fact that as the reaction of the pile to driving increases, the reaction to the ram increases correspondingly. That is, when the pile encounters considerable resistance, the rebound of the ram is higher and the energy is increased automatically. The operator is required to observe the driving operations closely to identify changing driving conditions and compensate for such changes with his controls to avoid damaging the pile.

Diesel hammers can be used on all types of piles and they are best suited to jobs where mobility or frequent relocation of the pile driving equipment is necessary.

PILE CHART

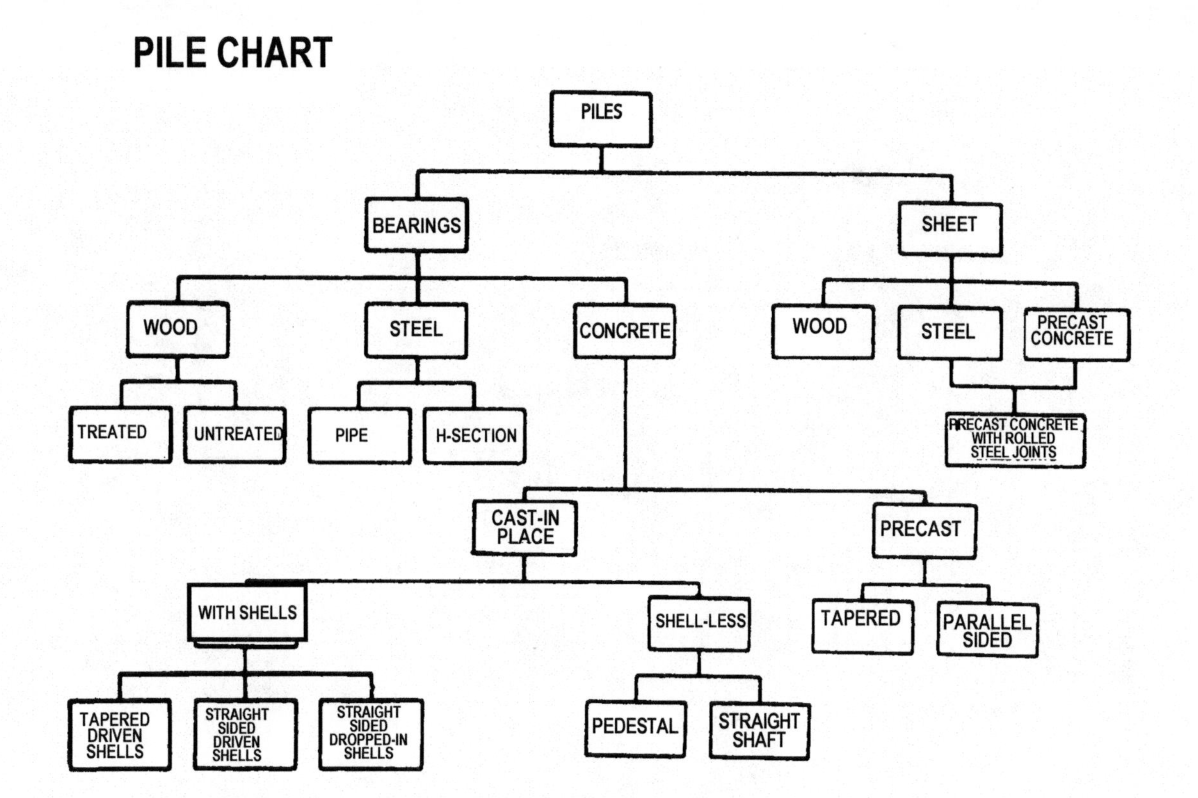

CONCRETE SHEET PILING

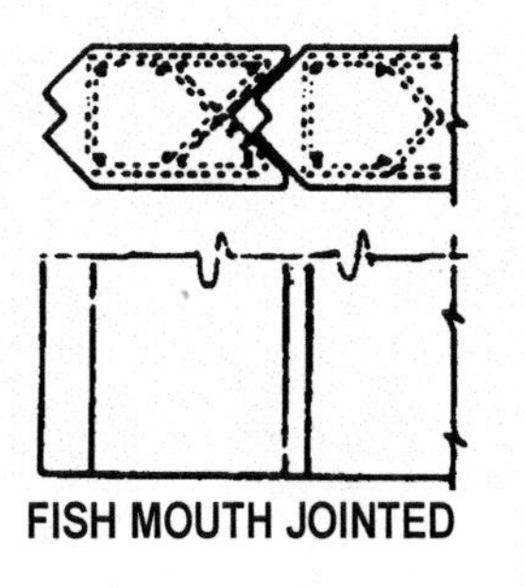
FISH MOUTH JOINTED

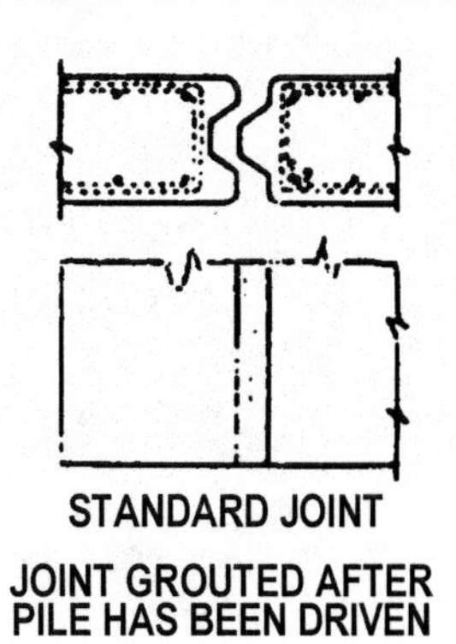
STANDARD JOINT

JOINT GROUTED AFTER PILE HAS BEEN DRIVEN

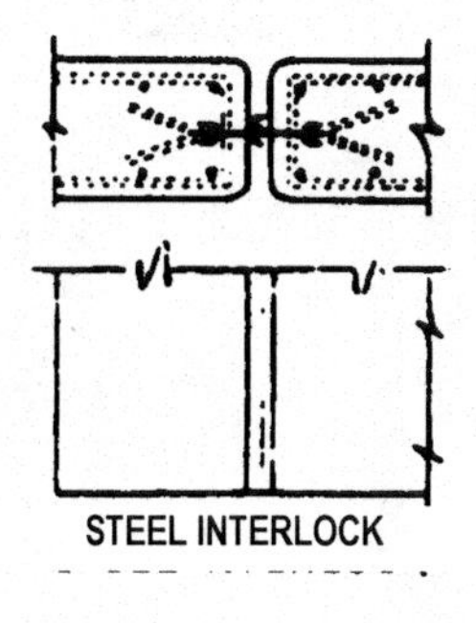
STEEL INTERLOCK

STEEL SHEET PILING

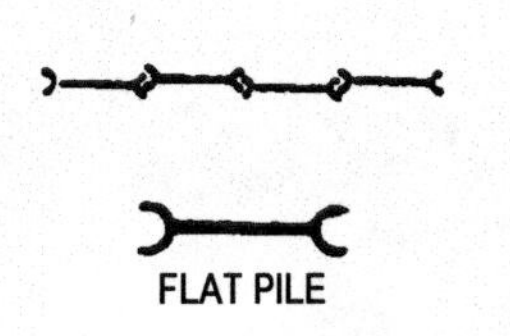
FLAT PILE

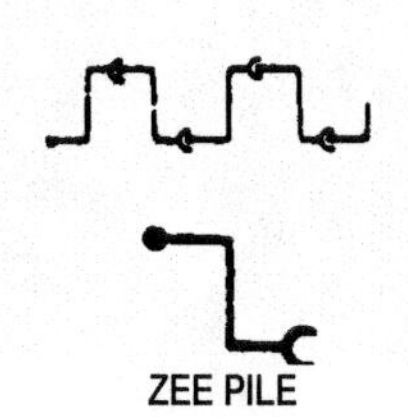
ZEE PILE

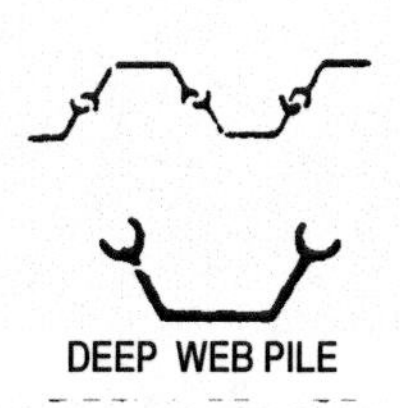
DEEP WEB PILE

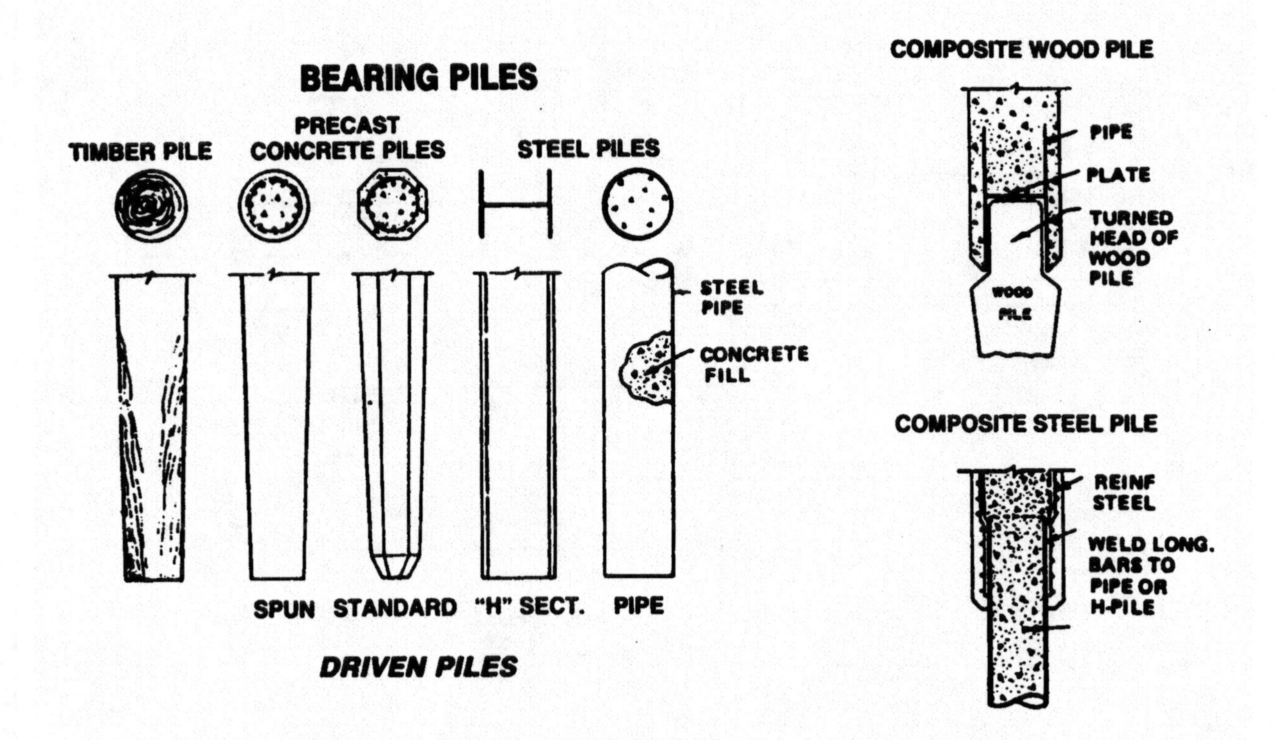

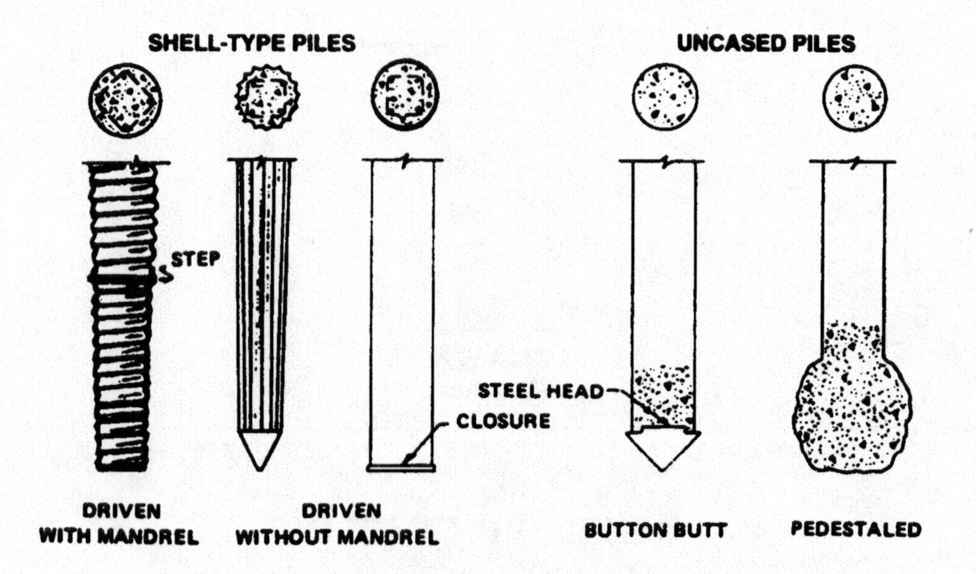

CAST-IN-PLACE PILES

CONCRETE MASONRY PAVING UNITS

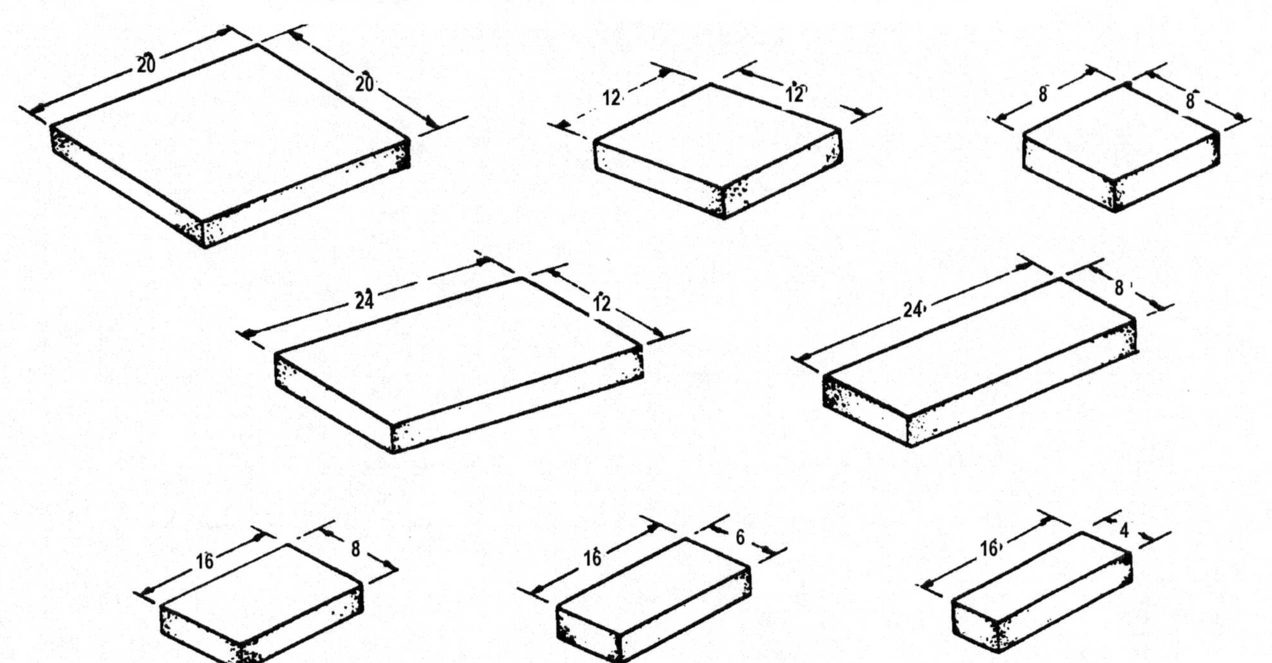

NOTE: Sizes are nominal and will vary by manufacture.

HEXAGON PAVER UNITS
Various Sizes Available

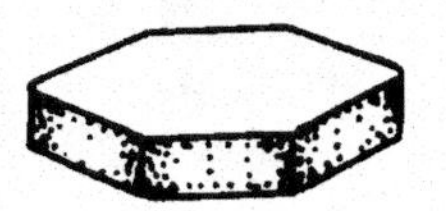

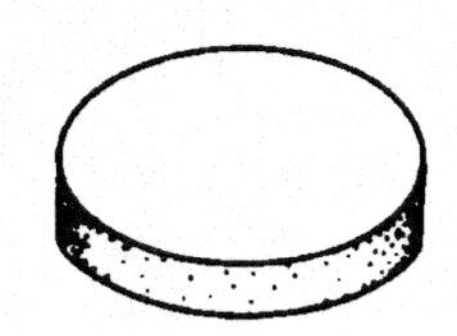

ROUND PAVING UNITS
Various Sizes Available

VEHICULAR PAVING UNITS

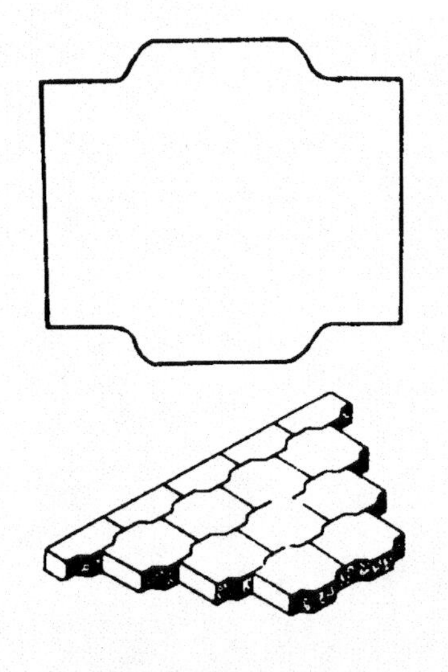

INTERLOCKING PAVER
7¼" X 3" X 8½"

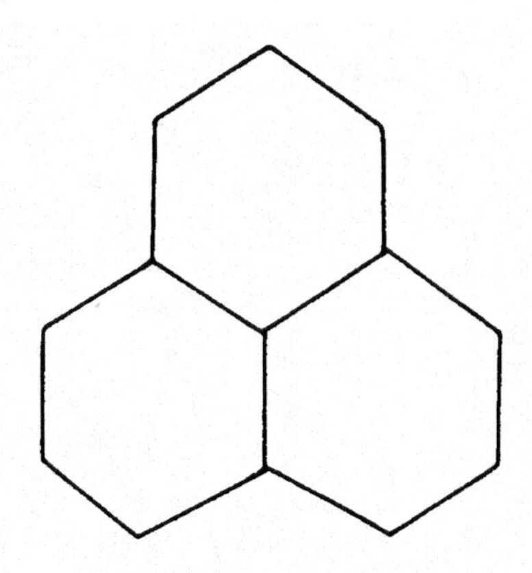

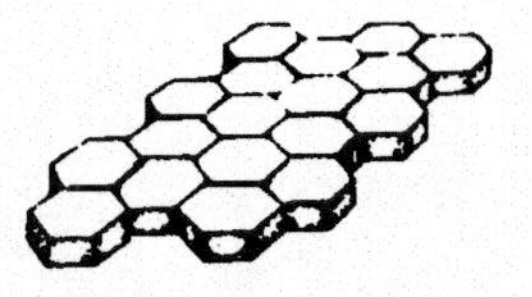

INTERLOCKING PAVER
12" X 3⅝" X 12"

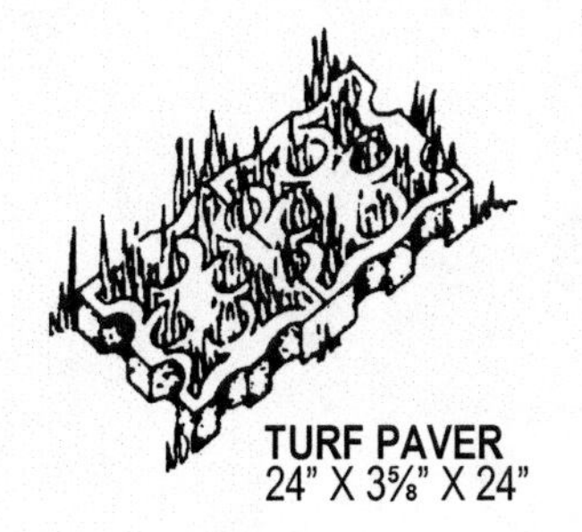

TURF PAVER
24" X 3⅝" X 24"

CAPACITIES FOR SEPTIC TANKS SERVING AN INDIVIDUAL DWELLING

No. of bedrooms	Capacity of tank (gals.)
2 or less	750
3	900
4	1,000

CONCRETE / FORMWORK

FORM NOMENCLAURE

FALSEWORK NOMENCLATURE

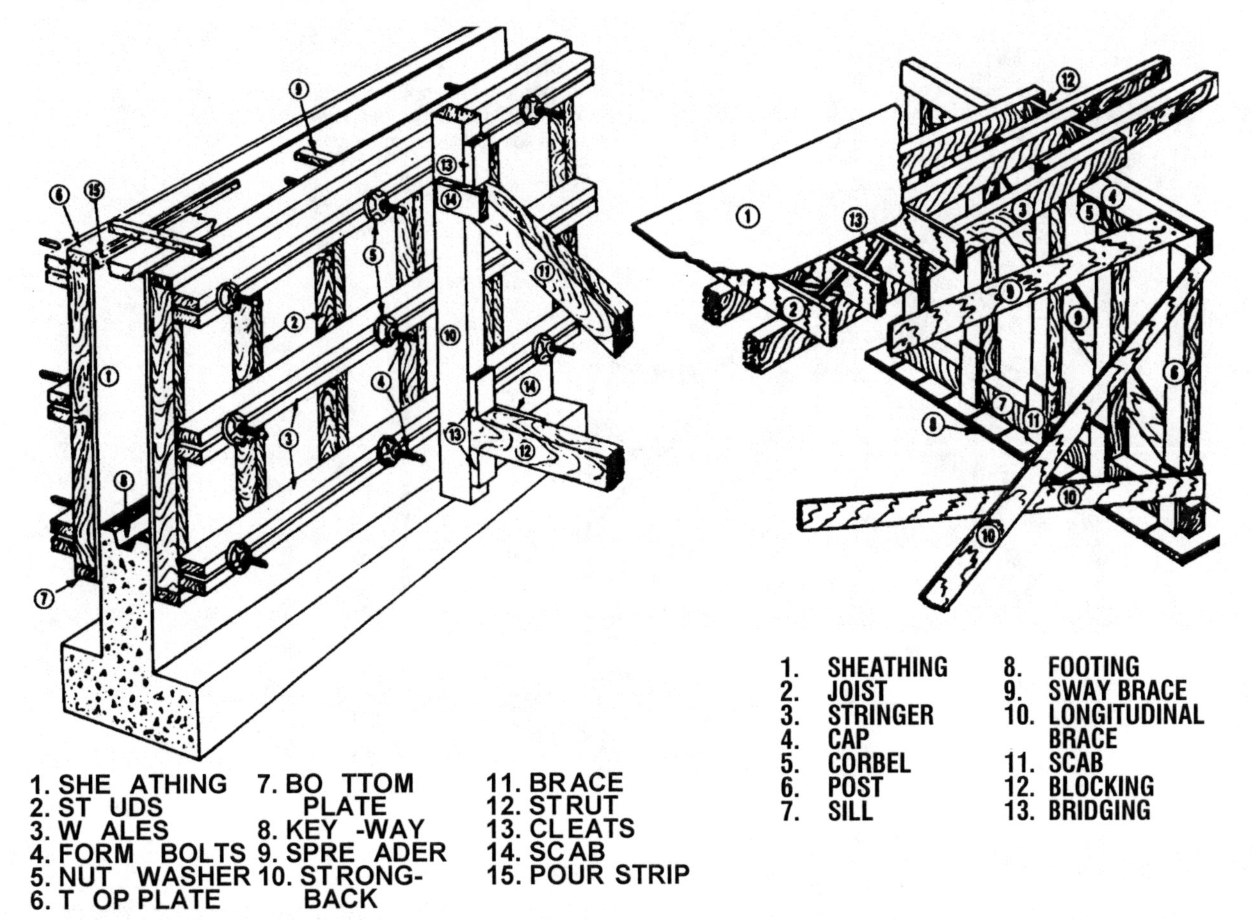

TYPICAL PAN-JOIST FORM CONSTRUCTION

3 FOOT SINGLE TAPERED END FORMS
3 FOOT INTERMEDIATE STEEL FORMS
END CAPS
WOOD SOFFIT PLANK
END CAP
STRINGER
HEADER FOR TEE-HEADED BEAM
BEAM FORM

TYPICAL WAFFLE SLAB FORM CONSTRUCTION

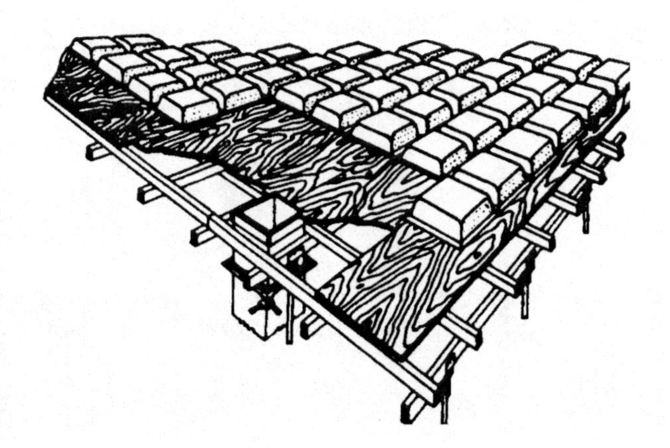

COMMON TYPES OF STEEL REINFORCEMENT BARS

ASTM specifications for billet steel reinforcing bars (A 615) require identification marks to be rolled into the surface of one side of the bar to denote the producer's mill designation, bar size and type of steel. For Grade 60 and Grade 75 bars, grade marks indicating yield strength must be show. Grade 40 bars show only three marks (no grade mark) in the following order:
1st — Producing Mill (usually an initial)
2nd — Bar Size Number (#3 through # 18)
3rd — Type (N for New Billet)

NUMBER SYSTEM — GRADE MARKS

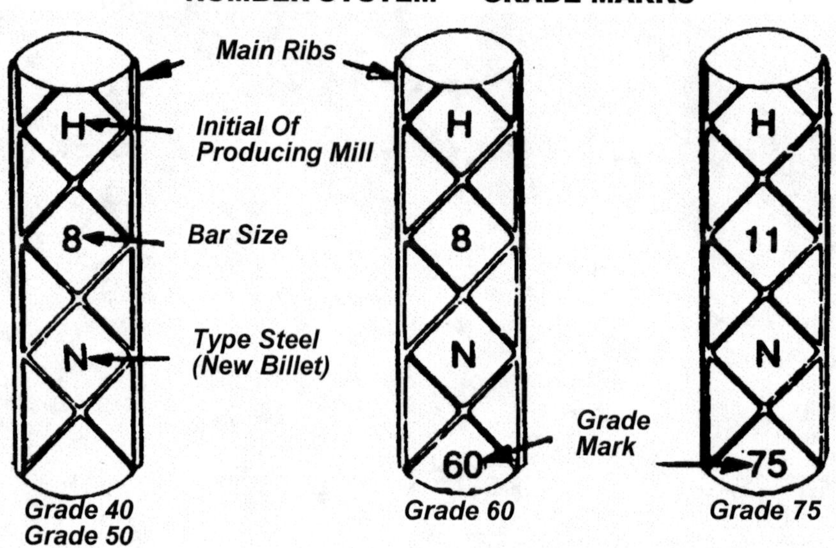

LINE SYSTEM — GRADE MARKS

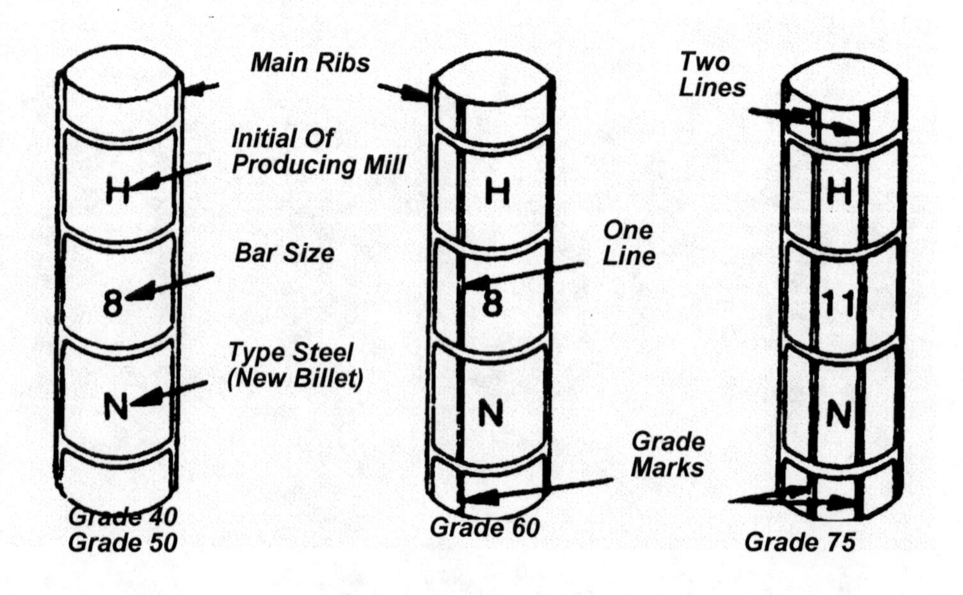

CONCRETE / REINFORCING STEEL 03210

STANDARD SIZES OF STEEL REINFORCEMENT BARS

STANDARD REINFORCEMENT BARS				
		Nominal Dimensions		
Bar Designation Number*	Nominal Weight lb. per ft.	Diameter, in.	Cross Sectional Area, sq. in.	Perimeter, in.
3	0.376	0.375	0.11	1.178
4	0.668	0.500	0.20	1.571
5	1.043	0.625	0.31	1.963
6	1.502	0.750	0.44	2.356
7	2.044	0.875	0.60	2.749
8	2.670	1.000	0.79	3.142
9	3.400	1.128	1.00	3.544
10	4.303	1.270	1.27	3.990
11	5.313	1.410	1.56	4.430
14	7.65	1.693	2.25	5.32
18	13.60	2.257	4.00	7.09

*The bar numbers are based on the number of 1/8 inches included in the nominal diameter of the bar.

Type of Steel & ASTM Specification No.	Size Nos. Inclusive	Grade	Tensile Strength Min., psi.	Yield (a) Min., psi
Billet Steel A 615	3-11	40	70,000	40,000
	3-11 14, 18	60	90,000	60,000
	11, 14, 18	75	100,000	75,000

CONCRETE / WELDED WIRE FABRIC 03220

Style Designation	Steel Area sq. in per ft.		Weight Approx. lbs. per 100 sq. ft.
	Longit.	Transv.	
Rolls			
6 x 6 — W1.4 x W1.4	.03	.03	21
6 x 8 — W2 x W2	.04	.04	29
6 x 6 — W2.9 x W2.9	.08	.06	42
6 x 6 — W4 x W4	.08	.08	58
4 x 4 — W1.4 x W1.4	.04	.04	31
4 x 4 — W2 x W2	.06	.06	43
4 x 4 — W2.9 x W2.9	.09	.09	62
4 x 4 — W4 x W4	.12	.12	
Sheets			
6 x 6 — W2.9 x W2.9	.06	.06	42
6 x 6 — W4 x W4	.08	.08	58
6 x 6 — W5.5 x W5.5	.11	.11	80
4 x 4 — W4 x W4	.12	.12	86

Insofar as is possible, the moisture content should be kept uniform to avoid problems in determining the proper amount of water to be added for mixing. Mixing water must be reduced to compensate for moisture in the aggregate in order to control the slump of the concrete and avoid exceeding the specified water-cement ratio.

Handling Concrete by Pumping Methods. Transportation and placement of concrete by pumping is another method gaining increased popularity. Pumps have several advantages, the primary one being that a pump will high-lift concrete without the need for an expensive crane and bucket. Since the concrete is delivered through pipe and hoses, concrete can be conveyed to remote locations in buildings, in tunnels, to locations otherwise inaccessible on steep hillside slopes for anchor walls, pipe bedding or encasement, or for placing concrete for chain link fence post bases. Concrete pumps have been found to be economical and expedient in the placement of concrete, and this has promoted the use and acceptance of this development. The essence of proper concrete pumping is the placement of the concrete in its final location without segregation.

Modern concrete pumps, depending on the mix design and size of line, can pump to a height of 200 feet or a horizontal distance of 1,000 feet. They can handle, economically, structural mixes, standard mixes, low slump mixes, mixes with two-inch maximum size aggregate and light weight concrete. When a special pump mix is required for structural concrete in a major structure, the mix design must be approved by the Engineer and checked and confirmed by the Supervisor of the Materials Control Group. The Inspector should obtain the pump manufacturer's printed information and evaluate its characteristics and ability to handle the concrete mixture specified for the project.

If concrete is being placed for a major reinforced structure, it is important that the placement continue without interruption. The Inspector should be sure that the contractor has ready access to a back-up pump to be used in the event of a breakdown. In order to further insure the success of the concrete placement by the pumping method, the user should be aware of the following points:

(a) A protective grating over the receiving hopper of the pump is necessary to exclude large pieces of aggregate or foreign material.

(b) The pump and lines require lubrication with a grout of cement and water. All of the excess grout is to be wasted prior to pumping the concrete.

(c) All changes in direction must be made by a large radius bend with a maximum bend of 90 degrees. Wye connections induce segregation and shall not be used.

(d) Pump lines should be made of a material capable of resisting abrasion and with a smooth interior surface having a low coefficient of friction. Steel is commonly used for pump lines, because a chemical reaction occurs between the concrete and the aluminum. Aluminum pipe should not be used for pumping concrete and some of the new plastic or rubber tubing is gaining acceptance. Hydrogen is generated which results in a swelling of the concrete, causing a significant reduction in compressive strength. This reaction is aggravated by any of the following: abrasive coarse aggregate, non-uniformly graded sand, low-slump concrete, low sand-aggregate ratio, high-alkali cement or when no air-entraining agent is used.

(e) During temporary interruptions in pumping, the hopper must remain nearly full, with an occasional turning and pumping to avoid developing a hard slug of concrete in the lines.

(f) Excessive line pressures must be avoided. When this occurs, check these points as the probable cause: segregation caused by too low a slump or too high a slump; large particle contamination caused by large pieces of aggregate or frozen lumps not eliminated by the grating; poor gradation of aggregates or particle shape; rich or lean spots caused by improper mixing.

(g) Corrections must be made to correct excessive slump loss as measured at the transit-mixed concrete truck and as measured at the hose outlet. This may be attributable to porous aggregate, high temperature or rapid setting mixes.

(h) Two transit-mix concrete trucks must be used simultaneously to deliver concrete into the pump hopper. These trucks must be discharged alternately to assure a continuous flow of concrete as trucks are replaced.

(i) Samples of concrete for test specimens prepared to determine the acceptance of the concrete quality are to be taken as required for conventional concrete.

Sampling is done before the concrete is deposited in the pump hopper. However, it is suggested that, where possible, the effect of pumping on the compressive strength be checked by taking companion samples, so identified, from the end of the pump line at the same time. The Record of Test must be properly noted as being a special mix used for pumping purposes. This will enable the Materials Control Group to compile a complete history of mix designs and their respective compressive strengths.

The prudent use of pumped concrete can result in economy and improved quality. However, only the control exercised by the operator will assure continued high standards of quality concrete.

Pump lines must be properly fastened to supports to eliminate excessive vibration. Couplings must be easily and securely fastened in a manner that will prevent mortar leakage. It is preferable to use the flexible hose only at the discharge point. This hose must be moved in such a manner as to avoid kinks or sharp bends. The pump line should be protected from excessive heat during hot weather by water sprinkling or shade.

Admixture	Purpose	Effects on Concrete	Advantages	Disadvantages
Accelerator	Hasten setting.	Improves cement dispersion and increases early strength.	Permits earlier finishing, form removal, and use of the structure.	Increases shrink age, decreases sulfate resistance, tends to clog mixing and handling equipment.
Air-Entraining Agent	Increase workability and reduce mixing water.	Reduces segregation, bleeding and increases freeze-thaw resistance. Increases strength	Increases workability and reduces finishing time.	Excess will reduce strength and increase slump. Bulks concrete volume.
Bonding Agent	Increase bond to old concrete.	Produces a non-dusting, slip resistant finish,	Permits a thin topping without roughening old concrete, self-curing, ready in one day.	Quick setting and susceptible to damage from fats, oils and solvents.
Densifier	To obtain dense concrete.	Increased workability and strength.	Increases workability and increases waterproofing characteristics, more impermeable.	Care must be used to reduce mixing water in proportion to amount used.
Foaming Agent	Reduce weight.	Increases insulating properties.	Produces a more plastic mix, reduces dead weight loads.	Its use must be very carefully regulated — following instructions explicitly.
Retarder	Retard setting.	Increases control of setting.	Provides more time to work and finish concrete.	Performance varies with cement used — adds to slump. Requires stronger forms.
Water Reducer and Retarder	Increase compressive and flexural strength.	Reduces segregation, bleeding, absorption, shrinkage, and increases cement dispersion.	Easier to place work, provides better control.	Performance varies with cement. Of no use in cold weather.
Water Reducer, Retarder and Air-Entraining Agent	Increases workability.	Improves cohesiveness. Reduces bleeding and segregation.	Easier to place and work.	Care must be taken to avoid excessive air entrainment.

ARCHITECTURAL WALL PATTERNS (BONDS)

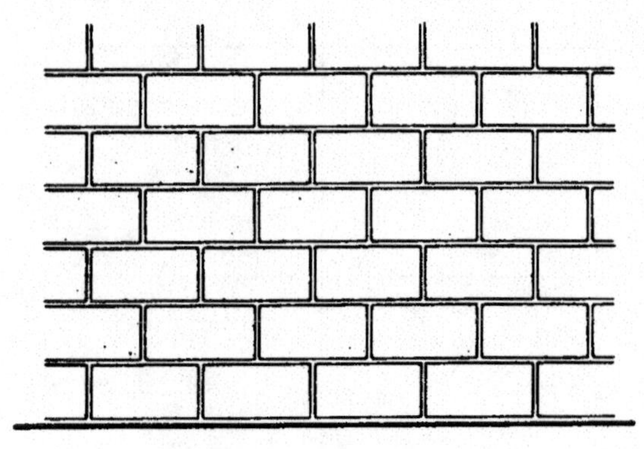

COMMON BOND
8" x 16" UNITS

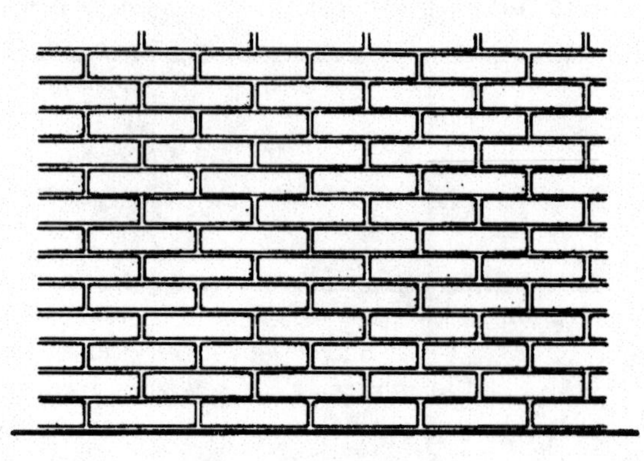

COMMON BOND
4" x 16" UNITS

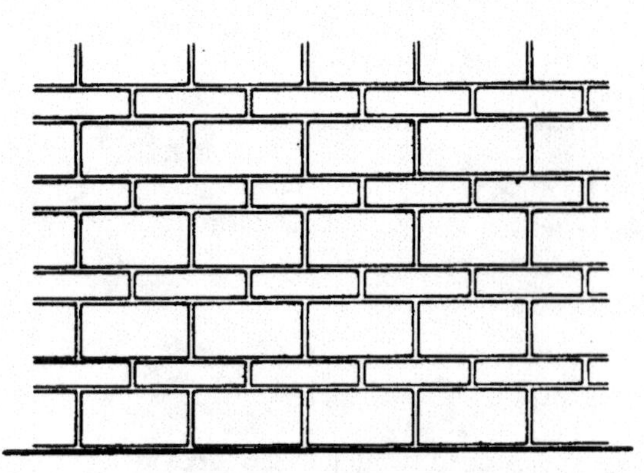

COURSED ASHLER
8" x 16" & 4" x 16" UNITS

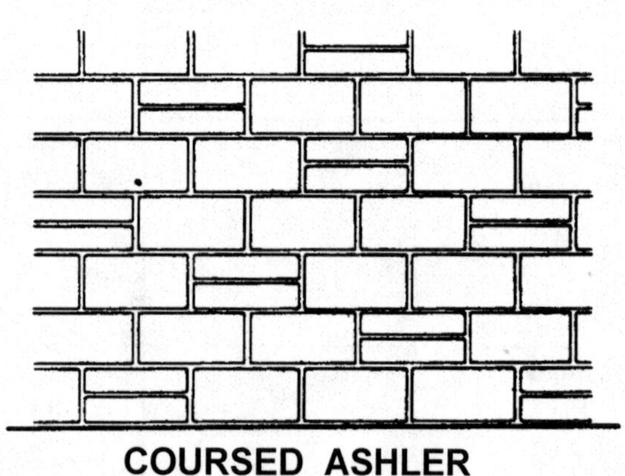

COURSED ASHLER
8" x 16" & 4" x 16" UNITS

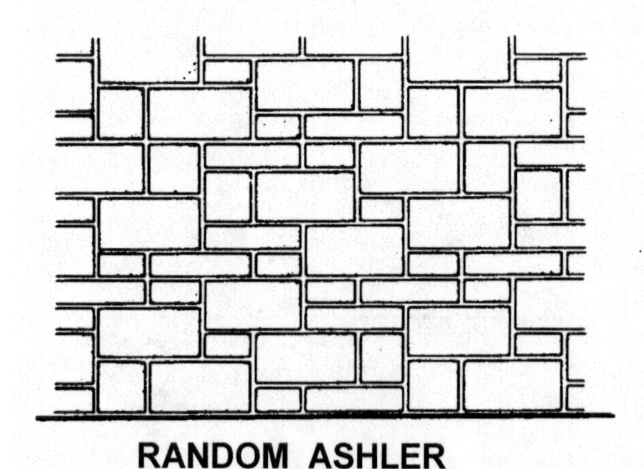

RANDOM ASHLER
8" x 16" & 4" x 16" UNITS
AND 4" X 8" UNITS

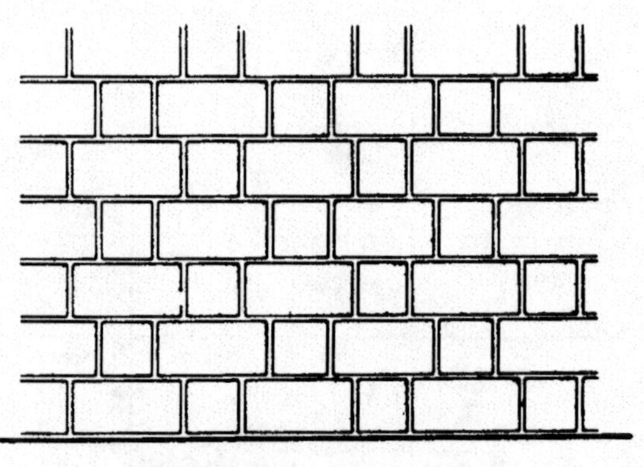

COURSED ASHLER
8" x 16" & 8" x 8" UNITS

ARCHITECTURAL WALL PATTERNS (BONDS) — (Continued)

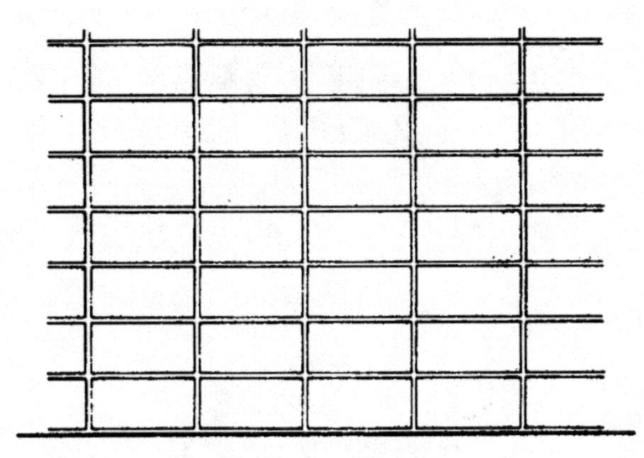

STACKED BOND

8" x 16" UNITS

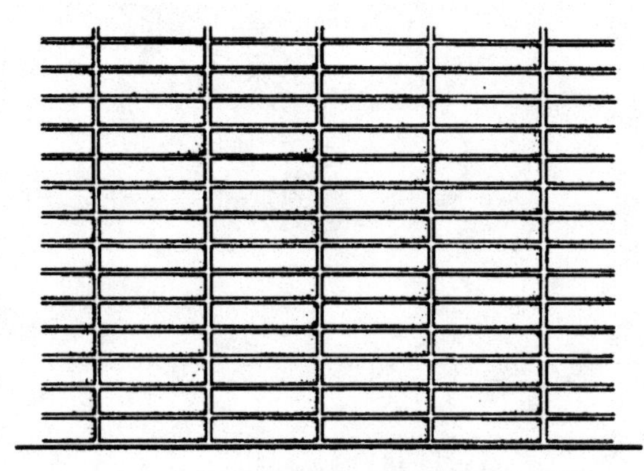

STACKED BOND

8" x 16" & 4" x 16" UNITS

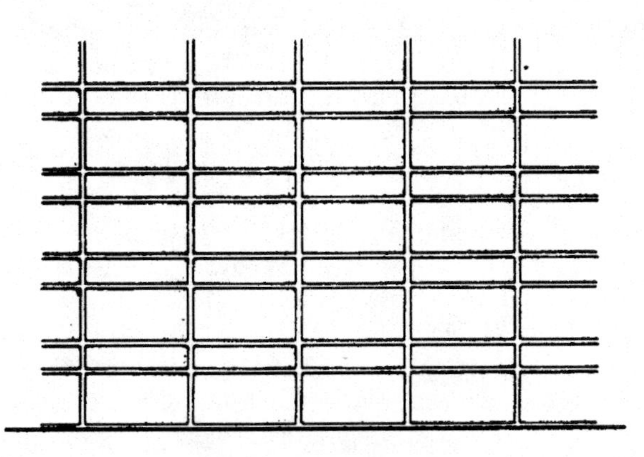

STACKED BOND

8" x 16" & 4" x 16" UNITS

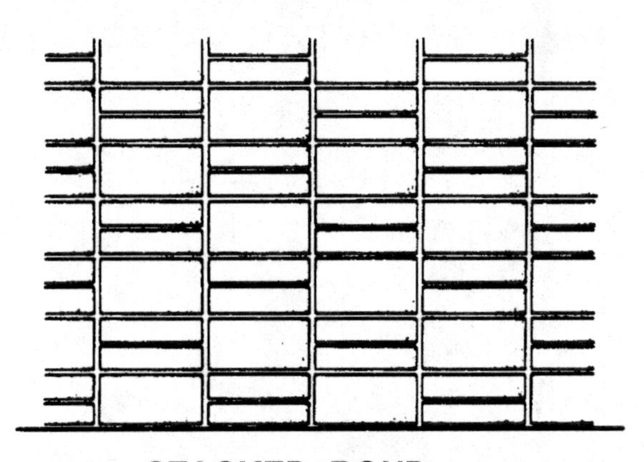

STACKED BOND

8" x 16" & 4" x 16" UNITS

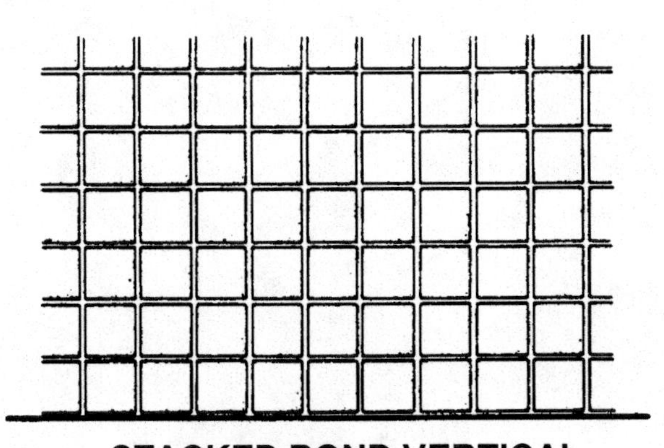

STACKED BOND VERTICAL SCORED UNITS

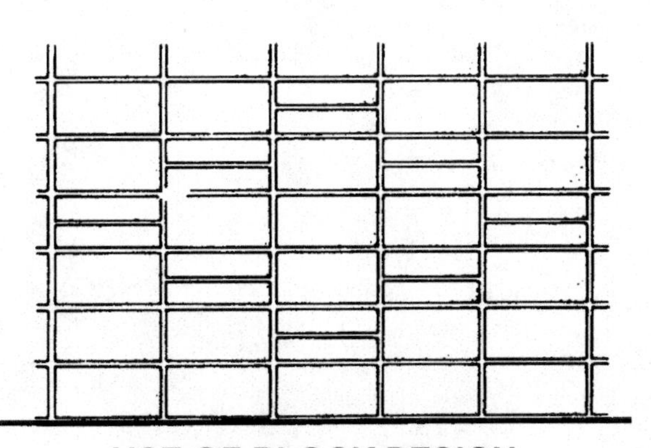

USE OF BLOCK DESIGN IN STACKED BOND

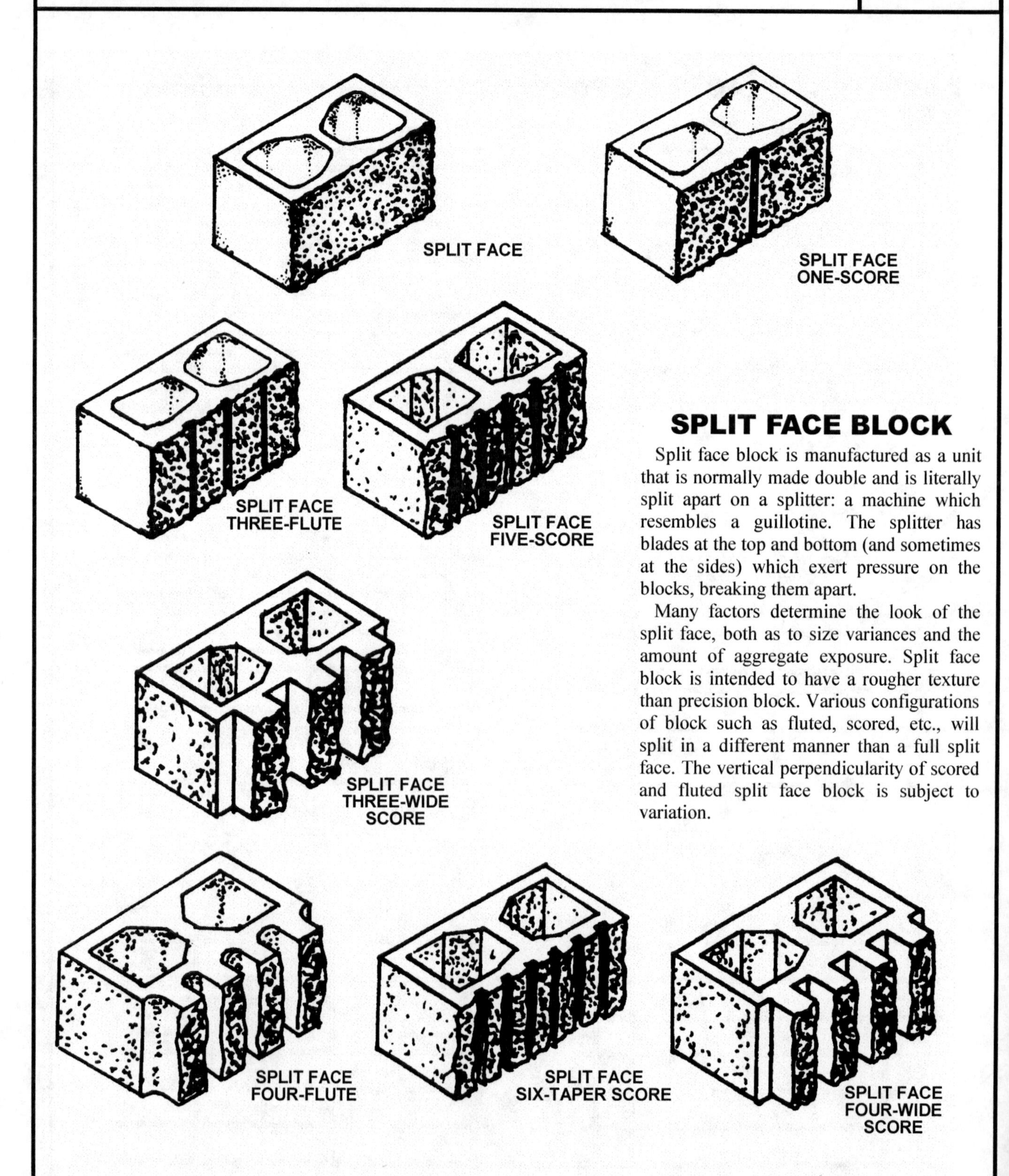

SPLIT FACE BLOCK

Split face block is manufactured as a unit that is normally made double and is literally split apart on a splitter: a machine which resembles a guillotine. The splitter has blades at the top and bottom (and sometimes at the sides) which exert pressure on the blocks, breaking them apart.

Many factors determine the look of the split face, both as to size variances and the amount of aggregate exposure. Split face block is intended to have a rougher texture than precision block. Various configurations of block such as fluted, scored, etc., will split in a different manner than a full split face. The vertical perpendicularity of scored and fluted split face block is subject to variation.

NOTE: Split face units shown in this manual are a small sampling of the broad range of concrete masonry architectural units available from the industry on special order. Depths and widths of scores vary. Consult a local manufacturer for specific information.

TYPICAL DETAILS — LINTELS AND BOND BEAMS

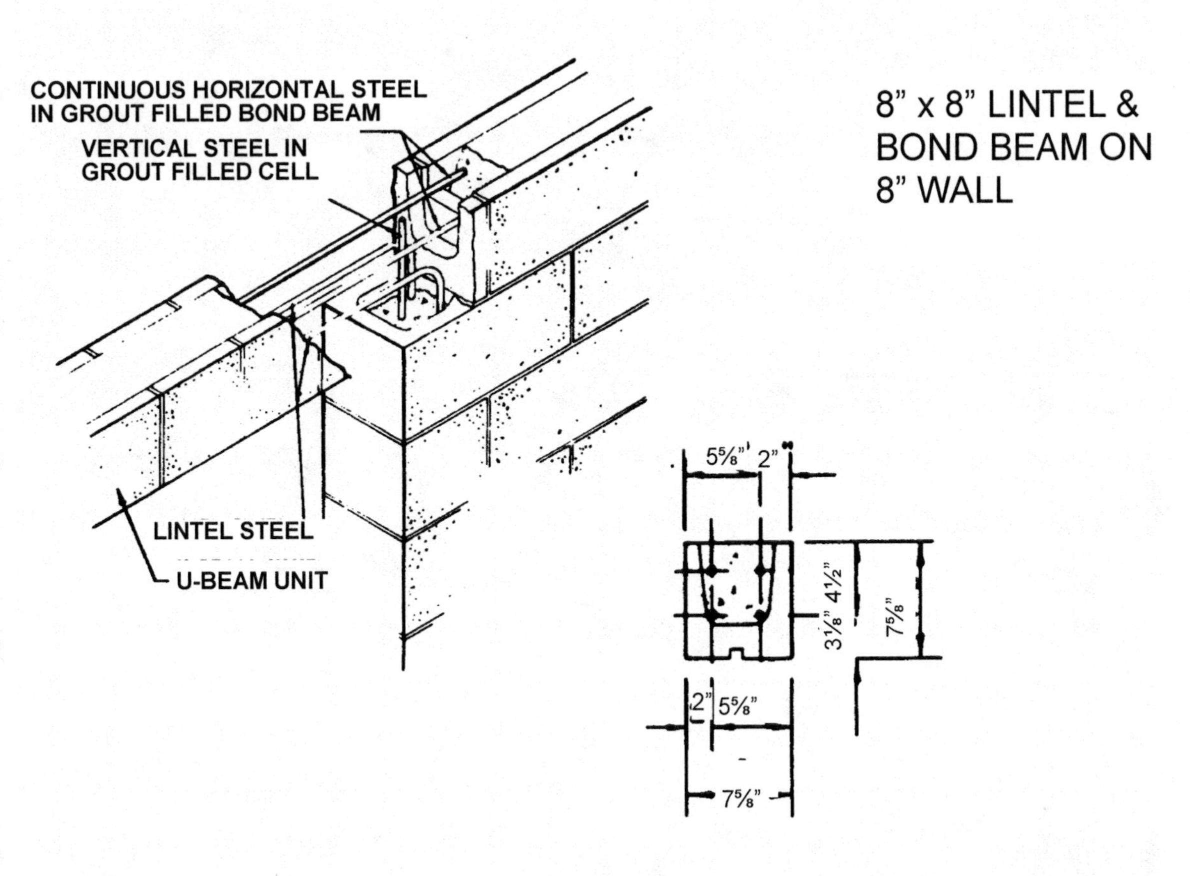

8" x 16"
BOND BEAM ON
8" WALL

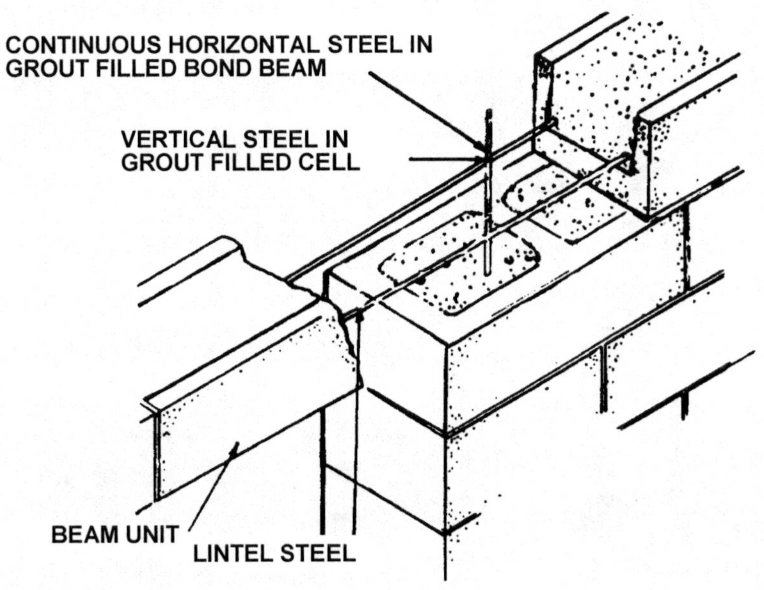

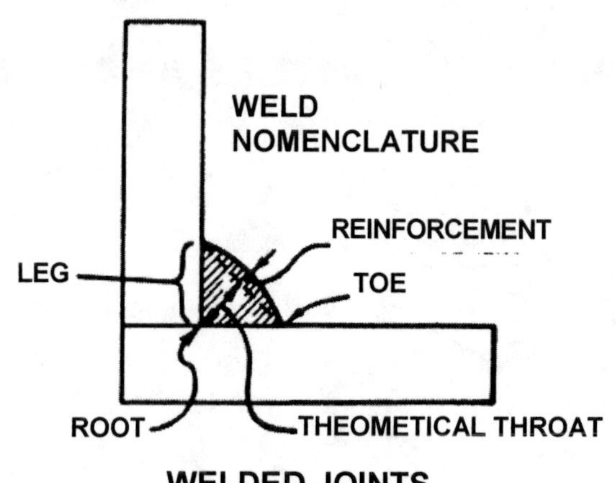

WELDED JOINTS

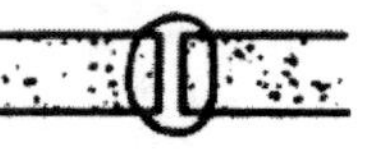
SQUARE BUTT

SINGLE VEE BUTT

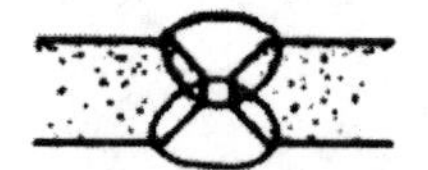
DOUBLE VEE BUTT

SINGLE U BUTT

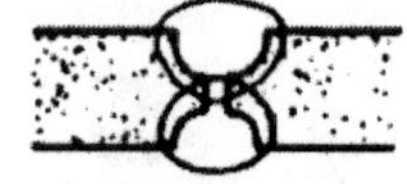
DOUBLE U BUTT

SINGLE FILLET LAP

DOUBLE FILLET LAP

STRAP JOINT

SINGLE BEVEL TEE

DOUBLE BEVEL TEE

SINGLE J TEE

SQUARE TEE

DOUBLE J TEE

CLOSED CORNER (FLUSH) JOINT

HALF OPEN CORNER JOINT

WELDING POSITIONS

FLAT (F)

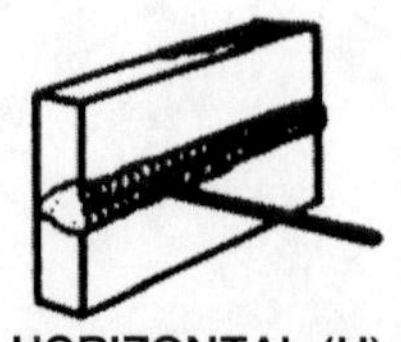
HORIZONTAL (H)

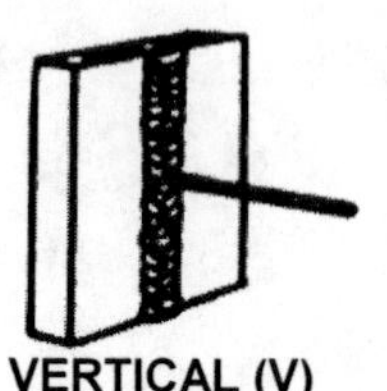
VERTICAL (V)

OVERHEAD (OH)

BOLTS IN COMMON USAGE

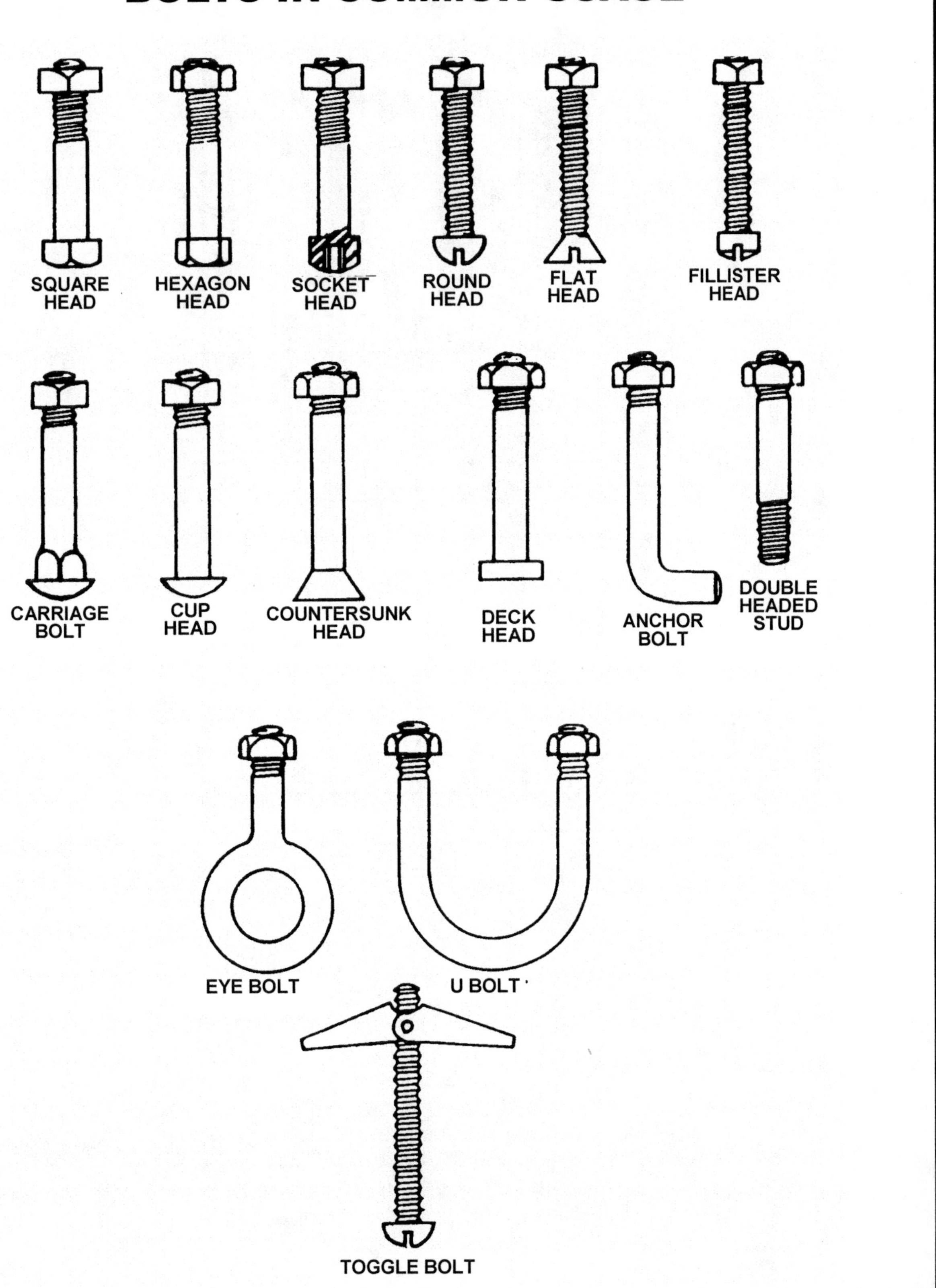

COMMON WIRE NAILS (ACTUAL SIZE)

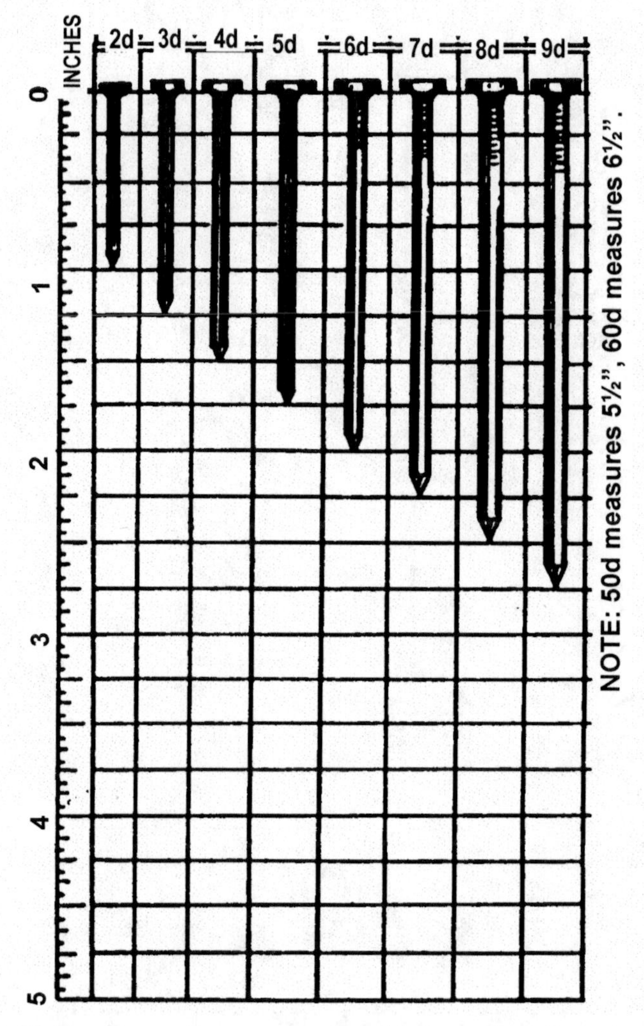

Cut Nails. Cut nails are angular-sided, wedge-shaped with ablunt point.

Wire Nails. Wire nails are round shafted, straight, pointed nails, and are used more generally than cut nails. They are stronger than cut nails and do not buckle as easily when driven into hard wood, but usually split wood more easily than cut nails. Wire nails are available in a variety of sizes varying from two penny to sixty penny.

Nail Finishes. Nails are available with special finishes, Some are galvanized or cadmium plated to resist rust. To increase the resistance to withdrawal, nails are coated with resins or asphalt cement (called cement coated). Nails which are small, sharp-pointed, and often placed in the craftsman's mouth (such as lath or plaster board nails) generally blued and sterilized.

COMMON WIRE NAILS (ACTUAL SIZE) (Cont.)

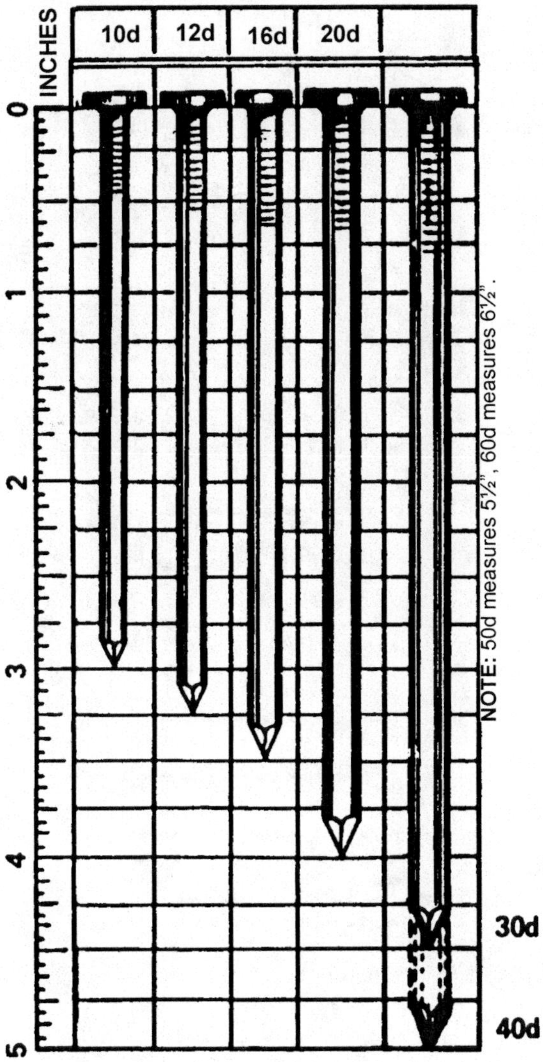

INCHES
0
1
2
3
4
5

10d
12d
16d
20d
30d
40d

NOTE: 50d measures 5½", 60d measures 6½".

COMMON WIRE
CONCRETE 
PLASTER BOARD 
SMOOTH BOX
SCAFFOLD, (DUPLEX HD) 
ROOFING 
CASING
SHINGLE
FINISHING
SLATING 
BLUED LATH 
CUT

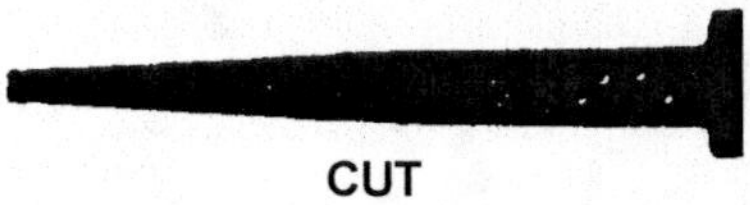

PLYWOOD — BASIC GRADE MARKS
AMERICAN PLYWOOD ASSOCIATION (APA)

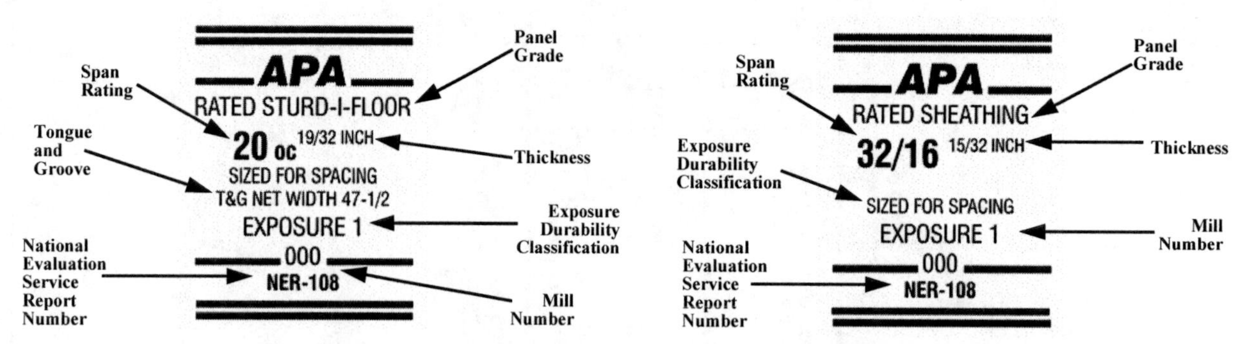

The American Plywood Association's trademarks appear only on products manufactured by APA member mills. The marks signify that the product is manufactured in conformance with APA performance standards and/or U.S. Product Standard PS 1-83 for Construction and Industrial Plywood.

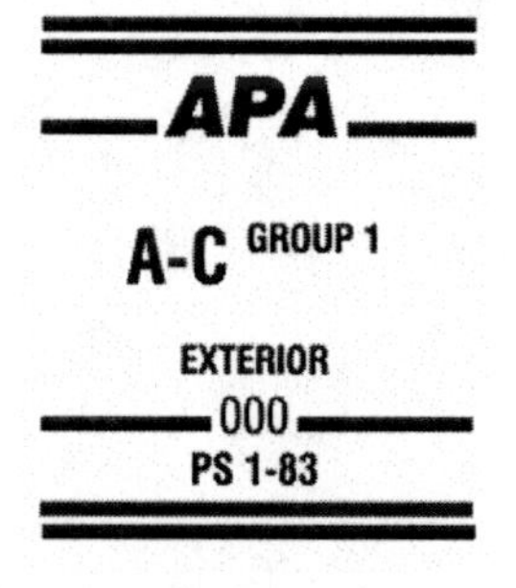

APA A-C
For use where appearance of one side is important in exterior applications such as soffits, fences, structural uses, boxcar and truck linings, farm buildings, tanks, trays, commercial refrigerators, etc. **Exposure Durability Classification: Exterior. Common Thicknesses:** ¼, 11/32, ¾, 15/32, ½, 19/32, 5/8, 23/32, ¾.

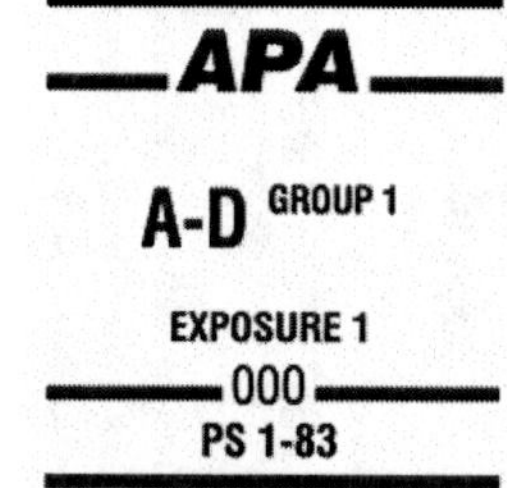

APA A-D
For use where appearance of only one side is important in interior applications, such as paneling, built-ins, shelving, partitions, flow racks, etc. **Exposure Durability Classifications: Interior, Exposure 1. Common Thicknesses:** ¼, 11/32, 3/8, 15/32, ½, 19/32, 5/8, 23/32, ¾.

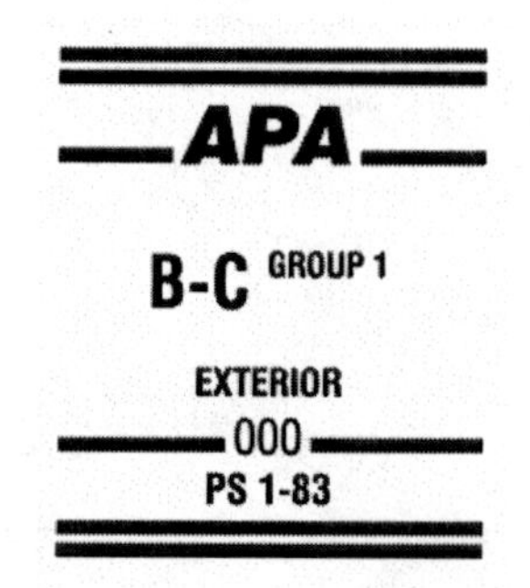

APA B-C
Utility panel for farm service and work buildings, boxcar and truck linings, containers, tanks, agricultural equipment, as a base for exterior coatings and other exterior uses. **Exposure Durability Classification: Exterior. Common Thicknesses:** ¼, 11/32, ¾, 15/32, ½, 19/32, 5/8, 23/32, ¾.

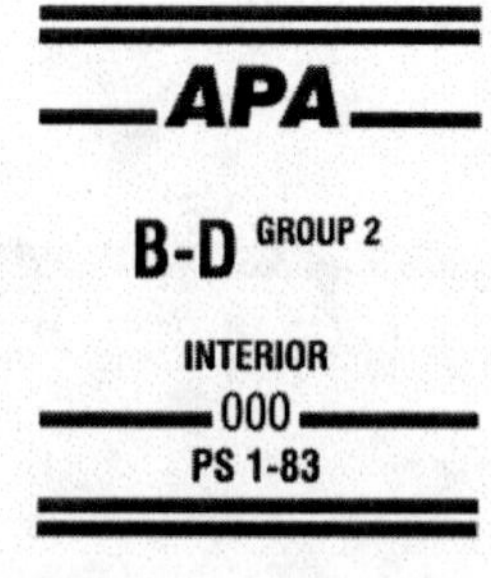

APA B-D
Utility panel for backing, sides or built-ins, industry shelving, slip sheets, separator boards, bins and other interior or protected applications. **Exposure Durability Classifications: Interior, Exposure 1. Common Thicknesses:** ¼, 11/32, 3/8, 15/32, ½, 19/32, 5/8, 23/32, ¾.

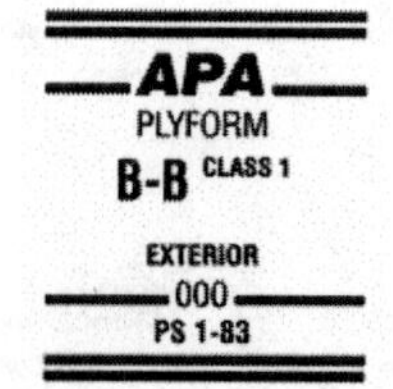

APA proprietary concrete form panels designed for high reuse. Sanded both sides and mill-oiled unless otherwise specified. Class I, the strongest, stiffest and more commonly available, is limited to Group 1 faces, Group 1 or 2 crossbands, and Group 1, 2, 3 or 4 inner plies. Class II is limited to Group 1 or 2 faces (Group 3 under certain conditions) and Group 1, 2, 3 or 4 inner plies. Also available in HDO for very smooth concrete finish, in Structural I, and with special overlays. **Exposure Durability Classification: Exterior. Common Thicknesses:** 19/32, ¾, 23/32, ¼.

Plywood panel manufactured with smooth, opaque, resin-treated fiber overlay providing ideal base for paint on one or both sides. Excellent material choice for shelving, factory work surfaces, paneling, built-ins, signs and numerous other construction and industrial applications. Also available as a 303 Siding with texture-embossed or smooth surface on one side only and Structural I. **Exposure Durability Classification: Exterior. Common Thicknesses:** 11/32, 3/8, 19/32, ½, 19/32, 5/8, 23/32, ¾.

SPECIALTY PANELS

HDO • A • A • G-1 • EXT APA • 000 • PS1 83

Plywood panel manufactured with a hard, semi-opaque resin-fiber overlay on both sides. Extremely abrasion resistant and ideally suited to scores of punishing construction and industrial applications, such as concrete forms, industrial tanks, work surfaces, signs, agricultural bins, exhaust ducts, etc. Also available with skid-resistant screen-grid surface and in Structural I. ***Exposure Durability Classification:*** **Exterior.** ***Common Thicknesses:*** **3/8, ½, 5/8, 3/4**

MARINE• A • A • EXT APA • 000 • PS1 83

Specialty designed plywood panel made only with Douglas fir or western larch, solid jointed cores, and highly restrictive limitations on core gaps and faces repairs. Ideal for both hulls and other marine applications. Also available with HDO or MDO faces. ***Exposure Durability Classification:*** **Exterior.** ***Common Thicknesses:*** **1/4, 3/8, ½, 5/8, 3/4.**

Unsanded and touch-sanded panels, and panels with "B" or better veneer on one side only, usually carry the APA trademark on the panel back. Panels with both sides of "B" or better veneer, or with special overlaid surfaces (such as Medium Density Overlay), carry the APA trademark on the panel edge, like this:

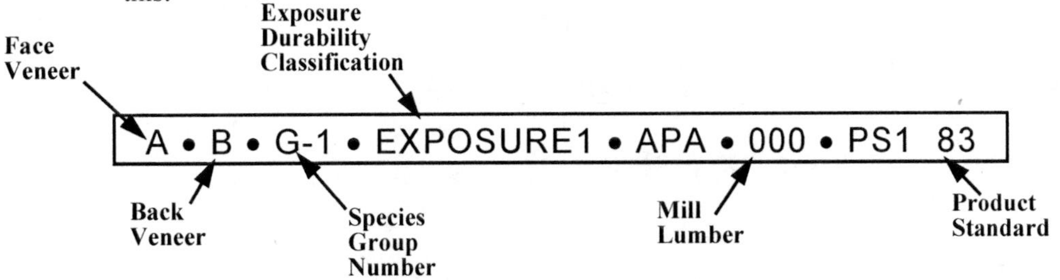

GLOSSARY OF TERMS

Some of the words and terms used in the grading of lumber follow:

Bow. A deviation flatwise from a straight line drawn from end to end of the piece. It is measured at the point of greatest distance from the straight line.

Checks. A separation of the wood which normally occurs across the annual rings and usually as a result of seasoning.

Crook. A deviation edgewise from a straight line drawn from end to end of the piece. It is measured at the point of greatest distance from the straight line.

Cup. A deviation from a straight line drawn across the piece from edge to edge. It is measured at the point of greatest distance from the straight line.

Flat Grain. The annual growth rings pass through the piece at an angle of less than 45 degrees with the flat surface of the piece.

Warp. Any deviation from a true or plane surface, including crook, cup, bow or any combination thereof.

Mixed Grain. The piece may have vertical grain, flat grain or a combination of both vertical and flat grain.

Pitch. An accumulation of resin which occurs in separations in the wood or in the wood cells themselves.

Shake. A separation of the wood which usually occurs between the rings of annual growth.

Splits. A separation of the wood due to tearing apart of the wood cells.

Vertical Grain. The annual growth rings pass through the piece at an angle of 45 degrees or more with the flat surface of the piece.

Wane. Bark or lack of wood from any cause, except eased edges (rounded) on the edge or corner of a piece of lumber.

LUMBER GRADING

GRADING-MARK ABBREVIATIONS

GRADES

(Listed alphabetically — not by quality)

COM	Common
CONST	Construction
ECON	Economy
No. 1	Number One
SEL-MER	Select Merchantable
SEL-STR	Select Structural
STAN	Standard
UTIL	Utility

ALSC TRADEMARKS

CLIS	California Lumber Inspection Service
NELMA	Northeastern Lumber Mfrs. Assoc., Inc.
NH&PMA	Northern Hardwood & Pine Mfrs. Assoc., Inc.
PLIB	Pacific Lumber Inspection Bureau
RIS	Redwood Inspection Service
SPIB	Southern Pine Inspection Bureau
TP	Timber Products Inspection
WCLB	West Coast Lumber Inspection Bureau
WWP	Western Wood Products Association

SPECIES GROUPINGS

AF	Alpine Fir
DF	Douglas Fir
HF	Hem Fir
SP	Sugar Pine
PP	Ponderosa Pipe
LP	Lodgepole Pine
IWP	Idaho White Pine
ES	Engelmann Spruce
WRC	Western Red Cedar
INC CDR	Incense Cedar
L	Larch
LP	Lodgepole Pine
MH	Mountain Hemlock
WW	White Wood

MOISTURE CONTENT

S-GRN	Surfaced at a moisture content of more than 19%.
S-DRY	Surfaced at a moisture content of 19% or less.
MC-15	Surfaced at a moisture content of 15% or less.

FRAMING ESTIMATING RULES OF THUMB

For 16" O.C. stud partitions figure 1 stud for every L.F. of wall; add for top and bottom plates.

For any type of framing, the quantity of basic framing members (in L.F.) can be determined based on spacing and surface area (S.F.):

12" O.C.	1.2 L.F./S.F.
16" O.C.	1.0 L.F./S.F.
24" O.C.	0.8 L.F./S.F.

(Doubled-up members, bands, plates, framed openings, etc., must be added.)

Framing accessories, nails, joist hangers, connectors, etc., should be estimated as separate material costs. Installation should be included with framing. Rule of thumb allowance is 0.5 to 1.5% of lumber cost for rough hardware. Another is 30 to 40 pounds of nails per M.B.F.

BOARD FEET/LINEAR FEET FOR LUMBER

Nominal Size	Actual Size	Board Feet Per Linear Foot	Linear Feet Per 1000 Board Feet
1 x 2	¾ x 1 ½	.167	6000
1 x 3	¾ x 2 ½	.250	4000
1 x 4	¾ x 3 ½	.333	3000
1 x 6	¾ x 5 ½	.500	2000
1 x 8	¾ x 7 ¼	.666	1500
1 x 10	¾ x 9 ¼	.833	1200
1 x 12	¾ x 11 ¼	1.0	1000
2 x 2	1 ½ x 1 ½	.333	3000
2 x 3	1 ½ x 2 ½	.500	2000
2 x 4	1 ½ x 3 ½	.666	1500
2 x 6	1 ½ x 5 ½	1.0	1000
2 x 8	1 ½ x 7 ¼	1.333	750
2 x 10	1 ½ x 9 ¼	1.666	600
2 x 12	1 ½ x 11 ¼	2.0	500

Redwood. Redwood is a fairly strong and moderately lightweight material. The heartwood is red but the sapwood is white. One of the principal advantages of redwood is that the heartwood is highly resistant (but not entirely immune) to decay, fungus and insects. Standard Specifications require that all redwood used in permanent installations shall be "select heart." Grade marking shall be in accordance with the standards established in the California Redwood Association. Grade marking shall be done by, or under the supervision of the Redwood Inspection Service.

Redwood is graded for specific uses as indicated in the following table:

REDWOOD GRADING

Type of Lumber	Grade	Typical Use
Grades for Dimension Only Listed Here	Clear All Heart	Exceptionally fine, knot free, straight-grained timbers. This grade is used primarily for stain finish work of high quality.
	Clear	Same as Clear All Heart except that this grade may contain sound sapwood and medium stain.
	Select Heart	**This grade only is to be used in Agency work, unless otherwise specified in the plans or specifications.** It is sound, live heartwood free from splits or streaks with sound knots. It is generally used where the timber is in contact with the ground, as in posts, mudsills, etc.
	Select Construction Heart	Slightly less quality than Select Heart. It may have some sapwood in the piece. Used for general construction purposes when redwood is needed.
	Construction Common	Same requirement as Construction Heart except that it will contain sapwood and medium stain. Its resistance to decay and insect attack is reduced.
	Merchantable	Used for fence posts, garden stakes, etc.
	Economy	Suitable for crating, bracing and temporary construction.

DOUGLAS FIR GRADING

Type of Lumber	Grade	Typical Use
Select Structural Joists and Planks	Select Structural	Used where strength is the primary consideration, with appearance desirable.
	No. 1	Used where strength is less critical and appearance not a major consideration.
	No. 2	Used for framing elements that will be covered by subsequent construction.
	No. 3	Used for structural framing where strength is required but appearance is not a factor.
Finish Lumber	Superior	For all types of uses as casings, cabinet, exposed members, etc., where a fine appearance is desired.
	Prime	
	E	
Boards (WCLIB)* * Grading is by West Coast Lumber Inspection Bureau rules, but sizes conform to Western Wood Products Assn. rules. These boards are still manufactured by some mills.	Select Merchantable	Intended for use in housing and light construction where a knotty type of lumber with finest appearance is required.
	Construction	Used for sub-flooring, roof and wall sheathing, concrete forms, etc. Has a high degree of serviceability.
	Standard	Used widely for general construction purposes, including subfloors, roof and wall sheathing, concrete forms, etc. Seldom used in exposed construction because appearance.
	Utility	Used in general construction where low cost is a factor and appearance is not important. (Storage shelving, crates, bracing, temporary scaffolding etc.)

BOARD FEET CONVERSION TABLE

Nominal Size (In.)	ACTUAL LENGTH IN FEET								
	8	10	12	14	16	18	20	22	24
1 x 2		1 2/3	2	2 1/3	2 2/3	3	3 ½	3 2/3	4
1 x 3		2 ½	3	3 ½	4	4 ½	5	5 ½	6
1 x 4	2 ¾	3 1/3	4	4 2/3	5 1/3	6	6 2/3	7 1/3	8
1 x 5		4 1/6	5	5 5/6	6 2/3	7 ½	8 1/3	9 1/6	10
1 x 6	4	5	6	7	8	9	10	11	12
1 x 7		5 5/8	7	8 1/6	9 1/3	10 ½	11 2/3	12 5/6	14
1 x 8	5 1/3	6 2/3	8	9 1/3	10 2/3	12	13 1/3	14 2/3	16
1 x 10	6 2/3	8 1/3	10	11 2/3	13 1/3	15	16 2/3	18 1/3	20
1 x 12	8	10	12	14	16	18	20	22	24
1¼ x 4		4 1/6	5	5 5/6	6 2/3	7 ½	8 1/3	9 1/6	10
1¼ x 6		6 ¼	7 ½	8 ¾	10	11 ¼	12 ½	13 ¾	15
1¼ x 8		8 1/3	10	11 2/3	13 1/3	15	16 2/3	18 1/3	20
1¼ x 10		10 5/12	12 ½	14 7/12	16 2/3	18 ¾	20 5/6	22 11/12	25
1¼ x 12		12 ½	15	17 ½	20	22 ½	25	27 ½	30
1½ x 4	4	5	6	7	8	9	10	11	12
1½ x 6	6	7 ½	9	10 ½	12	13 ½	15	16 ½	18
1½ x 8	8	10	12	14	16	18	20	22	24
1½ x 10	10	12 ½	15	17 ½	20	22 ½	25	27 ½	30
1½ x 12	12	15	18	21	24	27	30	33	36
2 x 4	5 1/3	6 2/3	8	9 1/3	10 1/3	12	13 1/3	14 2/3	16
2 x 6	8	10	12	14	16	18	20	22	24
2 x 8	10 2/3	13 1/3	16	18 2/3	21 1/3	24	26 2/3	29 1/3	32
2 x 10	13 1/3	16 2/3	20	23 1/3	26 2/3	30	33 1/3	36 2/3	40
2 x 12	16	20	24	28	32	36	40	44	48
3 x 6	12	15	18	21	24	27	30	33	36
3 x 8	16	20	24	28	32	36	40	44	48
3 x 10	20	25	30	35	40	45	50	55	60
3 x 12	24	30	36	42	48	54	60	66	72
4 x 4	10 2/3	13 1/3	16	18 2/3	21 1/3	24	26 2/3	29 1/3	32
4 x 6	16	20	24	28	32	36	40	44	48
4 x 8	21 1/3	26 2/3	32	37 1/3	42 2/3	48	53 1/3	58 2/3	64
4 x 10	26 2/3	33 1/3	40	46 2/3	53 1/3	60	66 2/3	73 1/3	80
4 x 12	32	40	48	56	64	72	80	88	96

MOISTURE PROTECTION / ROOFING — 07000

SLOPE AREA CALCULATIONS

Rise and Run	Multiply Flat Area by	LF of Hips or Valleys per LF of Common Run
2 in 12	1.014	1.424
3 in 12	1.031	1.436
4 in 12	1.054	1.453
5 in 12	1.083	1.474
6 in 12	1.118	1.500
7 in 12	1.158	1.530
8 in 12	1.202	1.564
9 in 12	1.250	1.600
10 in 12	1.302	1.641
11 in 12	1.357	1.685
12 in 12	1.413	1.732

MOISTURE PROTECTION / DOWNSPOUTS — 07600

DOWNSPOUT/VERTICAL LEADER CALCULATIONS

Roof Type	Slope	S.F. Roof/ Sq. In. Leader
Gravel	Less than ¼" per foot	300
Gravel	Greater than ¼" per foot	250
Metal or Shingle	Any	200

Alternate calculations:

$$\text{Diameter of downspout/leader} = 1.128\sqrt{\frac{\text{Area of drainage}}{\text{SF Roof/Sq. Inch}}}$$

TYPICAL MINIMUM SIZE OF VERTICAL CONDUCTORS AND LEADERS

Size of leader or conductor (Inches)	Maximum projected roof area (Square feet)
2	544
2 ½	987
3	1,610
4	3,460
5	6,280
6	10,200
8	22,000

TYPICAL MINIMUM SIZE OF ROOF GUTTERS

	MAXIMUM PROJECTED ROOF AREA FOR GUTTERS OF VARIOUS SLOPES			
Diameter gutter (Inches)	1/16 in. Ft. slope (Sq. ft.)	1/8 in. per Ft. slope (Sq. ft.)	1/4 in. per Ft. slope (Sq. ft.)	1/2 in. per Ft. /slope (Sq. ft.)
3	170	240	340	480
4	360	510	720	1,020
5	625	880	1,250	1,770
6	960	1,360	1,920	2,770
7	1,380	1,950	2,760	3,900
8	1,990	2,800	3,980	5,600
10	3,600	5,100	7,200	10,000

STUDLESS SOLID PARTITION

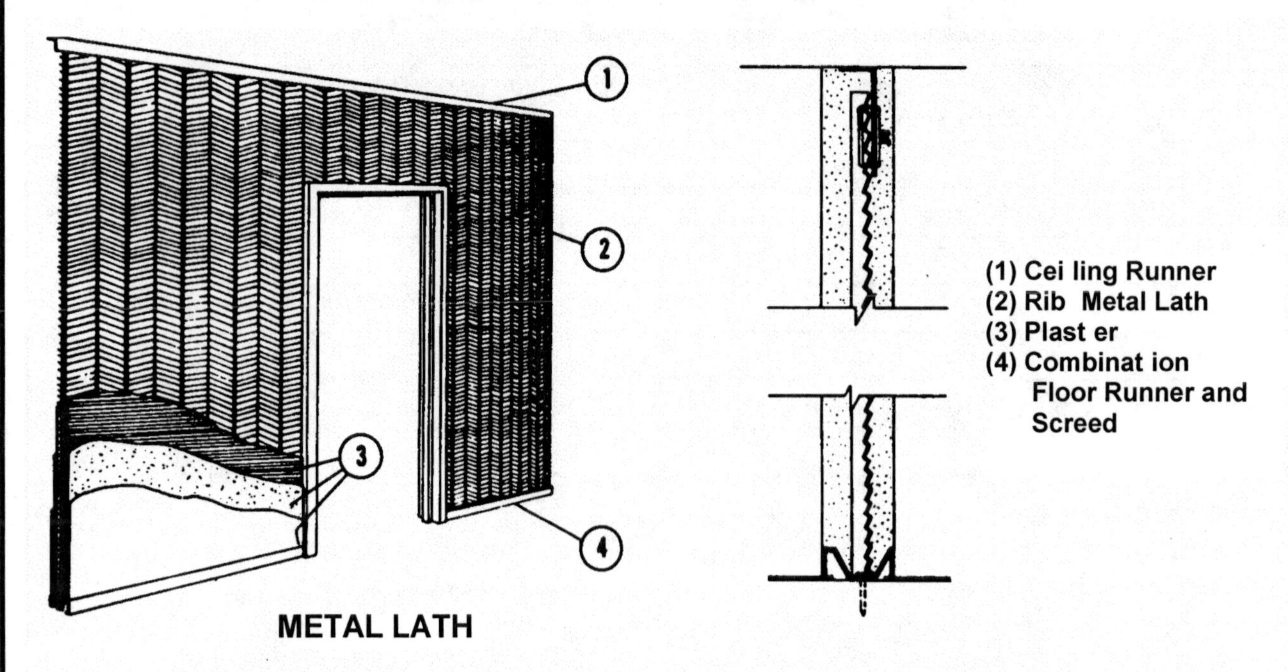

STUDLESS SOLID PARTITION

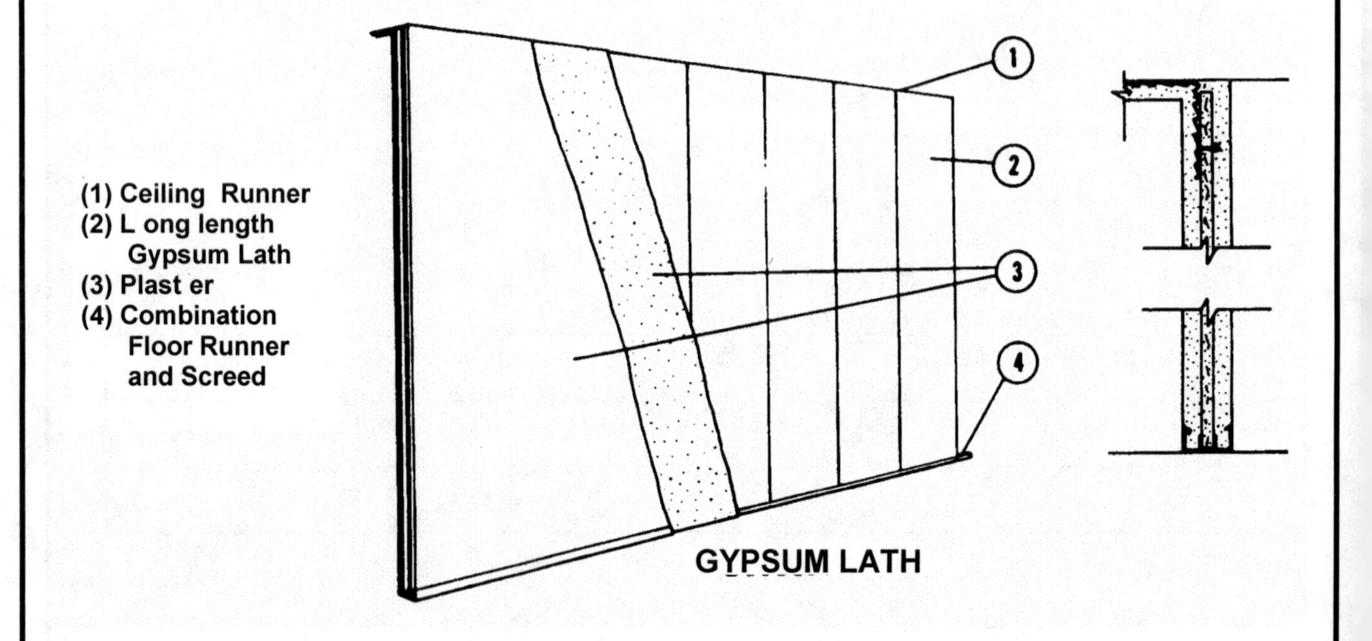

STEEL STUD

Hollow Partition

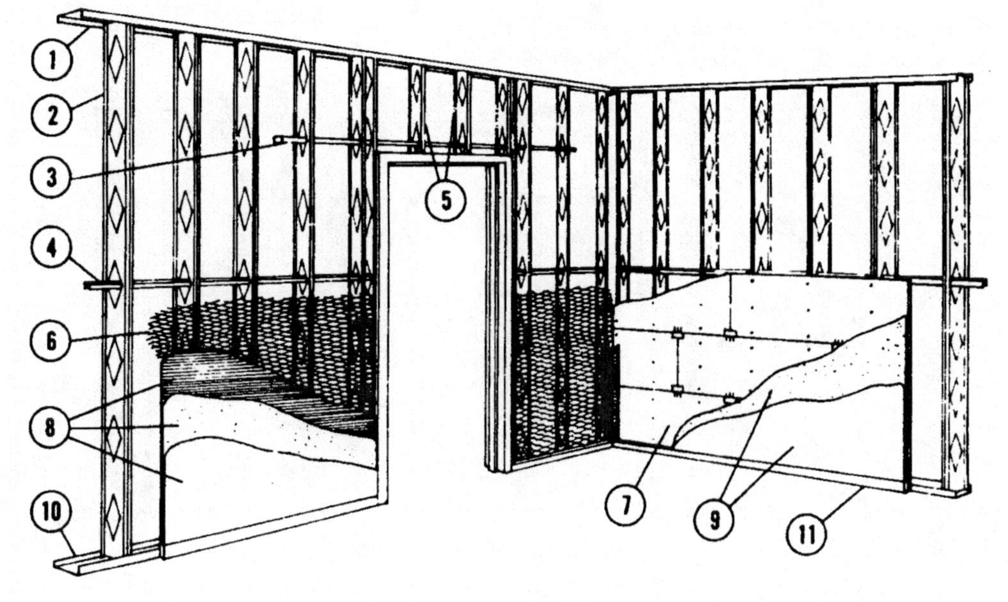

(1) Ceiling Runner Track
(2) Nailable Stud
(3) Door Opening Stiffener
(4) Partition Stiffener
(5) Jack Studs
(6) Metal or Wire Fabric Lath (screwed or wire tied)
(7) Gypsum Lath (nailed, clipped or screwed)
(8) Three Coats of Plaster (Scratch, Brown, Finish)
(9) Two Coats of Plaster (Brown, Finish)
(10) Floor Runner Track
(11) Flush Metal Base

SCREW STUD

Hollow Partition

(1) Ceiling Runner Track
(2) Screwed Stud
(3) Door Opening Stiffener
(4) Partition Stiffener
(5) Jack Studs
(6) Metal or Wire Fabric Lath (screwed on)
(7) Gypsum Lath (screwed on)
(8) Three Coats of Plaster (Scratch, Brown, Finish)
(9) Two Coats of Plaster (Brown, Finish)
(10) Floor Runner Track
(11) Flush Metal Base

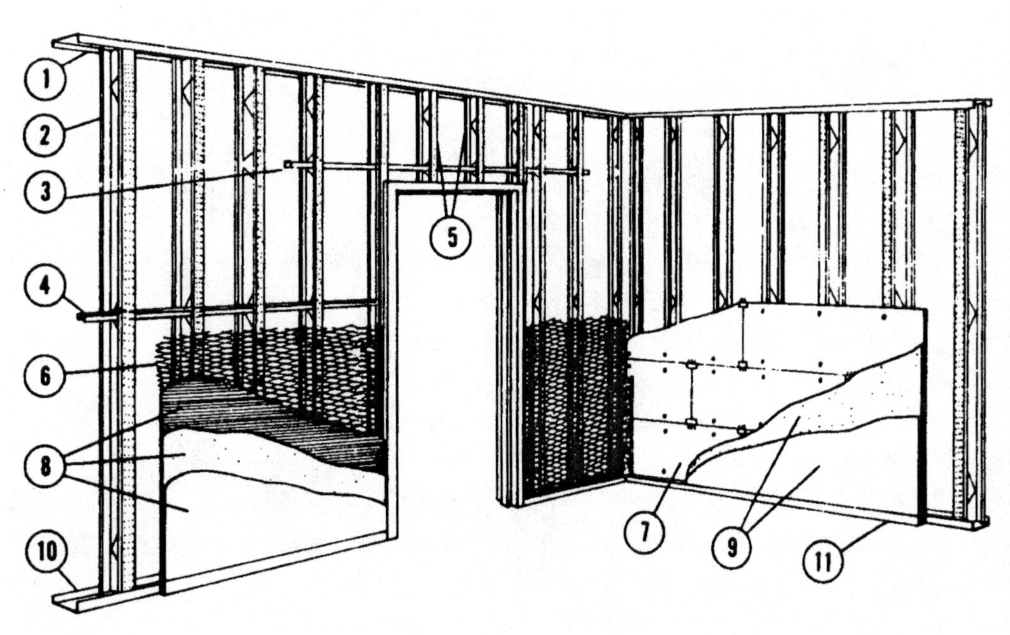

LOAD-BEARING HOLLOW PARTITION

Structural Stud

(1) Ceiling Runner Track
(2) Structural Studs (prefabricated)
(3) Structural Stud (nailable)
(4) Jack Studs
(5) Partition Stiffener (bridging)
(6) Metal or Wire Fabric Lath (wired-tied, nailed or stapled)
(7) Gypsum Lath (nailed or stapled)
(8) Three Coats of Plaster (Scratch, Brown, Finish)
(9) Two Coats of Plaster (Brown, Finish)

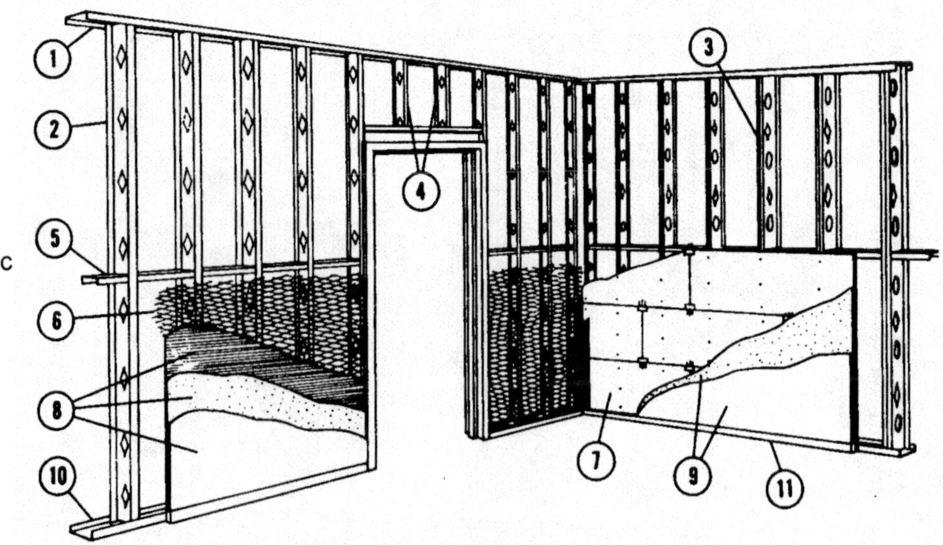

(10) Floor Runner Track
(11) Flush Metal Base

VERTICAL FURRING

With Studs

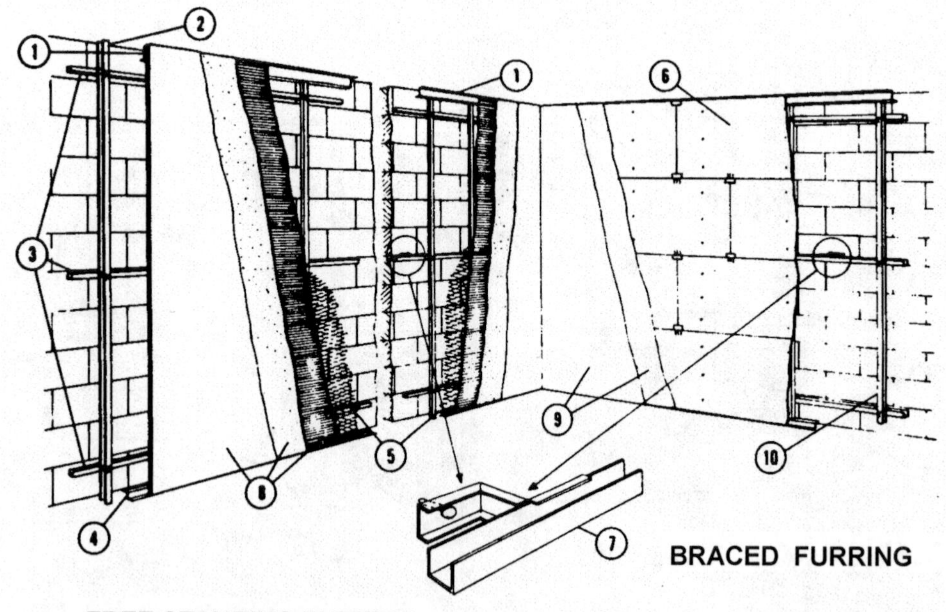

BRACED FURRING

FREE STANDING FURRING

(1) Ceiling Runner Track
(2) Channel Studs
(3) Horizontal Stiffener
(4) Floor Runner
(5) Metal or wire Fabric Lath
(6) Gypsum Lath
(7) Bracing
(8) Three Coats of Plaster (Scratch, Brown, Finish)
(9) Two Coats of Plaster (Brown, Finish)
(10) Screw Channel Studs

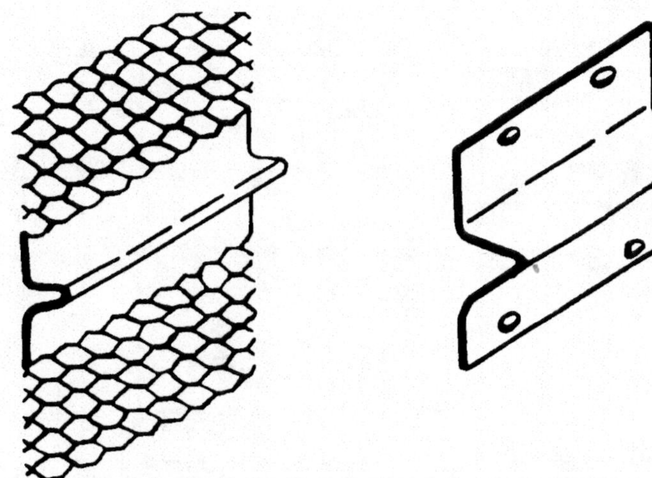

BASE OR PARTING SCREEDS

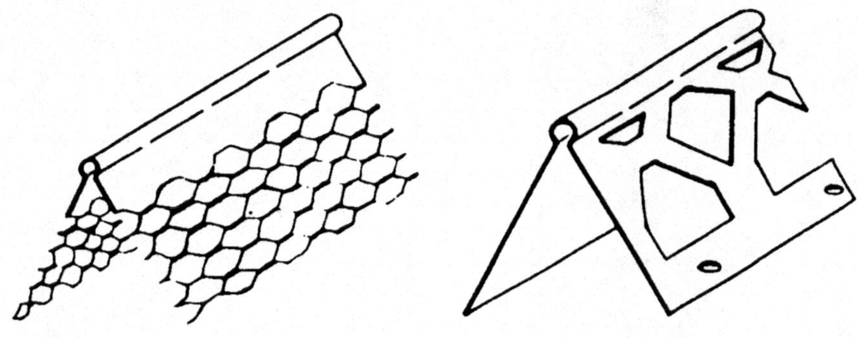

SMALL NOSE CORNER BEADS

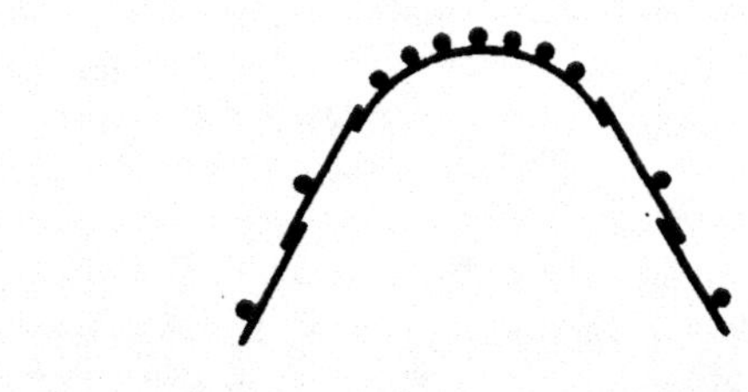

WIRE BULL NOSE CORNER BEADS

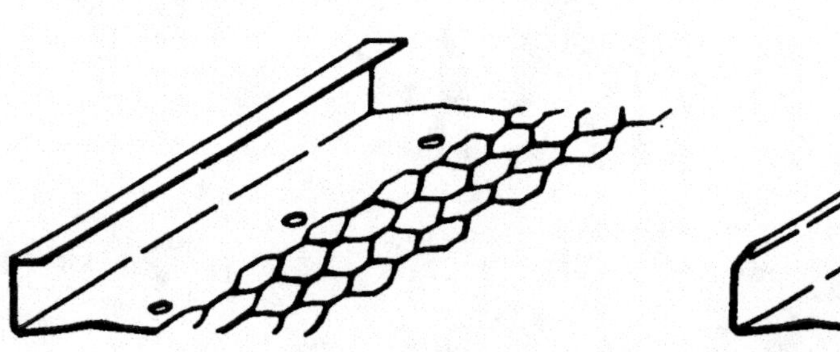

SQUARE CASING BEADS

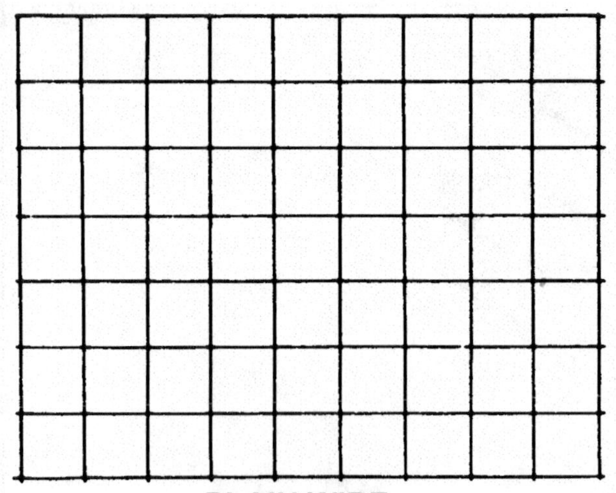

PLAIN WIRE
FABRIC LATH

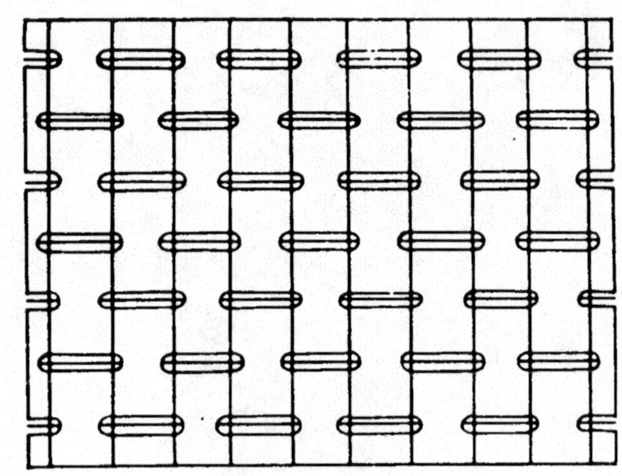

SELF-FURRING
WIRE FABRIC LATH

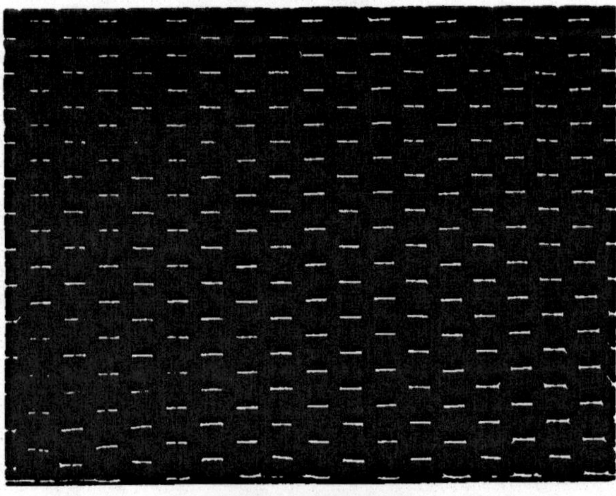

PAPER BACKED
WOVEN WIRE
FABRIC LATH

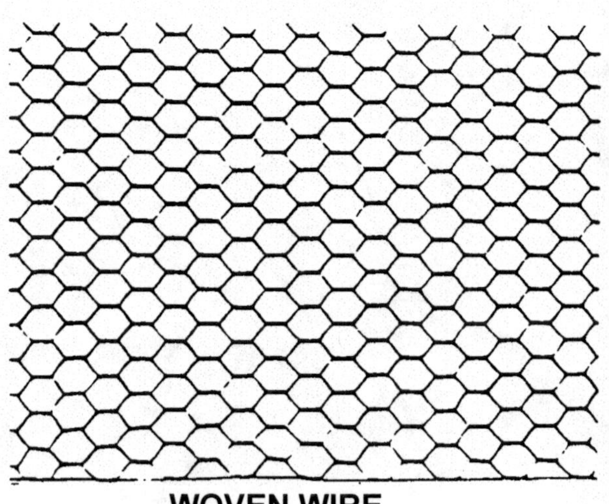

WOVEN WIRE
FABRIC LATH
(Also Available Self-Furred)

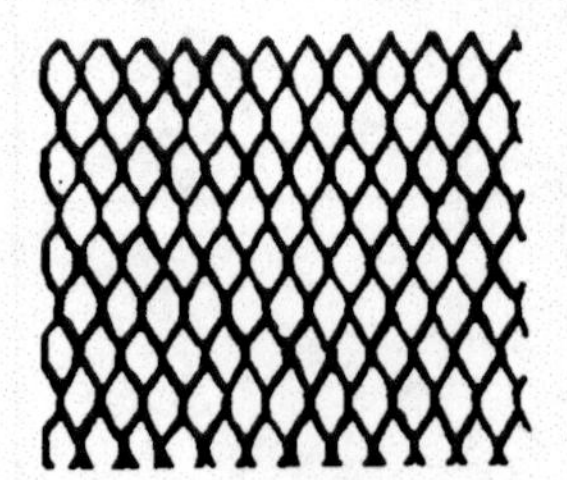

FLAT
DIAMOND MESH
METAL LATH

SELF-
FURRING
METAL LATH

FLAT RIB
METAL LATH

RIB
METAL LATH

RIB
METAL LATH

STRESS RELIEF (CONTROL JOINTS)

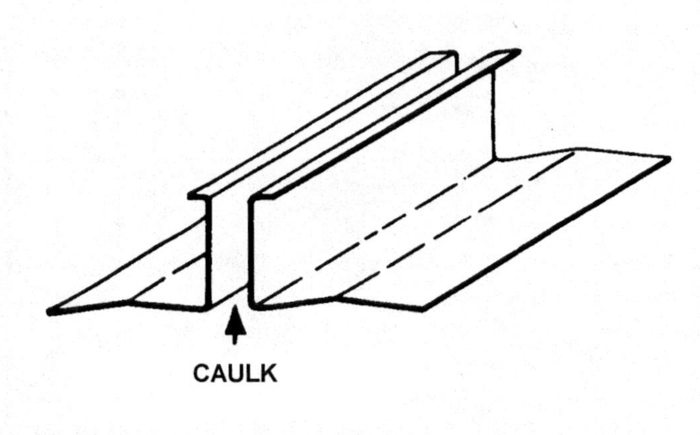

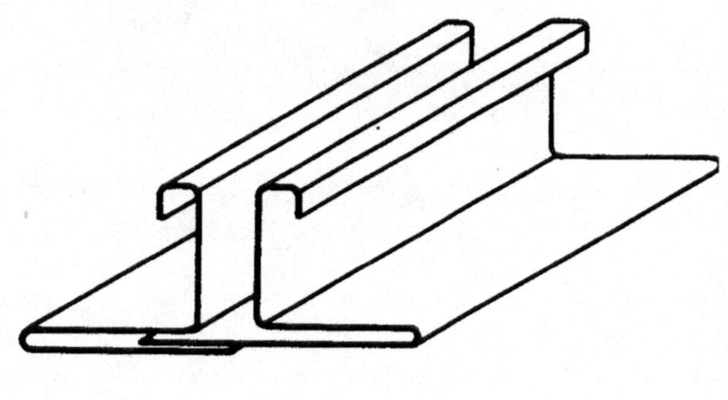

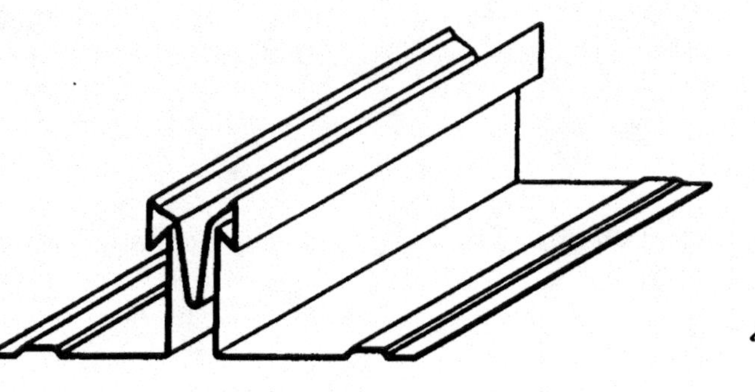

REVEALS

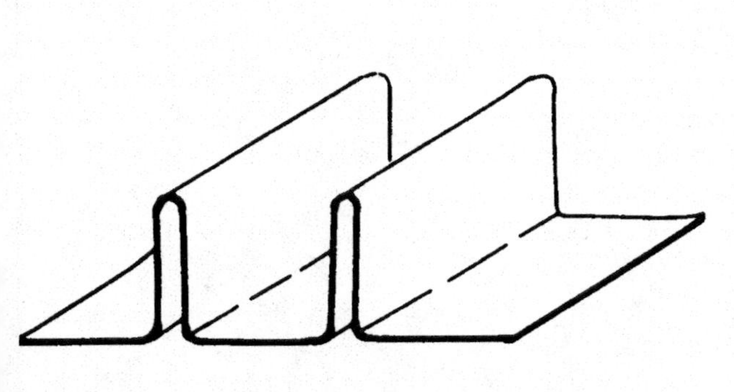

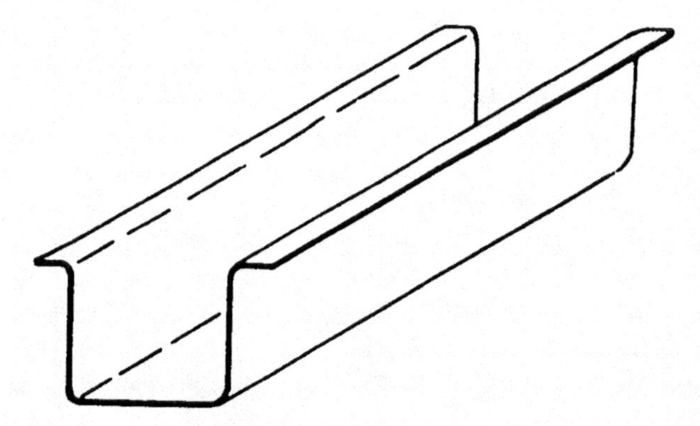

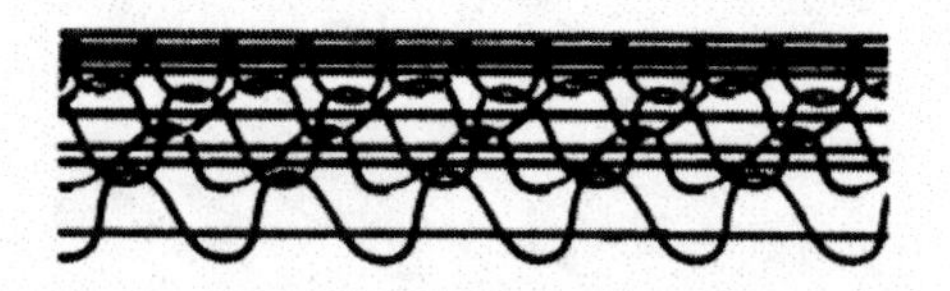

CORNER REINFORCEMENT
(EXTERIOR) WIRE

CORNER REINFORCEMENT
(EXTERIOR) EXPANDED METAL

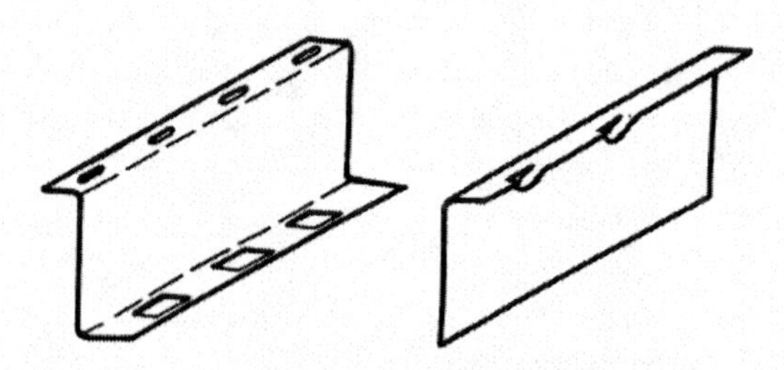

PARTITION RUNNERS
(Z AND L SHAPE)

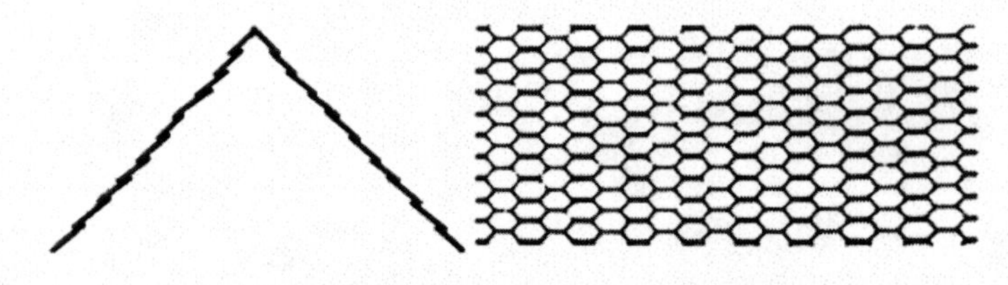

EXPANDED METAL CORNERITE

WIRE CORNERITE

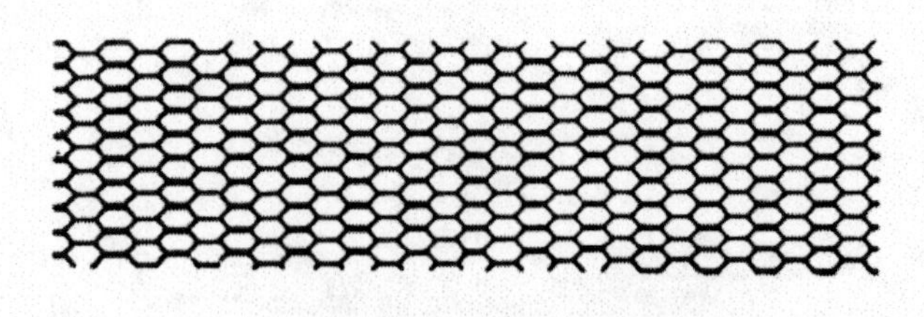

STRIP REINFORCEMENT
(EXPANDED METAL)

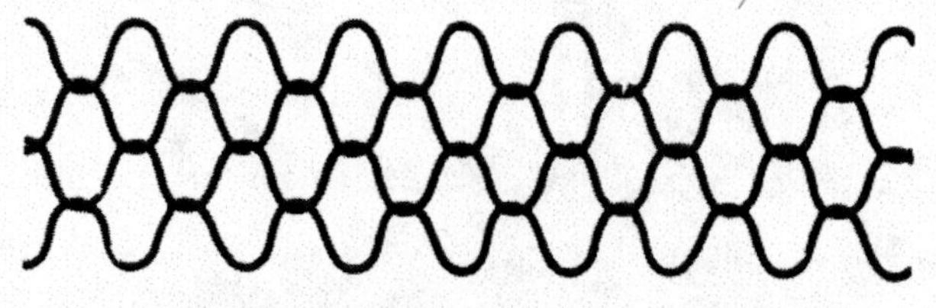

STRIP REINFORCEMENT (WIRE)

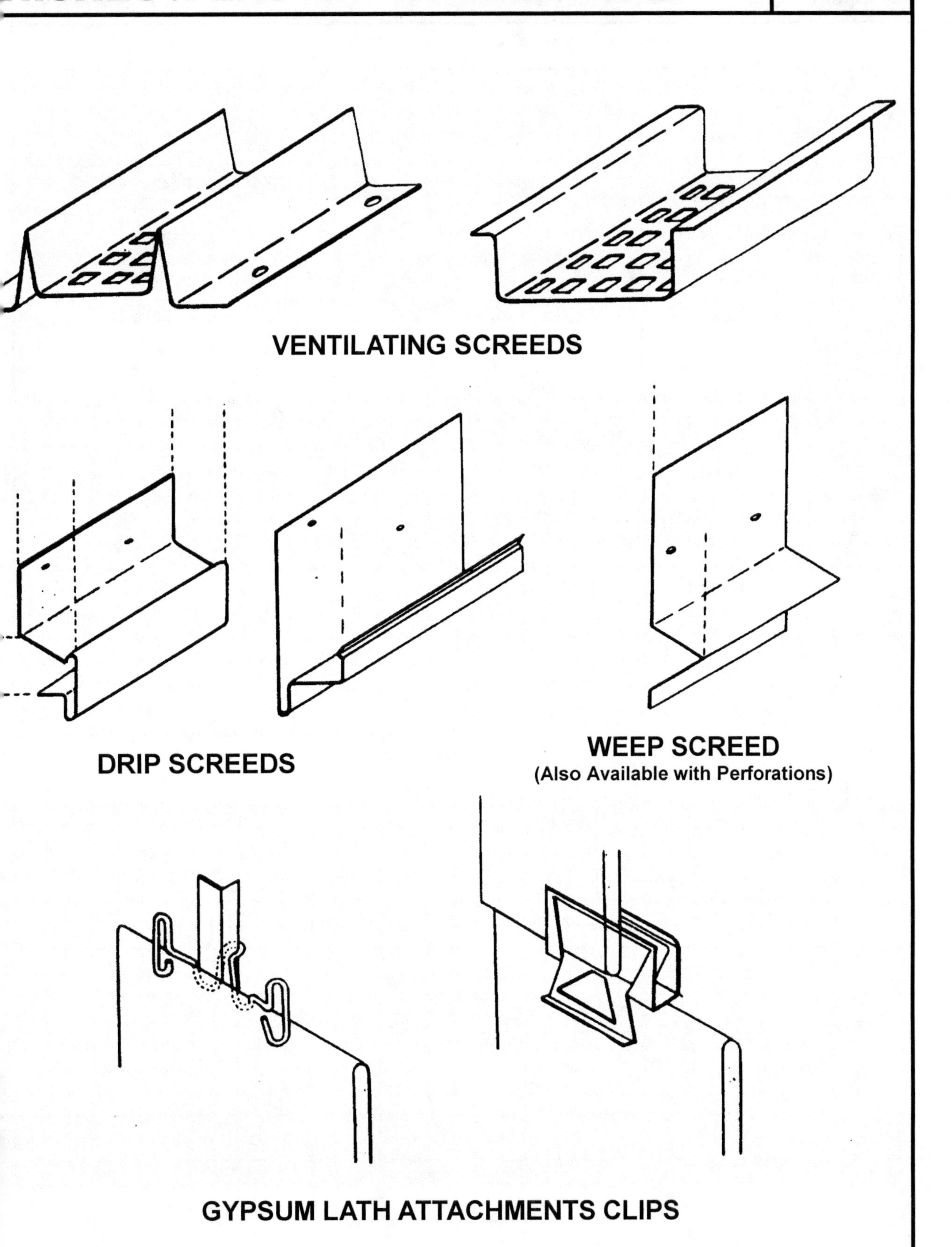

VENTILATING SCREEDS

DRIP SCREEDS

WEEP SCREED
(Also Available with Perforations)

GYPSUM LATH ATTACHMENTS CLIPS

PLASTERING TABLES

THICKNESS OF PLASTER

PLASTER	FINISHED THICKNESS OF PLASTER FROM FACE OF LATH, MASONRY, CONCRETE	
	Gypsum Plaster	Portland Cement Plaster
Expanded Metal Lath	5/8" minimum	5/8" minimum
Wire Fabric Lath	5/8" minimum	3/4" minimum (interior) 7/8" minimum (exterior)
Gypsum Lath	1/2" minimum	
Gypsum Veneer Base	1/16" minimum	1/2" minimum
Masonry Walls	1/2" minimum	7/8" maximum
Monolithic Concrete Walls	5/8" maximum	1/2" maximum
Monolithic Concrete Ceilings	3/8" maximum	

GYPSUM PLASTER PROPORTIONS

NUMBER OF COATS	COAT	PLASTER BASE OR LATH	MAXIMUM VOLUME AGGREGATE PER 100# NEAT PLASTER (CUBIC FEET)	
			Damp Loose Sand	Perlite or Vermiculite
Two-Coat Work	Basecoat	Gypsum Lath	2 1/2	2 1/2
	Basecoat	Masonry	3	3
Three-Coat Work	First Coat	Lath	2	2
	Second Coat	Lath	3	3
	First & Second Coat	Masonry	3	3

PORTLAND CEMENT PLASTER

COAT	VOLUME CEMENT	MAXIMUM WEIGHT (OR VOLUME) LIME PER VOLUME CEMENT	MAXIMUM VOLUME SAND PER VOLUME CEMENT	APPROXIMATE MINIMUM THICKNESS	MINIMUM PERIOD MOIST CURING	MINIMUM INTERVAL BETWEEN COATS
First	1	20 lbs.	4	3/8"	48 Hours	48 Hours
Second	1	20 lbs.	5	1st & 2nd Coats total 3/4"	48 Hours	7 Days
Finish	1	1	3	1st, 2nd & Finish Coats total 7/8"	-	

PORTLAND CEMENT - LIME PLASTER

COAT	VOLUME CEMENT	MAXIMUM WEIGHT (OR VOLUME) LIME PER VOLUME CEMENT	MAXIMUM VOLUME SAND PER VOLUME CEMENT	APPROXIMATE MINIMUM THICKNESS	MINIMUM PERIOD MOIST CURING	MINIMUM INTERVAL BETWEEN COATS
First	1	1	4	3/8"	48 Hours	48 Hours
Second	1	1	4 1/2	1st & 2nd Coats total 3/4"	48 Hours	7 Days
Finish	1	1	3	1st, 2nd & Finish Coats total 7/8"	-	

METAL STUD CONSTRUCTION

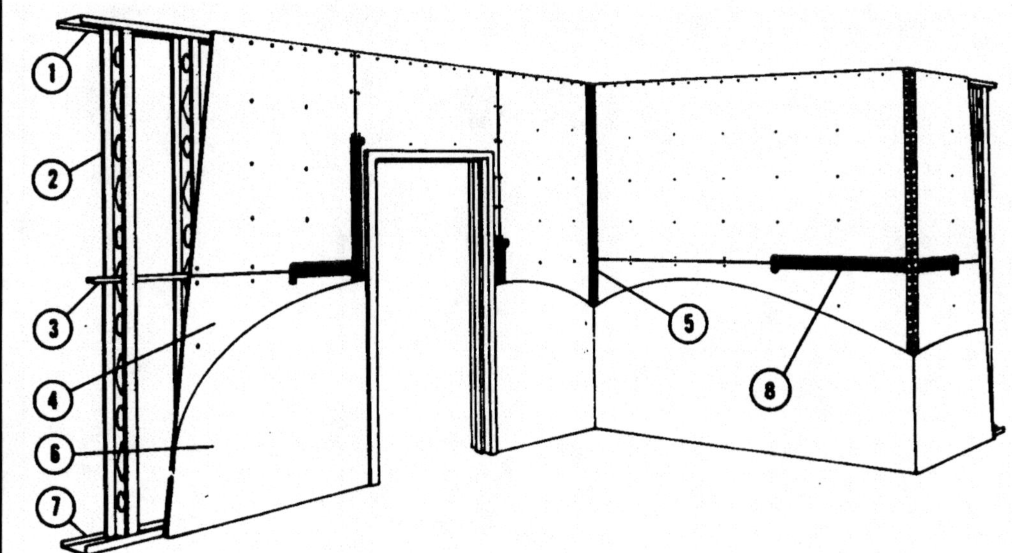

(1) Ceiling Runner Track
(2) Metal Stud (nailable or screw)
(3) Horizontal Stiffener
(4) Large Size Lath
(5) Angle Reinforcement
(6) Veneer Plaster 1/16 to 1/8 inch thick)
(7) Floor Runner Track
(8) Joint Reinforcement

WOOD STUD CONSTRUCTION

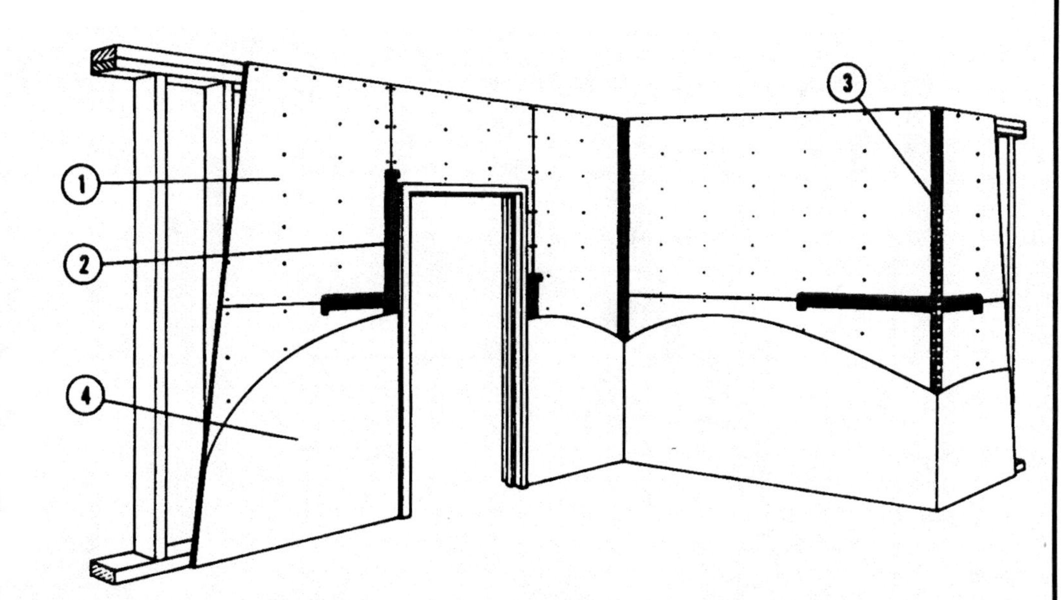

(1) Large Size Lath
(2) Joint Reinforcement
(3) Corner Bead
(4) Veneer Plaster (1/16 to 1/8 inch thick)

EXTERIOR LATH AND PLASTER

OPEN WOOD FRAME CONSTRUCTION

(1) Wire Backing
(2) Building Paper
(3) Wire Fabric Lath
(4) Approved Fasteners
(5) Weep Screed
(6) Three Coats of Plaster (Scratch, Brown, Finish)

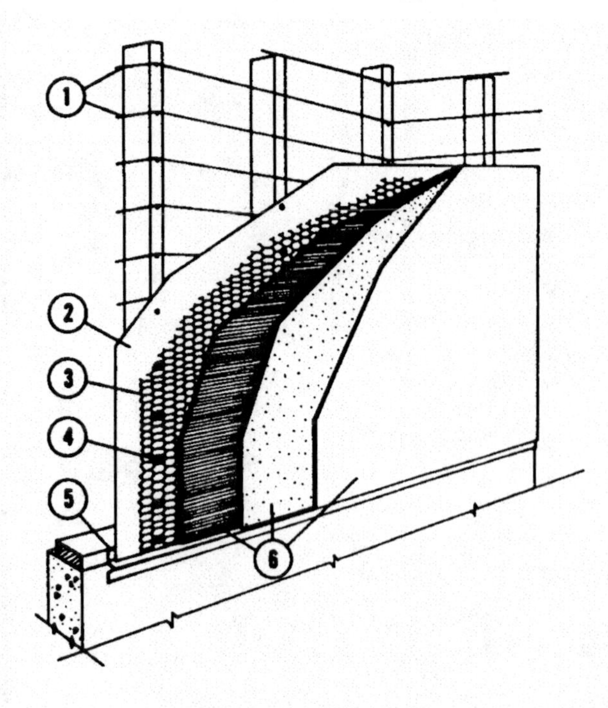

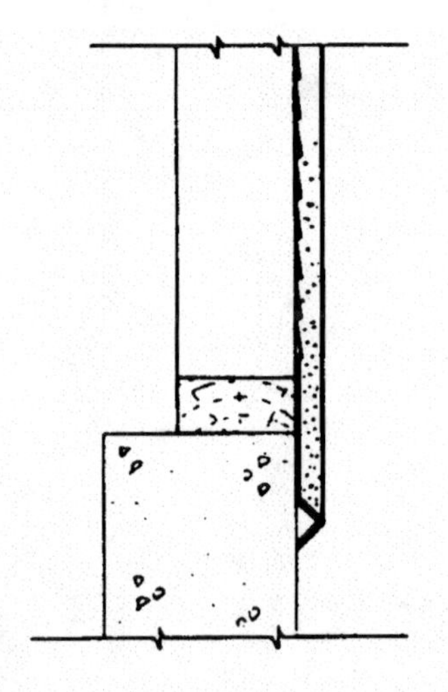

SHEATHED WOOD FRAME CONSTRUCTION

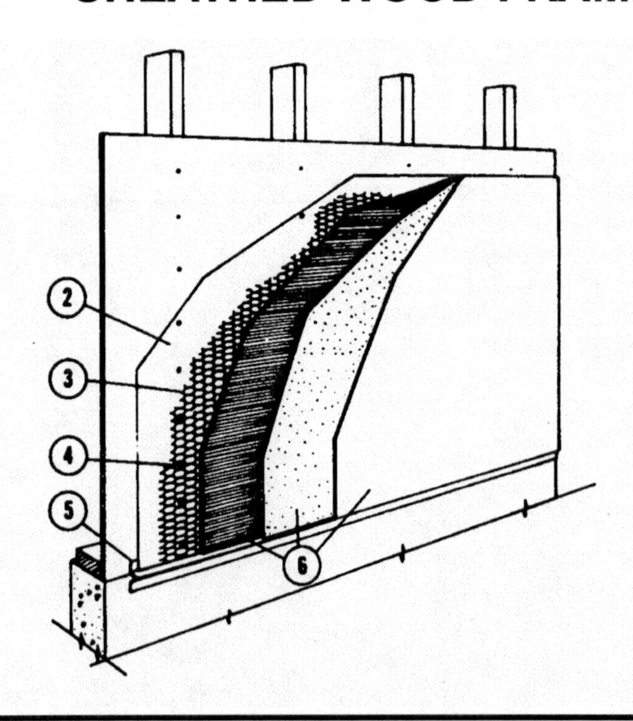

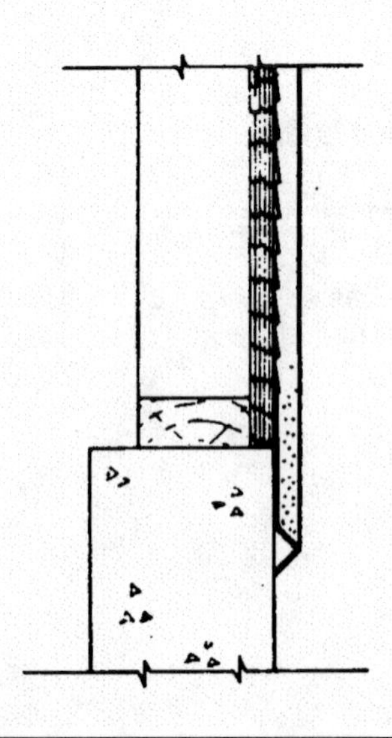

TYPICAL SIZES OF FIXTURE WATER SUPPLY PIPES

Fixture (inches)	Nominal pipe size
Bath tubs	1/2
Combination sink and tray	1/2
Drinking fountain	3/8
Dishwasher (domestic)	1/2
Kitchen sink, residential	1/2
Kitchen sink, commercial	3/4
Lavatory	3/8
Laundry tray, 1, 2 or 3 compartments	1/2
Shower (single head)	1/2
Sinks (service, slop)	1/2
Sinks flushing rim	3/4
Urinal (flash tank)	3/8
Urinal (direct flush valve)	1
Water closet (tank type)	3/8
Water closet (flush valve type)	1
Hose bibs	1/2
Wall hydrant	1/2

BNi® Building News

Geographic Cost Modifiers

The costs as presented in this book attempt to represent national averages. Costs, however, vary among regions, states and even between adjacent localities.

In order to more closely approximate the probable costs for specific locations throughout the U.S., this table of Geographic Cost Modifiers is provided. These adjustment factors are used to modify costs obtained from this book to help account for regional variations of construction costs and to provide a more accurate estimate for specific areas. The factors are formulated by comparing costs in a specific area to the costs as presented in the Costbook pages. An example of how to use these factors is shown below. Whenever local current costs are known, whether material prices or labor rates, they should be used when more accuracy is required.

Cost Obtained from Costbook Pages **X** Location Cost Adjustment Factor Divided by 100 **=** **Adjusted Cost**

For example, a project estimated to cost $125,000 using the Costbook pages can be adjusted to more closely approximate the cost in Los Angeles:

$$\$125{,}000 \times \frac{105}{100} = \$131{,}250$$

BNi® Building News

GEOGRAPHIC COST MODIFIERS		01025
State	Metropolitan Area	Multiplier
AK	ANCHORAGE	132
AL	ANNISTON	81
	AUBURN	82
	BIRMINGHAM	82
	DECATUR	84
	DOTHAN	83
	FLORENCE	84
	GADSDEN	82
	HUNTSVILLE	84
	MOBILE	86
	MONTGOMERY	81
	OPELIKA	82
	TUSCALOOSA	81
AR	FAYETTEVILLE	79
	FORT SMITH	79
	JONESBORO	78
	LITTLE ROCK	82
	NORTH LITTLE ROCK	82
	PINE BLUFF	80
	ROGERS	79
	SPRINGDALE	79
	TEXARKANA	79
AZ	FLAGSTAFF	94
	MESA	94
	PHOENIX	95
	TUCSON	93
	YUMA	94
CA	BAKERSFIELD	116
	CHICO	118
	FAIRFIELD	120
	FRESNO	118
	LODI	117
	LONG BEACH	119
	LOS ANGELES	119
	MERCED	118
	MODESTO	114
	NAPA	120
	OAKLAND	124
	ORANGE COUNTY	118
	PARADISE	114
	PORTERVILLE	116
	REDDING	114
	RIVERSIDE	116
	SACRAMENTO	118
	SALINAS	120
	SAN BERNARDINO	116
	SAN DIEGO	117
	SAN FRANCISCO	129

GEOGRAPHIC COST MODIFIERS		01025
State	**Metropolitan Area**	**Multiplier**
CA	SAN JOSE	126
	SAN LUIS OBISPO	113
	SANTA BARBARA	116
	SANTA CRUZ	120
	SANTA ROSA	121
	STOCKTON	117
	TULARE	118
	VALLEJO	120
	VENTURA	116
	VISALIA	118
	WATSONVILLE	118
	YOLO	118
	YUBA CITY	118
CO	BOULDER	103
	COLORADO SPRINGS	100
	DENVER	101
	FORT COLLINS	110
	GRAND JUNCTION	99
	GREELEY	108
	LONGMONT	103
	LOVELAND	110
	PUEBLO	105
CT	BRIDGEPORT	113
	DANBURY	113
	HARTFORD	112
	MERIDEN	113
	NEW HAVEN	113
	NEW LONDON	110
	NORWALK	117
	NORWICH	110
	STAMFORD	117
	WATERBURY	112
DC	WASHINGTON	105
DE	DOVER	105
	NEWARK	106
	WILMINGTON	106
FL	BOCA RATON	80
	BRADENTON	80
	CAPE CORAL	78
	CLEARWATER	81
	DAYTONA BEACH	75
	FORT LAUDERDALE	83
	FORT MYERS	78
	FORT PIERCE	81
	FORT WALTON BEACH	76
	GAINESVILLE	80
	JACKSONVILLE	78
	LAKELAND	78

GEOGRAPHIC COST MODIFIERS — 01025

State	Metropolitan Area	Multiplier
FL	MELBOURNE	75
	MIAMI	81
	NAPLES	79
	OCALA	79
	ORLANDO	77
	PALM BAY	75
	PANAMA CITY	77
	PENSACOLA	76
	PORT ST. LUCIE	81
	PUNTA GORDA	78
	SARASOTA	80
	ST. PETERSBURG	80
	TALLAHASSEE	75
	TAMPA	80
	TITUSVILLE	75
	WEST PALM BEACH	80
	WINTER HAVEN	78
GA	ALBANY	86
	ATHENS	89
	ATLANTA	92
	AUGUSTA	86
	COLUMBUS	79
	MACON	83
	SAVANNAH	87
HI	HONOLULU	138
IA	CEDAR FALLS	91
	CEDAR RAPIDS	102
	DAVENPORT	106
	DES MOINES	104
	DUBUQUE	95
	IOWA CITY	97
	SIOUX CITY	91
	WATERLOO	91
ID	BOISE CITY	102
	POCATELLO	102
IL	BLOOMINGTON	113
	CHAMPAIGN	109
	CHICAGO	125
	DECATUR	107
	KANKAKEE	113
	NORMAL	113
	PEKIN	111
	PEORIA	111
	ROCKFORD	113
	SPRINGFIELD	108
	URBANA	109
IN	BLOOMINGTON	102
	ELKHART	96

GEOGRAPHIC COST MODIFIERS — 01025

State	Metropolitan Area	Multiplier
IN	EVANSVILLE	99
	FORT WAYNE	100
	GARY	107
	GOSHEN	96
	INDIANAPOLIS	103
	KOKOMO	101
	LAFAYETTE	101
	MUNCIE	101
	SOUTH BEND	102
	TERRE HAUTE	100
KS	KANSAS CITY	120
	LAWRENCE	109
	TOPEKA	96
	WICHITA	87
KY	LEXINGTON	91
	LOUISVILLE	102
	OWENSBORO	101
LA	ALEXANDRIA	89
	BATON ROUGE	93
	BOSSIER CITY	90
	HOUMA	93
	LAFAYETTE	91
	LAKE CHARLES	93
	MONROE	89
	NEW ORLEANS	95
	SHREVEPORT	90
MA	BARNSTABLE	124
	BOSTON	128
	BROCKTON	118
	FITCHBURG	120
	LAWRENCE	121
	LEOMINSTER	120
	LOWELL	124
	NEW BEDFORD	118
	PITTSFIELD	118
	SPRINGFIELD	119
	WORCESTER	120
	YARMOUTH	124
MD	BALTIMORE	95
	CUMBERLAND	98
	HAGERSTOWN	90
ME	AUBURN	87
	BANGOR	87
	LEWISTON	87
	PORTLAND	88
MI	ANN ARBOR	119
	BATTLE CREEK	111
	BAY CITY	116

GEOGRAPHIC COST MODIFIERS		01025
State	Metropolitan Area	Multiplier
MI	BENTON HARBOR	111
	DETROIT	120
	EAST LANSING	117
	FLINT	116
	GRAND RAPIDS	112
	HOLLAND	112
	JACKSON	107
	KALAMAZOO	111
	LANSING	117
	MIDLAND	115
	MUSKEGON	112
	SAGINAW	116
MN	DULUTH	107
	MINNEAPOLIS	112
	ROCHESTER	107
	ST. CLOUD	105
	ST. PAUL	112
MO	COLUMBIA	114
	JOPLIN	103
	KANSAS CITY	118
	SPRINGFIELD	96
	ST. JOSEPH	117
	ST. LOUIS	115
MS	BILOXI	79
	GULFPORT	79
	HATTIESBURG	79
	JACKSON	79
	PASCAGOULA	79
MT	BILLINGS	96
	GREAT FALLS	90
	MISSOULA	91
NC	ASHEVILLE	73
	CHAPEL HILL	79
	CHARLOTTE	82
	DURHAM	81
	FAYETTEVILLE	75
	GOLDSBORO	80
	GREENSBORO	81
	GREENVILLE	79
	HICKORY	72
	HIGH POINT	81
	JACKSONVILLE	72
	LENOIR	72
	MORGANTON	72
	RALEIGH	80
	ROCKY MOUNT	72
	WILMINGTON	72
	WINSTON SALEM	77

GEOGRAPHIC COST MODIFIERS — 01025

State	Metropolitan Area	Multiplier
ND	BISMARCK	84
	FARGO	98
	GRAND FORKS	81
NE	LINCOLN	84
	OMAHA	91
NH	MANCHESTER	106
	NASHUA	106
	PORTSMOUTH	111
NJ	ATLANTIC CITY	126
	BERGEN	129
	BRIDGETON	125
	CAPE MAY	125
	HUNTERDON	128
	JERSEY CITY	130
	MIDDLESEX	129
	MILLVILLE	125
	MONMOUTH	129
	NEWARK	129
	OCEAN	130
	PASSAIC	130
	SOMERSET	128
	TRENTON	128
	VINELAND	125
NM	ALBUQUERQUE	91
	LAS CRUCES	91
	SANTA FE	91
NV	LAS VEGAS	109
	RENO	97
NY	ALBANY	119
	BINGHAMTON	116
	BUFFALO	118
	DUTCHESS COUNTY	119
	ELMIRA	118
	GLENS FALLS	120
	JAMESTOWN	112
	NASSAU	137
	NEW YORK	148
	NEWBURGH	119
	NIAGARA FALLS	121
	ROCHESTER	118
	ROME	109
	SCHENECTADY	119
	SUFFOLK	137
	SYRACUSE	118
	TROY	119
	UTICA	109

GEOGRAPHIC COST MODIFIERS		01025
State	Metropolitan Area	Multiplier
OH	AKRON	112
	CANTON	107
	CINCINNATI	105
	CLEVELAND	114
	COLUMBUS	115
	DAYTON	115
	ELYRIA	114
	HAMILTON	105
	LIMA	115
	LORAIN	114
	MANSFIELD	115
	MASSILLON	107
	MIDDLETOWN	115
	SPRINGFIELD	109
	STEUBENVILLE	115
	TOLEDO	109
	WARREN	111
	YOUNGSTOWN	111
OK	ENID	86
	LAWTON	86
	OKLAHOMA CITY	85
	TULSA	80
OR	ASHLAND	109
	CORVALLIS	112
	EUGENE	112
	MEDFORD	109
	PORTLAND	114
	SALEM	112
	SPRINGFIELD	112
PA	ALLENTOWN	118
	ALTOONA	110
	BETHLEHEM	118
	CARLISLE	113
	EASTON	118
	ERIE	112
	HARRISBURG	113
	HAZLETON	118
	JOHNSTOWN	104
	LANCASTER	93
	LEBANON	115
	PHILADELPHIA	134
	PITTSBURGH	116
	READING	119
	SCRANTON	116
	SHARON	112
	STATE COLLEGE	98
	WILKES BARRE	116
	WILLIAMSPORT	97

GEOGRAPHIC COST MODIFIERS		01025
State	**Metropolitan Area**	**Multiplier**
PA	YORK	113
PR	MAYAGUEZ	73
	PONCE	74
	SAN JUAN	75
RI	PROVIDENCE	122
SC	AIKEN	89
	ANDERSON	71
	CHARLESTON	76
	COLUMBIA	76
	FLORENCE	73
	GREENVILLE	76
	MYRTLE BEACH	73
	NORTH CHARLESTON	81
	SPARTANBURG	73
	SUMTER	76
SD	RAPID CITY	81
	SIOUX FALLS	85
TN	CHATTANOOGA	84
	CLARKSVILLE	83
	JACKSON	83
	JOHNSON CITY	83
	KNOXVILLE	80
	MEMPHIS	84
	NASHVILLE	83
TX	ABILENE	88
	AMARILLO	92
	ARLINGTON	87
	AUSTIN	89
	BEAUMONT	88
	BRAZORIA	88
	BROWNSVILLE	73
	BRYAN	86
	COLLEGE STATION	86
	CORPUS CHRISTI	84
	DALLAS	89
	DENISON	87
	EDINBURG	73
	EL PASO	81
	FORT WORTH	87
	GALVESTON	93
	HARLINGEN	73
	HOUSTON	88
	KILLEEN	77
	LAREDO	78
	LONGVIEW	78
	LUBBOCK	91
	MARSHALL	87
	MCALLEN	73

GEOGRAPHIC COST MODIFIERS — 01025

State	Metropolitan Area	Multiplier
TX	MIDLAND	87
	MISSION	73
	ODESSA	87
	PORT ARTHUR	88
	SAN ANGELO	87
	SAN ANTONIO	90
	SAN BENITO	73
	SAN MARCOS	89
	SHERMAN	87
	TEMPLE	77
	TEXARKANA	79
	TEXAS CITY	93
	TYLER	84
	VICTORIA	74
	WACO	77
	WICHITA FALLS	87
UT	OGDEN	95
	OREM	93
	PROVO	93
	SALT LAKE CITY	92
VA	CHARLOTTESVILLE	86
	LYNCHBURG	83
	NEWPORT NEWS	88
	NORFOLK	91
	PETERSBURG	78
	RICHMOND	90
	ROANOKE	76
	VIRGINIA BEACH	91
VT	BURLINGTON	97
WA	BELLEVUE	119
	BELLINGHAM	111
	BREMERTON	113
	EVERETT	117
	KENNEWICK	101
	OLYMPIA	113
	PASCO	100
	RICHLAND	101
	SEATTLE	119
	SPOKANE	98
	TACOMA	116
	YAKIMA	104

GEOGRAPHIC COST MODIFIERS 01025

State	Metropolitan Area	Multiplier
WI	APPLETON	113
	BELOIT	117
	EAU CLAIRE	113
	GREEN BAY	112
	JANESVILLE	117
	KENOSHA	118
	LA CROSSE	114
	MADISON	116
	MILWAUKEE	118
	NEENAH	113
	OSHKOSH	113
	RACINE	118
	SHEBOYGAN	112
	WAUKESHA	118
	WAUSAU	113
WV	CHARLESTON	113
	HUNTINGTON	113
	PARKERSBURG	113
	WHEELING	113
WY	CASPER	85
	CHEYENNE	85

- D -

- E -

- F -

BNi® Building News

Other Estimating References from BNi Building News

▪14 BNi General Construction Costbook

Over 12,000 unit prices provide you with cost estimates for all aspects of construction -- from sitework and concrete, to doors and painting.

The *BNi General Construction Costbook 2014* is broken down into material and labor costs, to allow for maximum flexibility and accuracy in estimating. You'll find this costbook data invaluable when preparing detailed estimates, bids, checking prices, and calculating the impact of change orders. You also get detailed man-hour tables that let you see the basis for the labor ꞏts based on standard productivity rates.

ludes a Free PDF download version you can customize!

8 pages, 8½ x 11, $99.95

▪14 BNi Facilities Manager's Costbook

Quickly and easily estimate the cost of renovations, repairs and new construction for all types of facilities and commercial buildings. The *BNi Facilities Managers Costbook 2014* is the first place to turn, whether you're preparing a preliminary estimate, evaluating a contractor's bid, or submitting a formal budget proposal. It provides accurate and up-to-date material and labor costs for thousands of cost items. Labor costs based on the prevailing rates for each trade and type of work, PLUS man-hour tables tied to each unit costs, you can clearly see exactly how the labor cost were calculated and ꞏke any necessary adjustments. Equipment costs -- including rental ꞏd operating costs. Square-foot tables based on the cost-per-square-ꞏt of hundreds of actual projects -- invaluable data for quick, ballpark ꞏtimates.

ꞏludes a Free PDF download version you can customize!

8 pages, 8½ x 11, $99.95.

▪14 BNi Mechanical/Electrical Costbook

From pipe to duct to receptacle, this detailed reference book provides extensive coverage of the most technical aspects of building construction. With thousands of current, reliable mechanical and electrical costs at your fingertips, you can estimate quickly and accurately. All data is organized clearly so you can find the costs you need ... fast! Geographic Cost Modifiers allow you to tailor your estimates to specific areas of the country.

Includes a Free PDF download version you can customize!

478 pages, 8½ x 11, $99.95

2014 BNi Public Works Costbook

Now you can quickly and easily estimate the cost of all types of public works projects involving roads, excavation, drainage systems and much more.

The *BNi Public Works Costbook 2014* is the first place to turn, whether you're preparing a preliminary estimate, evaluating a contractor's bid, or submitting a formal budget proposal. It provides accurate and up-to-date material and labor costs for thousands of cost items, based on the latest national averages and standard labor productivity rates.

Square-foot tables based on the cost-per-square-foot of hundreds of actual projects -- invaluable data for quick, ballpark estimates.

Includes a Free PDF download version you can customize!

478 pages, 8½ x 11, $99.95

2014 BNi Electrical Costbook

From meter to duct, conduit to receptacle, this unique guide provides extensive coverage of the most technical aspects of building construction. Quickly and easily estimate the cost of all types of electrical projects involving new construction and renovations in both commercial and residential buildings.

Provides accurate, up-to-date material and labor costs for thousands of cost items, based on the latest national averages and standard labor productivity rates. Includes detailed 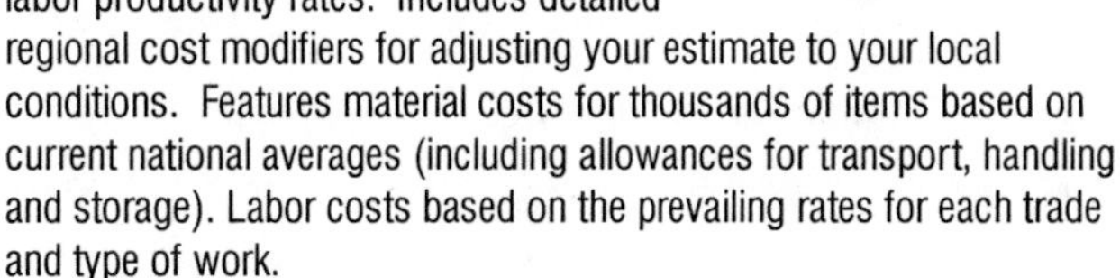
regional cost modifiers for adjusting your estimate to your local conditions. Features material costs for thousands of items based on current national averages (including allowances for transport, handling and storage). Labor costs based on the prevailing rates for each trade and type of work.

Includes a Free PDF download version you can customize!

354 pages, 8½ x 11, $93.95

2014 BNi Home Builder's Costbook

Here's the easy way to estimate the cost of all types of residential construction projects! Accurate and up-to-date material and labor costs for thousands of cost items, based on the latest national averages and standard labor productivity rates.

Includes detailed regional cost modifiers for adjusting your estimate to your local conditions. Material costs are included for thousands of items based on current national averages (including allowances for transport, handling and storage). Labor costs based on the prevailing rates for each trade and type of work.

Includes a Free PDF download version you can customize!

336 pages, 8½ x 11, $83.95

2014 BNi Home Remodeler's Costbook

This guide lets you quickly and easily estimate the cost of all types of home remodeling projects, including additions, new kitchens and baths, and much more.

It provides accurate and up-to-date material and labor costs for thousands of cost items, based on the latest national averages and standard labor productivity rates. Includes detailed regional cost modifiers for adjusting your estimate to your local conditions. Includes material costs for thousands of remodeling items as well as labor costs for each trade and type of work. Includes equipment costs including rental and operating costs and square-foot tables for quick, ballpark estimates.

Includes a Free PDF download version you can customize!

497 pages, 8½ x 11, $87.95

2014 BNi Remodeling Costbook

Now there's an EASY way to quickly and easily develop cost estimates that are right on target – everytime - no matter what kind of remodeling job you're handling.

The BNI *Remodeling Costbook* is like no other estimating tool you've ever used before. It's jam-packed with large, clear tables that give you key data at a glance. Whether you're developing a budget ... checking contractor's prices... working up a preliminary estimate...or submitting a formal bid, you'll have all the facts and figures you need for accurate, on-the-money estimates.

Includes a Free PDF download version you can customize!

525 pages, 8½ x 11, $99.95

2014 Architect's Square Foot Costbook

This manual presents detailed square foot costs for 65 buildings tailored specifically to meet the needs of today's architect. For each project you get a complete cost breakdown of the included systems, so you can easily calculate the impact of modifications and enhancements on your own projects. Several of the case studies presented feature significant green building strategies as indicated by the LEED Rating system, encompassing recycled construction waste, recycled material content and re-use, recycled rainwater, bio-retention and wetland storm management, indoor environmental air quality, energy efficient lighting, hybrid HVAC systems, green roofs, and native or adaptive vegetation.

This manual also features detailed costs to enhance the included case studies and allow the design professional to "mix and match" components to more closely follow their own design.

210 pages, 8½ x 11, $66.95

2014 Sweets Unit Cost Guide

Work with the most extensive, and respected database in the construction industry - developed by the experts at McGraw-Hill Construction and BNi Building News. Keep your estimates, bids, specifications, and cost checks right on target with precise, unit construction costs. Ensure cost accuracy to your region with Sweets 1,000+ location modifiers. This indispensable tool gives you the highest available level of precision in estimating unit costs for: Demolition, Sitework, Concrete & Masonry, Metals, Woods & Plastics, Thermal & Moisture Protection, Doors & Windows, Finishes, Specialties, Equipment, Furnishings, Special Construction, Conveying Systems, Mechanical Systems, Electrical Systems, and more!

558 pages, 8½ x 11, $99.95

ENR Square Foot Costbook 2014 Edition

As you know, square-foot costs can vary widely, making them difficult to use for estimating and budgeting. But the 2014 *ENR Square-Foot Costbook* eliminates this problem by giving you costs that are based on actual projects -- not hypothetical models.

For each building project you get a detailed narrative with background information on the specific project. This lets you put the cost data into context and make appropriate adjustments to your own projections.

Developed in partnership with Engineering News Record,Design Cost Data, and BNi Building News, this ready-reference costbook also features illustrations of each building type, a guide to 5-year cost trends for key building materials, Plus detailed unit-in-place costs for thousands of items -- from asphalt and anchor bolts, to vents and wall louvers. By purchasing this book, you will also receive the 2015 edition, on a no-risk free trial basis

192 pages, 8½ x 11, $63.95

2014 Sweets Green Building Square Ft. Costbook

A comprehensive collection of 59 recent LEED® and sustainable building projects along with their actual square foot costs, broken down by CSI MasterFormat section. All of the costs have been carefully adjusted for 2014. Included are government projects, commercial buildings, schools and libraries, medical facilities, residential buildings and recreational facilities. These projects all reflect the vast array of "green" materials and technologies being used today. For each building, you'll find a detailed narrative describing the major features of the building, and what steps were taken to minimize the environmental impact both in its construction and its operation.

198 pages, 8½ x 11, $57.95

ENR Interiors Square Foot Costbook 2014 Edition

The interior make up of a project can dramatically impact the overall cost. Interiors cover a wide range of disciplines: from carpentry to millwork ... drywall to flooring ... specialties to furnishings ...and more.

In this completely up-to-date estimating guide, Design & Construction Resources has teamed up with *Design Cost Data*™ (DCD) magazine to provide you with a reliable and easy-to-use way to evaluate the cost impact of modifications to interior plans for just about any type of building construction.

The ***2014 Interiors Square Foot Costbook*** covers new construction, addition, renovation, adaptive re-use and tenant build-out, with all costs precisely escalated to 2014. It features 65 "real life" cost studies, with each project clearly broken down by all of its interior components presented on a cost-per-square-foot basis. It itemizes the materials used along with their cost to assist in developing a conceptual estimate for interior construction.

224 pages, 8½ x 11, $64.95

ENR Mechanical/Electrical Square Foot Costbook 2014

Mechanical, Electrical and Plumbing (MEP) systems make up a significant part of a building's cost. Design & Construction Resources has teamed up with Design Cost Data™ (DCD) magazine to provide the first ever resource to take a full-blown look at the cost effect of these systems related to the total building cost.

Along with the cost studies, users have access to Unit-In-Place Costs for Mechanical and Electrical systems. These costs, coupled with the case tudies, give a complete estimating reference for MEP systems.

76 pages, 8½ x 11, $64.95

NR Remodeling Contracting Costbook 2014 Edition

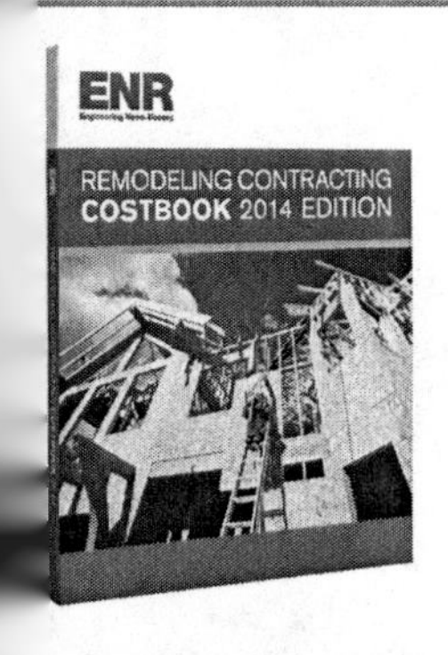

The ENR Remodeling Contracting Costbook solves the unique challenges of estimating remodeling jobs -- which isn't easy due to all the hidden factors that are not visible when estimating remodeling work -- yet become visible after you start the job.

The labor costs in the ENR Remodeling Contracting Costbook have all been adjusted to account for working during on-going operations or off-hours construction. And the material costs lect the fact that smaller order quantities are usually involved.

4 pages, 8½ x 11, $59.95

The Blue Book: Guide to Construction Costs 2014

Find thousands of thoroughly-researched construction material and installation costs you can use when making budgets, checking prices, calculating the impact of change orders or preparing bids. You'll find costs listed for scheduling, testing, temporary facilities, equipment, signage, and more. Quick square-foot costs are listed by division and building type. You'll find costs for every type of construction – from demolition and excavation to material and installation costs for finishes, flooring and much more.

160 pages, 8-1/2 x 11, $49.95

2014 NEC Costbook

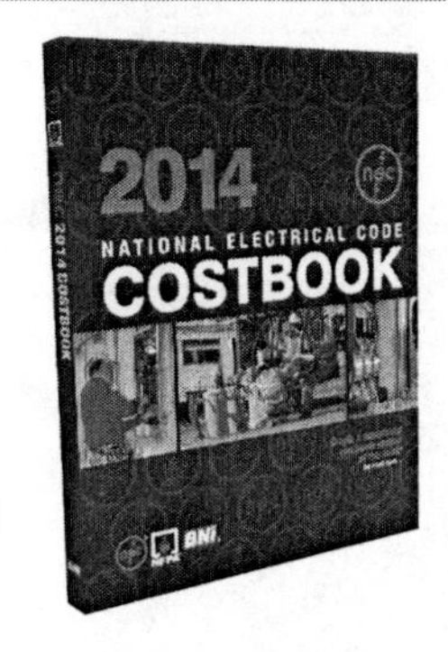

Developed by the NFPA and BNi Building News you'll find every electrical estimate for any type of job you handle — from residential remodeling to new commercial construction. You'll find detailed estimates of wireways, conduit, boxes, conductors, cabling, utility poles, transformers, safety switches, panelboards and just about every electrical component available today. Every type of electrical cost estimate you'll need is covered here.

210 pages, 8½ x 11, $61.95

ENR Residential Contracting Costbook 2014 Edition

Just for home builders! Here at last is a comprehensive cost estimating guide that presents you with all the data you need -- whether you're building a small single-family home or a townhouse ... using wood frame construction or masonry. It covers everything from doors, windows, siding, and brick to landscaping and all kitchen and bath materials.

The *2014 ENR Residential Contracting Costbook* has the reliable residential construction costs you need.

194 pages, 8½ x 11, $59.95

ENR General Contracting Costbook 2014 Edition

Up-to-date and easy-to-use construction costs tailored to fit the needs of today's general construction market. All prices in the ENR General Contracting Costbook are listed in CSI Masterformat for a clear easy-to-read style that helps you find the right costs for exactly what you're looking for.

This fully-indexed volume has ALL the data you need. Everything is clearly organized making it easy to pinpoint appropriate cost data. Includes a detailed city cost index for fine-tuning your estimate to practically any region of the country.

224 pages, 8½ x 11, $59.95

Construction Inspection Manual - 8th Edition

These checklists cover closeout procedures, facility operations, sitework, pre-cast concrete, metal fabrications, rough carpentry, waterproofing, thermal and moisture protection, doors, windows and roofs, raceways, water supplies, finishes, fire protection, and much more.

A detailed matrix is linked to each item on every checklist, showing if it pertains primarily to special inspections, architectural inspections, engineering inspections or other inspections. Within each of these categories, it indicates sub-groups such as testing laboratory, hazardous material, or safety (for special inspections), civil, structural, mechanical or electrical (for engineering inspections), and legal, government, fire protection, etc (for other inspections).

373 pages, 8½ x 11, $76.95

Marketing Handbook - 3rd Edition

These days, driving new business is the key to your survival, and this highly acclaimed marketing tool is your key to driving new business. With 64 articles by more than 70 contributors and 18 reviewers, the all-new and thoroughly revised Marketing Handbook covers every aspect of marketing in the design and construction industry from initial market research, to getting leads, to making presentations, to website development. If you're not a marketing professional, then you really this book.

In a clear, no-nonsense manner, *Marketing Handbook for the Design & Construction Professional* shows you how to:

Create winning proposals and presentations
Take full advantage of low-cost public relations activities
Develop a winning plan for getting the most from trade shows
Create a high-profile website on a budget

638 pages, 8½ x 11, $69.95

Business Letters for the Construction Industry

Here you'll find over 100 professionally-written model letters for virtually every situation. Put your business in the best possible light with a well-written letter — especially when you can use that letter to improve a bad situation. Use these letters over and over again to resolve disputes, win new clients, clarify proposals, coordinate with architects, subcontractors, owners, and insurers, schedule meetings and inspections, and to respond to complaints or difficult situations. Included are letters responding to threats of legal action, of commendation to workers, of job performance, apology for defective or delayed work; letters for justification of change orders and price increases; letters explaining your insurance liability, drug testing, injury at work, overtime, equipment use, and more.

376 pages, 8½ x 11, $59.95

Standard Estimating Practice - 8th Edition

This practical "how-to" reference presents a standard set of practices and procedures proven to create consistent construction estimates. Every step of estimating is covered in detail -- from spec and plan review to what to expect on bidding day. With the procedures in this book you'll see your estimating results become more accurate and more consistent. It includes practical checklists and forms to help you include everything in your bids -- including insurance, outside services, taxes, equipment rental and much more.

526 pages, 8½ x 11, $89.00

Profit, Risk & Leadership:

The Business of Construction & Design-Build

This new book takes you on a clear path through the maze of running a construction business: how to make money, how to manage risk, and how to plan, organize and lead toward the achievement of your goals.
This book teaches the business of construction, not methods of construction. It is written primarily for general building, institutional, governmental, commercial, industrial, entertainment, transportation, power, water/waste water and other civil projects. Essential reading for anyone who is in construction management or aspires to that role.

192 pages, 8½ x 11 $59.95

Design Management Guide

Design-build has shown to be a superior method of product delivery. Knowing the roles of each "player" in the design-build process is the key to putting together a successful design-build management team.

This guide is a handbook for putting together a design-build management team: practical suggestions - distilled from the experiences of facility owners, developers, and senior design and construction practitioners - and procedures that can be applied to any situation at hand. While the guide is intended for design managers in design-build organizations, these principles may be applied equally to other project-delivery methods.

This guide is intended specifically for the design-build of "vertical" structures: buildings such as multi-family residential units, office buildings, high-rises, municipal buildings, places of worship, medical facilities, and schools.

The Design Management Guide compiles thoughts and best practices from many people through the design-build industry. It is the result of three, in-person workshops, along with much consideration and discussion by the workshop participants.

52 pages, 8½ x 11, $29.95

2012 Work Area Traffic Control Handbook

Published under the authority of the Southern California Chapter of the American Public Works Association, the Work Area Traffic Control Handbook is the product of a dedicated, 10-member committee comprised of leading experts in traffic control and safety. This valuable pocket-sized handbook has become the leading source of information for traffic control in construction work areas on local and county roads. Now in its 12th Edition, it gives you the information on traffic control you need.

Best of all, this wealth of information is all packed into a convenient, 4½ x 7½ format! And with a price of just $10.95, every supervisor on a job site should carry a copy in his or her pocket!

4½ x 7½, $10.95

Public Works Contracting: Start To Finish

This book is a comprehensive reference for any construction company, including start-ups, that wants to improve operations and management. The methods outlined in this book are straightforward, general approaches to contracting encountered in civil and commercial construction.

This book is for owners, engineers and supervisors, either on the jobsite or on office management teams, since it covers all the major aspects of organizing and managing a civil or commercial construction company from start to finish.

411 pages, 8½ x 11, $62.95

Public Works Inspectors Manual, 7th Edition

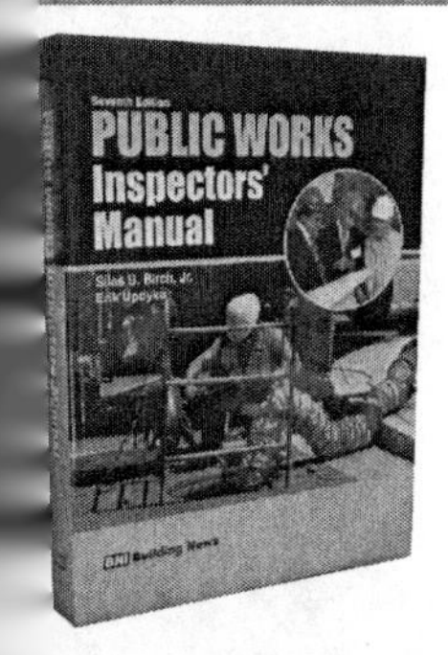

A complete operational technical guidebook for professionals charged with the responsibilities of inspecting all types of public works construction for city, county, state and federal agencies.

This comprehensive Manual has become the standard reference for public works professionals across the nation for over 50 years.

It clearly spells out, in-step-by-step detail, testing procedures and inspection guidelines for virtually every type of work and material - from asphalt and aggregates, to pipe and prestressed concrete.

More importantly, the richly illustrated *Public Works Inspector's Manual* shows you how to:

- *read and understand the contract documents*
- *ensure that required safety procedures are enforced*
- *minimize the cost of claims and disputes*
- *maintain accurate and complete records*
- *prepare thorough, well-documented reports*

The most comprehensive and authoritative text on public works inspection ever written and published. Of considerable value not only to inspection personnel, but also to contractors, engineers, architects and students.

74 pages, 8½ x 11, $74.95

2012 Greenbook: Standard Specifications for Public Works Construction

The *Standard Specifications for Public Works Construction,* popularly known as the "Greenbook", was originally published in 1967. The 2012 Edition is the 16th edition of this book, which is updated and republished every three years.

The Greenbook is designed to aid in furthering uniformity of plans and specifications accepted and used by those involved in public works construction and to take such other steps as are designed to promote more competitive bidding by private contractors.

The Greenbook provides specifications that have general applicability to public works projects -- from the thickness of asphalt to the type of glue used on road surface reflectors.

The Greenbook consists of five parts: General Provisions, Construction Materials, Construction Methods, Rock Products, and Pipeline System Rehabilitation.

It is used by many municipalities for gauging construction standards on public works construction. It is especially useful when comparing competitive bids among contractors, or for contractors in making sure their bids are to required standards.

530 pages, 8½ x 11, $84.50

Standard Plans for Public Works Construction 2012 Edition

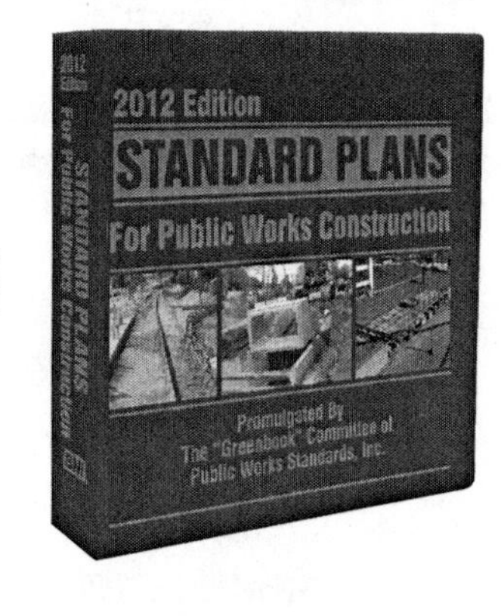

This essential resource is the "graphical companion" to the "Greenbook" (Standard Specifications for Public Works Construction).

Sometimes nothing explains construction standards like a clear illustration. This loose-leaf manual features hundreds of standardized drawings and dimensional details covering every aspect of public works construction; from street improvement and sewers and sanitation, to street lighting, traffic signals, retaining walls, landscaping and irrigation systems - now all in standard English units. Every dimension is clearly illustrated so that there are no questions on the requirements of the public works standards.

Standard Plans also features cross-sections that clearly illustrate the latest approved installation techniques, and exhaustive notes covering every detail - with clearly spelled out answers to any question you might have. Dimensions are listed in both feet and inches and metric measurements.

New in this 2012 version:

* *Traffic striping modeled after Caltrans Standard Plans*
* *Modified curb ramp plans for new construction and retrofits*
* *Updated notes on concrete boxes used in storm water conveyance*

500 pages, 8½ x 11, 3 ring binder $92.50

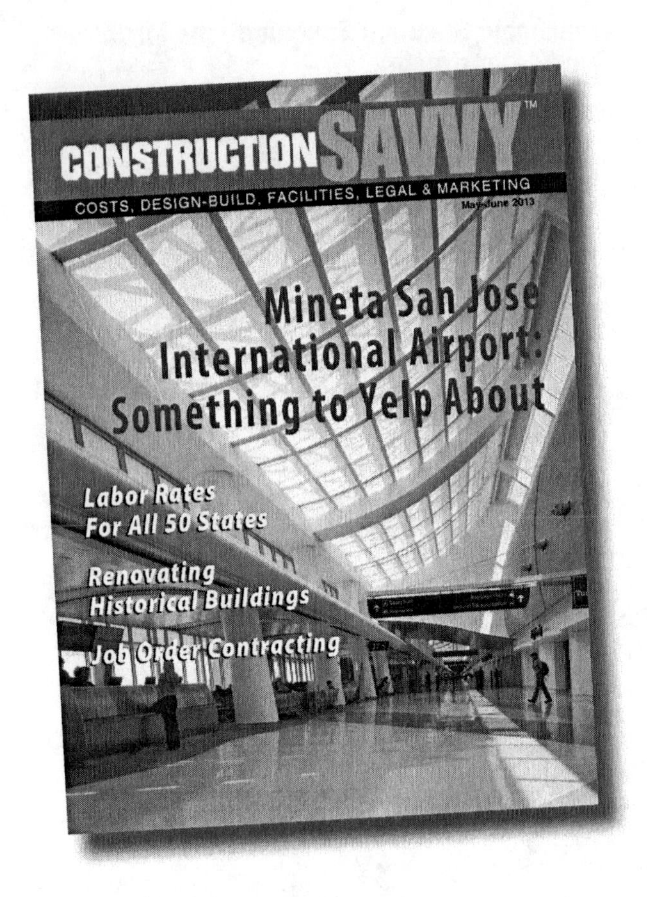

Have CONSTRUCTION SAVVY magazine delivered to your mailbox for one full year for only $29.95.

Each issue of Construction Savvy brings you important new ideas and cost information that you can apply immediately to the work you do.

For instance the May/June issue includes over 20 fact-filled articles that can help you with new ideas while providing practical techniques. For example you'll see:

* *Where to apply for public grants for restoring historical buildings*
* *How an apartment retrofit project reduce energy cost by 73%*
* *How chill beam technology can offer quiet, more effective HVAC in retrofits*
* *How to establish a baseline for go/no go decisions in your firm*
* *How a simple blog, if done right, can help you stand out from the competition*
* *Practical tips for avoiding accessibility lawsuits*
* *How using federal money for fire-sprinklers can stimulate economic growth.*
* *Find free square foot cost studies of actual building projects*

Savvy provides the knowledge, the practical advice, the tools and the construction costs professionals need.

Have each issue mailed directly to your door. One year subscription $29.95
Call today 1-888-264-2665

Construction Project Log Book

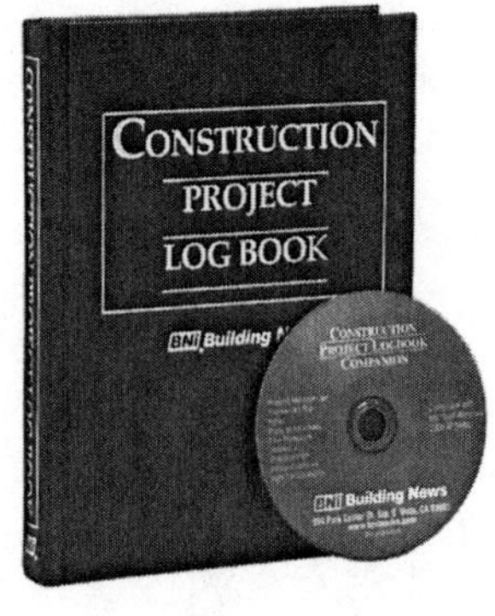

Proper documentation is perhaps the most important aspect of construction project management. In most states it is a legal requirement to keep daily records. With *Construction Project Log Book* that task is made as simple as possible.

The 365 Daily Work Log pages let you keep a detailed record for each day of the year. Additional forms such as Accident Reports and numerous checklists help make sure that you're covered.

Don't get caught without a written record ever again!

Document every shipment, machinery rental, delivery, delay, and weather condition -- all items that can effect productivity. With the interactive forms you can keep this information on your computer and enter new data daily into the interactive PDF forms. There are forms that actually do the math for you -- eliminating typical mistakes.

375 pages, 7 x 9 1/4, $39.95

Design Build Services: A Marketing and Business Development Handbook

The *Design Build Services: A Marketing and Business Development Handbook* is packed with ready-to-use forms and sample documents, laying out a complete "blueprint" for your success in design-build.

Written by an expert team of seasoned practitioners, consultants and university researchers, the *Design Build Services: A Marketing and Business Development Handbook* will quickly transform your firm into a competitive design-build powerhouse.

In clear, straightforward language, it covers how to conduct:

•*Market research* •Market planning •Communications planning
•Technology support •Networking for success
•Controlling costs and quality... and much more!

403 pages, 8½ x 11, $69.95

ORDER NOW WITH A FULL 30-DAY MONEY-BACK GUARANTEE

BNi Building News
990 Park Center Dr.
Suite E
Vista, CA 92081

☎ Phone Order Line
1-888-264-2665
Fax (760) 734-1540

In A Hurry?

We accept phone orders charged to your
❑ Visa, ❑ MasterCard, or ❑ American Express

Card# ____________________

Exp. date ____________ CVV2 ____________

Signature ____________________

Tax Deductible: Treasury regulations make these references tax deductible when used in your work. Save the canceled check or charge card statement as your receipt.

____
Name

____
e-mail address (for order tracking and special offers)

____
Company

____
Address

____
City/State/Zip ❑ This is a residence

Total enclosed ____________ (In California add 8.25% tax)

Order online: www.bnibooks.com
Find hundreds of construction references, forms and contracts to help you with your construction business.
Use promo code 841Y

Shipping via Fed Ex by dollar amount of order: $1 - $100 = $9.75, $101 - $150 = $15.25, $151 - $200 = $18.75.

Find Updated Davis Bacon wages and FREE estimating tools at www.constructionworkzone.com

Prices subject to change without notice

❑ 49.95 2014 BlueBook ACE Guide to Construction Costs
❑ 66.95 2014 Architect's Square Foot Costbook
❑ 99.95 2014 BNi Facilities Manager's Costbook
❑ 99.95 2014 BNi General Costbook
❑ 83.95 2014 BNi Home Builder's Costbook
❑ 87.95 2014 BNi Home Remodeler's Costbook
❑ 93.95 2014 BNi Electrical Costbook
❑ 99.95 2014 BNi Mechanical/Electrical Costbook
❑ 99.95 2014 BNi Public Works Costbook
❑ 99.95 2014 BNi Remodeling Costbook
❑ 99.95 2014 Sweets Unit Cost Guide
❑ 64.95 2014 ENR Mechanical/Electrical Square Costbook
❑ 63.95 2014 ENR Square Foot Costbook
❑ 64.95 2014 ENR Interiors Square Foot Costbook
❑ 59.95 2014 ENR Residential Contracting Costbook
❑ 59.95 2014 ENR Remodeling Contracting Costbook
❑ 59.95 2014 ENR General Contracting Costbook
❑ 61.95 2014 National Electrical Code Costbook
❑ 57.95 2014 Sweets Green Building Square Foot Costbook
❑ 89.00 Standard Estimating Practice
❑ 29.95 Design Management Guide
❑ 69.95 Profit Risk & Leadership: The Business of Construction & Design-Build
❑ 59.95 Business Letters for the Construction Industry
❑ 69.95 Marketing Handbook
❑ 76.95 Construction Inspection Manual
❑ 10.95 2012 Work Area Traffic Control Handbook
❑ 62.95 Public Works Contracting: Start to Finish
❑ 74.95 Public Works Inspectors Manual, 7th Edition
❑ 84.50 2012 Greenbook
❑ 92.50 2012 Standard Plans for Public Works Construction
❑ 69.95 Design Build Services
❑ 39.95 Construction Project Log Book
❑ 29.95 Construction Savvy